AF454294

Araucaria tree suspected of forming amber-bearing resin.

Photography © Peter M. Barczak
I. Rappsilber Photo: 107,108,113,
M. Barczak Photo: 108,110
Illustrations· K.Doszla Fig 3, Fig 12, Fig 13, Fig 14, Fig 53, Fig 56, Fig 60, Fig 61, Fig 102, Fig 110,
Illustrations: K.Korona Fig 23, Fig 24, Fig 38, Fig 39, Fig 118, Fig 119.
Cover: Shahadat Shakit
Translation: P.Rymarczyk
Proofreader: Chapter 1- Louise Kelly (Liverpool), all chapters Thalia Oosthuizen, Kleinmond, South Africa .
TRADEMARKS: Royal Brand logo is trademark and registered trademark.
ISBN: Print ISBN- 978-83-959437-7-5

FOR
MY SONS

Photo 1. Duxite. Petrified resin, preserved in a fossilized tree trunk. Lom Vršany u Mostu, Czechia. Found 2001.

ABOUT THE AUTHOR

Piotr M. Barczak developed the beginnings of his passion for geology and minerals while studying in one of the most important schools of modern Silesia, Poland – the Carolinum high school in Nysa. The first rector of this school was Christoph Scheiner, a world-famous Jesuit scholar; a mathematician and an astronomer whose observations contributed to the discovery of sunspots. Nobel laureate and American biochemist Konrad Bloch also graduated from the school. The ancestors of astronomer Nicolaus Copernicus, the creator of the heliocentric system, also lived near the town of Nysa. It was in this environment that the author became interested in biological sciences, chemistry, geography, and physics. At that time in the field of geography, a work was already created concerning the area of the nearby foothills at the junction of the Polish and Czech mountains, which was awarded in a regional competition of high school students. The author began the next period of education in the field of engineering and economics at the University of Economics in Wrocław, majoring in chemistry. During his studies, he had the opportunity to visit places rich in minerals in Lower Silesia and the Czech Republic. Since high school, he has also been collecting mineralogical specimens. He graduated from subsequent studies in Warsaw in the field of journalism and political science before he studied at the Warsaw University of Technology in the field of engineering.

Initially, the author worked as a journalist prior to starting his business, while constantly deepening his knowledge about minerals. Having his own company allowed him to develop a collection of minerals and fossils.

In 2017, Piotr M. Barczak obtained a PhD in economics in the field of management in the Warsaw School of Economics. During this period, as part of his business activity, he began researching amber and plant extracts. His need to find answers to the mystery of "living resin" was supported by funding from the European Union, which, through the National Center for Research and Development in Warsaw, approved the project entitled: "Development of Innovative Bio-nutraceuticals Based on Gypsywort and Natural Sources of Iodine". As a result, a business and scientific team was established. Its scientific directors included Prof. Habilitated Dr. Ewa Osińska from Warsaw University of Life Sciences (Department of Vegetables and Medicinal Plants) and Habilitated Dr. Daniel Załuski – head of Department of Pharmacognosy at the University of Toruń.

Piotr M. Barczak is the author of several works on amber, including the brochure: "Baltic Amber. Elixir of Immortality. Succinite Solution for Alzheimer's Disease and Dementia. Scientific Study" and "Amber. Elixir of Immortality: Health Properties of Compounds with Pharmacological Action". They have been written in collaboration with D. Załuski.

Many years of interest in amber and minerals sparked an impulse to write a book about Baltic amber. This is a difficult topic from a scientific point of view. It requires in-depth knowledge of many fields of science, including geology, minerology, botany, paleobotany, pharmacy, and history. Every aspect of scientific knowledge requires verification and checking, sometimes of individual words, hence it is a very tedious task. Ms. Alicja Pielińska from the Museum of the Earth, Polish Academy of Sciences in Warsaw has helped in writing the book by making available all possible scientific materials, books, and publications on amber.

Without her commitment, writing a book would be more difficult and its scientific value probably much lower. The author has been also supported by the Amber Museum in Gdańsk.

The author has used a number of scientific materials and over 100 photos and drawings, thanks to which the scientific content of the book may be more accessible to the reader. He has personally visited several countries, looking for answers to the mystery of the fossil resin. The author, having been at the Baltic Sea many times, has never personally found amber...

Photo 2. Amber- resin. Gablitz-Wienerwald, Austria. Photo by P. Barczak.

TABLE OF CONTENTS

CHAPTER VI — 289

CHAPTER VII — 307

PREFACE

altic amber fascinates thousands of tourists who come to the Baltic Sea. Small children and their parents are eager to dig in the sand while looking for precious resin. It is not easy to find, and every large piece of amber found in the "cucumber season" makes headlines. Another category of amber hunters are professionals. These are people who make a living from extracting amber of jewelry value, because each piece of fossilized resin has quite a lot of value. This group is quite hermetic, and the places and methods of extraction remain in the sphere of understatement and are shrouded in a certain mystery. The aforementioned group is particularly interested in the uniqueness of Baltic amber, created by "ancient" rivers, that was already traded by Persians, Vikings and Mycenaeans. From this group, wonderful jewelry products are created.

In this book, however, I deal with another aspect of Baltic amber. The initial idea was to analyze the knowledge of this fossilized resin in various aspects. The difficulty in getting to know amber is related to the multidimensionality of the issue. The problem is that science likes to specialize to be precise. This, however, limits knowledge of the issue of amber. If we deal with the question of the formation of the resin, we touch upon issues in the field of geology. This field is enough to fill many years of research and work of a scientist. You can look back a hundred thousand years ago, or into the past millions of years or tens of millions of years, which means that we delve into such vast knowledge that the life of one person may be too short to study the issue well.

Then, what do we find in amber? Insects. Thousands of insects. Every geological epoch, each geological era means different families and different species of preserved insects, all of which are to be obtained, dissected, and described in scientific articles. Again, work for a lifetime, infinitely long, because we reach millions of years into the past, and not even in one specific place on the planet – we need to look through the entire planet, because it has been changing over millions of years.

What about the plants? After all, they created succinite at different stages of development. What were they like? Are those plants we know today the same ones that existed millions of years ago? Further, some species have become extinct, some have evolved, some grow today in China and others in New Zealand or New Caledonia. Then there is chemistry, the compounds found in amber... Terpenes have already been counted; 40,000. And they can be found in amber. Maybe not in that number, but – since amber was found in peat, brown coal, surrounded by minerals, sediments, and rocks – it could have absorbed the catalog of chemical compounds occurring in the world of plants, trees and shrubs, and also essential oils.

Another topic is pharmacology and medicine. Again, you need to demonstrate detailed knowledge, and the topic is problematic because we are talking about human health. Anyone can misinterpret the research, which is I have avoided the words "heal" and "cure" in this book. These words are reserved for medical professionals. However, one can look at scientific works from the point of view of a PhD in economics in the field of management and assess the role of terpenes contained in Baltic amber and their possible effects on the body.

Therefore, while implementing innovative amber extracts, we conducted a study on liver cancer cells, determining the

toxicity range of the terpenes it contained in it. Interestingly, at a certain concentration of amber extract, half of the cancer cells were destroyed. Therefore, the research is a very important aspect of theoretical analyzes, although full human studies of amber extracts have not yet been conducted. In the book, I also touch upon the topic of amber teethers, about which there were some doubts from the US, Canadian and Australian markets. Perhaps the composition of terpenes explains the satisfaction of young children, nibbling on amber beads to ease the pain of erupting milk teeth, and frequent purchases of such teethers by their parents. Most of the terpenes found in the study have bactericidal and anti-inflammatory properties.

In the book, I put forward a controversial hypothesis that Baltic amber is a marketing term. There was no single Eridan river that created Baltic amber deposits in the Gdańsk region, and it seems that glaciers mixed amber in Central Europe in such a way that some of these ambers come from the Cretaceous era, not from the Eocene. I have also visited New Caledonia, New Zealand, and Tasmania, where the old amber forest has survived. It seems that scientists in the third millennium should take a broader view of world geography than the explorers of the eighteenth century. Some theories from that period are still used in theoretical works on the history of amber formation.

My private collection of minerals that I have assembled over 40 years also allowed me to conduct several experiments and observations related to amber. The search for the truth about amber was also an interesting adventure and a fascinating task for me. Perhaps the book will be an interesting read for you. Thank you for your interest.

CHAPTER
1

1. INTRODUCTION

SUMMARY

Amber of the Baltic Sea has been known for centuries. There are an enormous number of articles and books on the fossilized resin which originated millions of years ago. One of the places which holds the largest collection of amber literature is the Museum of the Earth, Polish Academy of Sciences in Warsaw, Poland, with over 2,600 books, brochures, and articles. Looking at this rich collection, however, it can be stated that the healing properties of amber have never been described in detail, and modern scientific research, as a matter of fact, only mentions this issue briefly.

Amber is a fossilized plant resin from millions of years ago, a mixture of organic and inorganic compounds. It can be classified as a mineraloid, i.e. a natural substance different from minerals. It contains groups of various elements and mixtures of chemical compounds. Mineraloids include pearls, jet, obsidian, opal. Baltic amber, also referred to as succinite, was probably created by damage to trees, as a result of which resin, containing compounds protecting tree structure, leaked.

Baltic amber is a complex molecular structure which, as a result of millions of years of transformations, has acquired a unique structure, additionally enriched with plant compounds occurring on earth over millions of years. Succinite because that is how Baltic amber is described in the scientific language, shows supramolecular structures and each amber study reveals new compounds known from plants.

Amber is a cross-linked polymer- this is chemical substance composed of many elements of chemical compounds interconnected into complex spatial structures contained in one piece of amber. That contains molecules or macromolecules interacting with each other. They easily migrate when we heat Baltic amber.

Succinite (the experts' name for Baltic amber) contains succinic acid. It has been suggested that succinic acid enters human cells, modifying their activity. Through such actions, intracellular processes are regulated, and the organism's homeostasis is restored.

Photo 3. Rumenite Ukraine. Photo by P. Barczak

Baltic amber is a fossilized plant resin, known from jewelry, but also from insects and air bubbles found in it. It forms over millions of years. The beginning of its formation is the geological era of the Eocene, which started about 50 million years ago. During this time, the resin hardened as a result of fossilization. Immersed in sands and sediments, without access to air, it acquired specific, unique features. The temperature and the changing pressure of the environment underground, where the fossilizing trees lay, triggered specific chemical processes in this polymerized resin. Chemical reactions, however, may still occur in Baltic amber, and the chemical compounds contained within this matter are still active and look for possible reactions. It is not known exactly how succinite is formed. Concepts and hypotheses regarding this question are largely based on conjecture and in the following book we will try to summarize this issue based on the existing scientific works from various fields.

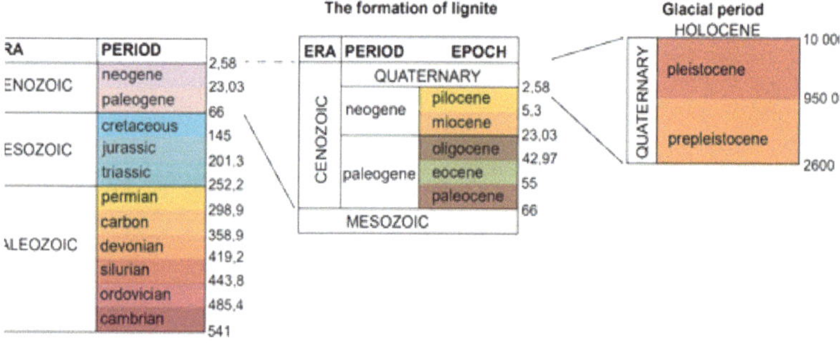

Fig. 1. Graphics depicting the period of millions of years when amber and the modern world were created. Based: Koźma, J. (2016) "Anthropogenic Landscape Changes Related to the Former Lignite Mining on the Example of the Polish Part of the Area of Łuk Mużakowa". Górnictwo Odkrywkowe, 57.

More and more amber deposits are becoming known in the world. Amber has become popular, so its deposits are exploited in various countries, from Congo, through Peru, Ethiopia, India, and New Zealand. The oldest ambers come from the Triassic Era, 230 million years ago. They have been found in the US and Italy. Baltic amber is younger – it comes from the Eocene period, which is assumed to have lasted 22 million years, occurring from 56 million to 33.9 million years ago. However, there are problems with chronological differentiation of individual pieces of Baltic amber because, for millions of years, fossil resins in the region of Central Europe were being mixed up by glaciers which, quite recently, around 10,000 years ago, devastated the entire geological system of this part of the world. Generally, however, all the succinites, i.e., amber from the Baltic Sea region, Poland, Germany, Ukraine, Kaliningrad in Russia, and Belarus, are similar in terms of their chemical composition and features. The deposits currently located in the Baltic Sea region (the Gulf of Gdańsk) have probably been pushed by glaciers from the southern regions of Europe.

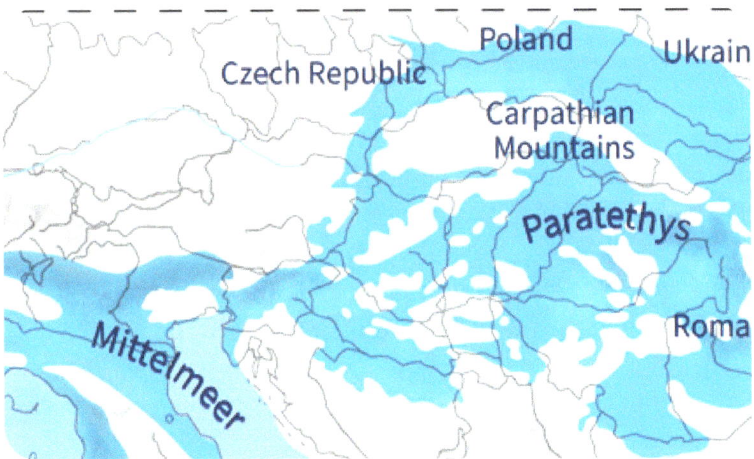

Fig. 2. A map of a hypothetical sea from around 23 million years ago. Based on: Rögl, F. (1999). "Mediterranean and Paratethys. Facts and Hypotheses of an Oligocene to Miocene Paleogeography (Short Overview)". Geologica carpathica, 50 (4), 339-349.

Rivers flowing from south to north may also have contributed to the "recent" warming around 10,000 years ago, when dissolved masses of ice formed deltas around the ancient sea -today, they are mostly Central European lands. Part of the water flowed down the river valleys, the sands of which have been enriched with amber to this day.

Fig. 3. Eocene – the Paleogene Period, lasting about 22 million years (from 56.0 to 33.9 million years ago). Painted by K. Doszla.

Let us go back to the Eocene times, 56 million years ago when the so-called Paleocene-Eocene Thermal Maximum (PETM) occurred on Earth. The temperature of the oceans increased by 6 degrees Celsius and in the Arctic the surface waters warmed up to 24 degrees Celsius, and tropical organisms and plants grew in the circumpolar zones. In the early Eocene (about 50 million years ago) annual temperatures were around 25 degrees Celsius, which means that the entire planet was very warm. The vegetation was thriving. Tropical and subtropical ecosystems

were characterized by vast species diversity. Cypress trees and redwoods grew in Antarctica. Palm trees grew in Alaska, Canada, and Greenland. Over time, changes in the layout of the continents resulted in the absorption of the accumulated CO_2 and, since 49 million years ago, the climate has been cooling down gradually. This led to the destruction of forests, which in turn led to the formation of peat bogs and vast areas with proto-amber. About 33 million years ago, the colder climate caused a change in the plant cover; the period of mixed forests in Central Europe began and the flora was similar to that of the modern day. Some plant species from this stage of development have survived in various parts of the planet, unchanged to this day. Swamp trees flourished in the warm climate of Central Europe. Inflowing warm air masses from the Paratethys Sea (a sea formed about 34 million years ago, separated from the Tethys Ocean) caused development of plants and the formation of extensive lignite deposits. Over the course of millions of years, the Paratethys Sea dried up due to another warming. Deposits of salt and gypsum were formed in Central Europe. The prevailing Mediterranean climate preserved vegetation similar to that which grows today. Approximately 5 million years ago, the Mediterranean Sea separated from the Atlantic Ocean and the water temperature reached 35 degrees Celsius. The climate in today's Poland and Ukraine began to become desert-like and forests were replaced by grasses [Słodkowska, Kasiński 2016]. The former tropics looked like today's desert parts of Australia.

Fig. 4. The hypothetical location of the Fennoscandia continent during the Eocene. Source: based on, Szwedo, J.; Sontag, E., "The Flies (Diptera) Say That Amber from the Gulf of Gdańsk, Bitterfeld and Rovno is the Same Baltic amber", Polish Journal of Entomology, 2013.

There has been adopted a concept that amber from Germany (Bitterfeld), Ukraine, and the Gdańsk region come from the Eocene period and were created around the same time. Baltic amber could have made quite a long journey. It was moved from the south of Europe, from the area of the former Paratethys Sea, which covered a large part of the south of the European continent. The common origin of Ukrainian, Belarusian, and Polish deposits, as well as Russian ones, is confirmed by the inclusions found in the fossil resin. The same insects have been noted in Bitterfeld amber from Germany and that from the Baltic coast, although separate species have also been found. For example, 10% of the ants found in the ambers are of the same type throughout that area. The majority of insects are the same in all the ambers from the mentioned countries. For example, from 88 species of Ceratopogonidae (a family of flies), as many as 23 are known from Bitterfeld and 26 from the Rovno deposit in Ukraine [Sontag, Szadziewski 2011].

However, despite the existence of numerous works, succinite (Baltic amber) – has remained mysterious. The type of trees that produced this unique resin have not yet been established unequivocally and

Photo 4. White Baltic amber with tarnish. Photo by P. Barczak.

definitively. Work on this topic has been going on for over 130 years – that is, from the time of the first papers written about trees that produce amber.

Discoveries related to amber, however, have a much longer history. It is said that in 1800 BC the Phoenicians reached the North Sea, where there were "tin islands" somewhere near Great

Photo 5. Amathus is one of the oldest ancient royal cities of Cyprus, founded around 1100 BC.

Britain. They also reportedly sailed to the Baltic Sea to the "electron's area". There, they loaded amber onto ships and then sold it to the Egyptians, Greeks, and Hebrews. The Carthaginians also wrote about a mysterious sea where the resin was obtained, but no one knew where the succinum (the Latin name for amber) they were selling had come from.

Amber was a priceless stone, expensive, accessible to few. Ancient Roman Emperor Nero admired how much of a valuable gift of nature amber was and collected large amounts of it. During the time of Emperor Marcus Aurelius, a small amber figurine reached the price of a slave [Chętnik 1952]. Calistratus, an ancient writer, noted that amber was a medicine and successfully treated confusion of the senses. An even earlier sage Flavius Philostratus, who lived in Alexander's time, wrote that the "poplar electrum" were pieces of a substance carried by the waves of Eridianus to the lands inhabited by barbarians. In light of the long history of amber, the Roman times are, however, almost "modern". The oldest known fragment written about amber is in the British Museum. Written in cuneiform script on an obelisk from Persia from the 10th century BC, it states that amber is mined in the north.

Amber was also described in Arab countries. The differences between amber and other resins were noticed in 1048 by an encyclopedist and researcher Al-Biruni, in his book "Collecting Information for Knowledge About Jewels". He mentioned that amber is a different substance from resins of the known trees, as it reacts differently and there are sometimes frozen insects in it.

In Europe, the first known written references to amber appear in 1405. In the old books of the Teutonic Order, it is mentioned that it was obtained on the Vistula Spit and the Sambia Peninsula. In Lochstädt, in a castle built by the Teutonic Order, there was the Amber Office (German: Bernsteinamt) from 1327. The largest excavations of fossilized resin took place in the western part of the Sambia coast in the villages of Sinyavino (German Gross Hubnicken), Donskoye (German: Gross Dirschheim), Lesnoye (German: Warnicken), and Strobschnee, Iantarnyi (German: Palmnicken) [Hartmann 1677].

From the 14th century, the Teutonic Order took control of the amber deposits on the Baltic Sea in the area of today's Gdańsk and Kaliningrad. The so-called Lords of Amber (Bersteinsherrn), who supervised the beaches, were appointed. Originally, amber did not arouse great emotions and only when the Teutonic Knights took over this area it became a valuable find. Order among the local population was brought about by terror. The peasants were threatened with flogging or imprisonment for going to the sea without an overseer [Elditt 1868]. Amber could not be collected or sold to anyone. In the 15th century, the death penalty for collecting amber was introduced, and the sentence was carried out without trial, by hanging the "unfortunate" one on the nearest tree. In order to hinder the trade in amber, administration of the Teutonic Order did not issue permits for amber turners to settle in their country. Stocks of amber in monastic castles were growing. Most of the amber was sold by the Order Knights to craftsmen in guilds in Bruges and Lübeck (now Belgium and Germany), where mainly rosaries were made of it. For the amber sold, the Order sought ginger, spices, and cloth [Różański 1959].

Fig. 5. The Teutonic Order introduced strict laws. Those unlawfully going to the shores of the Baltic Sea were threatened with the death penalty.

Over time, it started to be considered a precious raw material with valuable healing properties. It was called – using Latin names – lapis marinus or lapis qui burnestein vu lgariter nuncupatur (that is, burning stone). Originally, it was used as a raw material for making rosaries, but over time its healing properties were appreciated. After the abolition of the Teutonic Order, the Prussian Duke Albert took power. He supervised the procurement of amber all over the coast. On his behalf, the beaches were controlled by an amber master (Bernsteinmeister) with a few horse guards (Strandreiter). All white amber, which was particularly valued for its healing properties, went to Duke Albert, because using it relieved pain. The duke even sent pieces of this amber to Martin Luther to relieve nephrolithiasis he suffered from. White amber was the most valued. It was called "Hauptstein". Pieces of this type of amber had to be larger than the thumb, they were white, even transparent, in the color of white cabbage or whey. It was the most expensive type of amber. Another kind was Weisenstein: of white color but overgrown with other colors. It also had to be given to the duke-elector of Prussia, Albert. since it was possible to earn good money on amber and the number of transactions increased, the duke subcontracted the trade to an intermediary. Trade in this raw material was conducted exclusively by Paulus Jaske and his family.

Photo 6. Amber from Lithuania. Photo by P. Barczak.

Duke Albert continuously lacked money and that is why the punishments invented by the Teutonic Knights returned in 1513. The person who was not entitled to dig amber yet done so, was sentenced to death. The duke's warehouses became full again. All legally obtained amber was sent to the warehouse which was supervised by the Prussian state treasury. The supervisor had to be a nobleman. If an authorized person found amber near his home, he had to put it in a special bag and return it to the warehouse the next morning. The theft of a piece of amber from the duke's land, not just one found on the coast, was also punishable by death. In 1770, 391 barrels of amber were collected in Prussia. Effective restrictions allowed the rulers of Prussia to get rich. The strict regulations were not abolished until 1833. At that time, amber mining on one's own land was allowed. International trade also developed. The greatest amounts of amber from the Baltic region were shipped to Turkey, Egypt, Persia, and India [Ibid.]. In the 15th century, resourceful merchants even established amber trading posts all over the world, in Turkey, Persia, India, China, Japan, and in European cities in Livorno, Paris, London, Vienna and Budapest.

In 1551, a book by Andreas Aurifaber entitled "Succini Historia" was published. The book was first published in Prussia. It describes simple methods of obtaining amber, which involved, inter alia, hammering poles into the sand. Water incoming along the rod caused leaching of a fossilized resin lighter than water – of Baltic amber. "Yellow gold" was mostly found in the form of "nests" in clays, so after exhausting the deposit exploitation in a given place was ended and new places were sought. In the swampy areas, however, no mines were dug and a rather primitive method of trampling the wet peat mass was used, as a result of which, amber, being lighter than water, was poured to the surface.

Over time, new works on amber were printed. In 1534, the oldest herbarium in Poland, written by Stefan Falmierz and entitled "O ziołach y moczy ich" ("On Herbs and Their Powers"), was published in Kraków. The author wrote that amber is credited with the power of the "heart". It can cheer up, drive away the heartworm, and strengthen the spirit. Thanks to the books, it was possible to archive many recipes for medicines from amber which are known to this day.

Photo 7. A petrified tree trunk found by the author by the Baltic Sea. Photo by P. Barczak.

Another author, Hartmann [1677], writes in his book about the existence of mines in the vicinity of Elbląg near Gdańsk and about mining as much as 318 kg of amber there in 1641. Tylkowski (1680), in turn, mentions the extraction of amber at the mouth of the Vistula and from the vicinity of the Narew (see: Popiołek 2006). In Klebs' study (1883) it was written that, in 1728, amber was dug in Sambia in Primorie (German: Gross Kuhren), Filino (German: Klein Kuhren), and Svetlogorsk (German: Rauschen). Amber was a valued and precious raw material. Jewelers and craftsmen readily used it. Figures, rosaries, and rings were made of it. There is also a description how amber was used to make a piece of furniture for Augustus II the Strong, the Polish King.

Exploited amber deposits usually occurred in sandy lands, locally. Amber was mined in today's Gdańsk county in the villages of Kleszczewko, Różyny, Łęgowo, Bąkowo, Kokoszki (currently a district of Gdańsk), Brzeźno (Gdańsk disict), and Klukowo (now a housing estate in Gdańsk). In Klukowo, amber was mined from the beginning of the eighteenth century [Kaunhowen 1914].

The prehistoric amber forest has been a mystery since the earliest years. The first scientific hypothesis concerning the trees from which Baltic amber was made was developed by a German professor G. Goeppert. In a book published in Wrocław (Breslau), he assumed that the amber deposits were created by the pine Pinites succinifer. He suspected, however, that modern trees are different from those existing 50 million years ago [Goeppert 1883]. Interestingly, this concept also appears in today's studies.

In 1972, a group of English scientists led by J. Mills published a paper on chemical compounds contained in Baltic amber. It was based on specific chemical composition studies. With new analytical devices, 124 compounds contained in amber were detected, of which 71 were identified. Comparing the infrared spectra, a team of British scientists concluded that Baltic amber is similar to the resin of the tree growing in Australia and New Zealand, *Agathis* from the *Araucariaceae* family [Mills, White, Gough 1984]. The concept of H. Conwentz, who believed that the original amber forest was evergreen and diverse, with a significant admixture of tropical plants, including sequoia, pines, laurels, myrtles, and magnolias, started to seem realistic. Amber pines, according to him, belonged to four species of trees: *Pinus baltica, Pinus cembrifolia, Pinus parufora, and Pinus silvatica* [Conwentz 1890].

Regardless of the unsolved problem of the plants that make up amber, the content of succinic acid in Baltic amber and the abundance of resin secreted by trees remain mysteries. Fossil tree resins found all over the world and described in scientific literature usually do not contain succinic acid. The exception is Baltic amber, which contains a relatively large amount of it. On this basis, some geologists concluded that in the Eocene era, the sea in which "amber trees" grew was contaminated with hydrogen sulfide. The forests where Baltic amber formed were flooded with hydrogen sulfide water [Maidanovich 1988]. One can agree with this type of concept. The Bitterfeld deposit was probably shaped differently than the rest of Baltic amber.

German amber has been known since 1731, when King Augustus II the Strong was informed about its existence in the Bitterfeld region in Saxony, Germany [Rappsilber, Krumbiegel, Wimmer 2013]. In the 1950s and 1960s, significant amounts of it appeared in lignite mines, although the deposit has its renaissance at the present time, when amber has become an important jewelry product. Amber from this deposit was probably created in the area of former river deltas and lagoons. The trees from Bitterfeld underwent a metamorphosis related to the climate, which changed from warm to cool and, finally, to arctic. Bitterfeld amber differs from succinite, however, because it does not contain succinic acid in such a form as succinite does and there is no, for example, communic acid, whereas Baltic amber is rich in these chemical compounds.

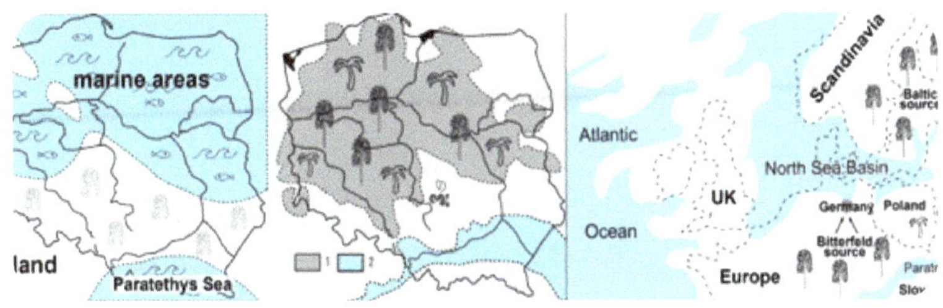

Fig. 6. Left figure: Eocene (56.0 million to 33.9 million years ago), there is a visible land in the region of Podkarpacie and Ukraine and there are areas flooded with water (blue). Central figure: a visible land and the Paratethys sea (blue). Right figure: Miocene geological era (23.03 million to 5.333 million years ago). You can see the Paratethys Sea and the deposition site of German amber in Bitterfeld. Based on: https://zywaplaneta.pl/ and [Wolfe et al. 2016]. "Bitterfeld Amber is not Baltic Amber: Three Geochemical Tests and Further Constraints on the Botanical Affinities of Succinite". Review of Palaeobotany and Palynology, 225, 21-32.

Searches for forest trees which produced amber are still ongoing. However, the resins of the trees *Agathis australis*, *Cupressaceae* (*Metasequoia glyptostroboides*), *Pinaceae* (*Pinus contorta*), *Podocarpaceae* (*Podocarpus totara*), and *Sciadopityaceae* (*Sciadopitys verticillata*), which have been subjected to tests, did not have the identical chemical composition as Baltic or Bitterfeld amber [Wolfe et al. 2016]. So, what tree formed Baltic amber? As part of one of the concepts of Baltic amber creation, it was assumed that proto-amber was formed in a mixture of bitumen (a natural thick liquid sometimes called asphalt), waxes, resins, and brown coal sediments. Such a mixture fossilized in paleo-peats, then in sea waters, and finally in the so-called blue earth-glauconite. Most of the Baltic amber was created in the surroundings of this matter. Therefore, succinite is a mixture of resins and chemical compounds – not a particular resin of one tree.

Photo 8. Fossilized tree from Slovakia. Photo by P. Barczak.

Baltic amber is a kind of a macerate, which was "pickling" underground for millions of years. A mixture of resins, oils, tree remnants, and geological compounds permeated from the environment to the interior of this "macerate".

The process could have happened as follows. In the first stage, resin oils evaporated underground, accumulated in the forest litter and soil, polycondensation and partial oxidation of the resins took place. Such mixtures shifted into different hollows. As a result of geological changes, the matter turned into bogs in vast areas flooded with water. Similarly, in such bogs, young ambers – copals – are formed: fossil resins of trees growing in tropical forests. It is assumed that in a few million years they will turn into ambers. In a similar way, another amber was formed – aikait – which is found in deposits of lignite in Hungary. It is brittle and does not contain succinic acid. The plants that formed it - about 89-90 million years ago -are found in Hungary today, but also in the fossils of Lake Tanganyika in Africa, at Lake Baikal, in Japan, Australia, and South America [Matsui 2010]. However, the area of a lignite mine in Hungary where aikait occurs has never been subject to glaciation. The environment of the Hungarian amber was made of clays, not glauconite – blue earth – usually found with succinite. Perhaps that is why aikait is a worthless fossilized resin of purely hobbyist – and no jewelry – value, and succinite is a jewelry rarity and a unique source of terpenoids from trees existing in the past.

Photo 9. Aikait, Csinger Tal, Ajka, Hungary. Photo by P. Barczak.

Succinite is found throughout Central Europe. It has been found on the present-day Polish coast, near Łeba, i.e., over 100 km from Gdańsk, and in Bory Tucholskie – a region near the Vistula pre-valley, over 140 km from Gdańsk. In the Brda and Wda river basins, numerous shallow open-pit mines have been created. Similar places of amber extraction were created in Kurpie, a place more than 300 km away from Gdańsk. Therefore, it should be presumed that the amber deposits on the Baltic Sea did not arise from the sediments of the hypothetical Eridan delta flowing north to south, from the ancient mainland of Fennoscandia. Rather, amber was transported by a glacier from the regions in the south of the country to the north. It underwent chemical changes pressed by the glacier.

Fig. 7. Map of the occurrence of amber (red dots) in Poland, according to R. Borzęcki. http://www.redbor.pl/

The sediments in which amber is found today often contain the minerals called andalusite and tourmaline. These types of minerals are found in Ukraine, Belarus, and Podkarpacie, Poland, i.e., hypothetical places of amber resin formation. In Ukraine, the place where amber occurs is Klesov, in Belarus it is Gatcha-Osova. It has been proved that amber deposits were transported during the ice age. Secondary sludge containing amber lies in the Polish Lowlands and Belarus at a depth of approx. 5m.

These are secondary sediments, the so-called placers. In the Quaternary (0.1–1 million years ago), melting glaciers carried ambers to the areas of the present Baltic Sea, mainly through the Vistula valley but also through other rivers. Paleogene deposits (66 million to 56 million years ago), however, lie at a depth of 50m underground.

Photo 10. Baltic amber and its varieties. Photo by P. Barczak.

The largest transport of amber-bearing delta sediments took place in the Pleistocene (from 2.58 million to 11.7 thousand years ago), when moving ice sheets torn fragments of amber-bearing delta sediments from the ground and carried them over a distance of hundreds of kilometers. Amber was transported in the form of glacial rafts, i.e., whole packets of sediments mechanically detached from the parent rock outcrops by the ice sheet. After the ice sheet subsided, the glacial rafts were settled and covered by younger sediments. Another type of transport for succinite deposits were fluvioglacial waters. Amber was dragged by the waters flowing out from under the melting ice sheets and was accumulated, along with sand and gravel, in the outwash fans in front of the head of the ice sheet. The later redeposition of amber nuggets is associated with the development of the Baltic Sea in the Holocene and the deposition of amber, washed out by waves from the Eocene and Pleistocene sediments, in the fossil and modern Baltic beaches changing their range [Małka, Kramarska 2013]. The water level in the Gulf of Gdańsk at that time was 30m lower than it is today.

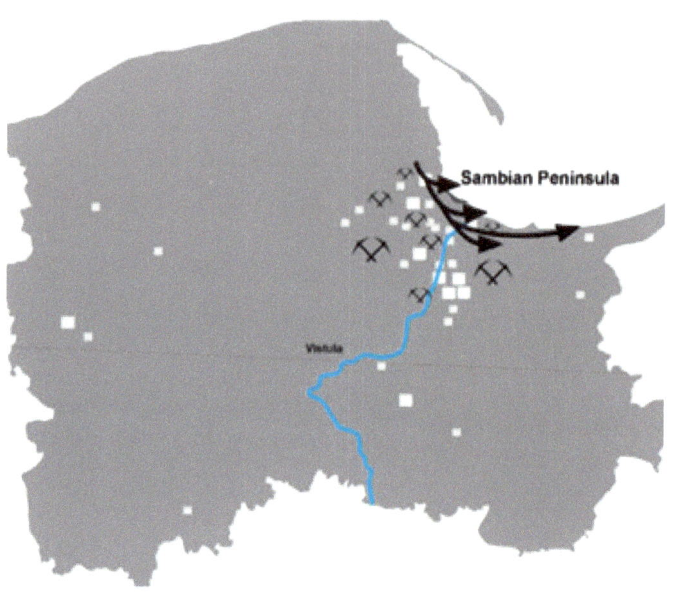

Fig. 8. Area of the Gulf of Gdańsk and directions of amber transport by sea currents. Places of extraction of Eocene amber deposits. Source: Szwedo, J.; Sontag, E., "The Flies (Diptera) Say That Amber from the Gulf of Gdańsk, Bitterfeld and Rovno is the Same Baltic Amber", Polish Journal of Entomology, 2013. Małka, A., "In the Footsteps of Amber Mining in the Pomeranian Voivodeship", Acta Universitatis Lawodsis Folia Geographica Socio-Oeconomica 2015.

Amber-bearing ice glacial rafts that were moved by the glacier have been found in Zielnów near Grudziądz, but also in Możdżanów near Ustka. However, amber was extracted from a depth of 120m, according to chronicles from the eighteenth century. The glacier also formed outwash sediments accumulated by the waters flowing out of the emerging tunnels from the glacier. Such a place is the Kurpie region in the central Narew basin in Poland, where amber deposits settled in the lowland area. Subsequent actions of glaciers took place in the Holocene, in which river regions existing to this day were formed, and amber sometimes accumulated in their sediments.

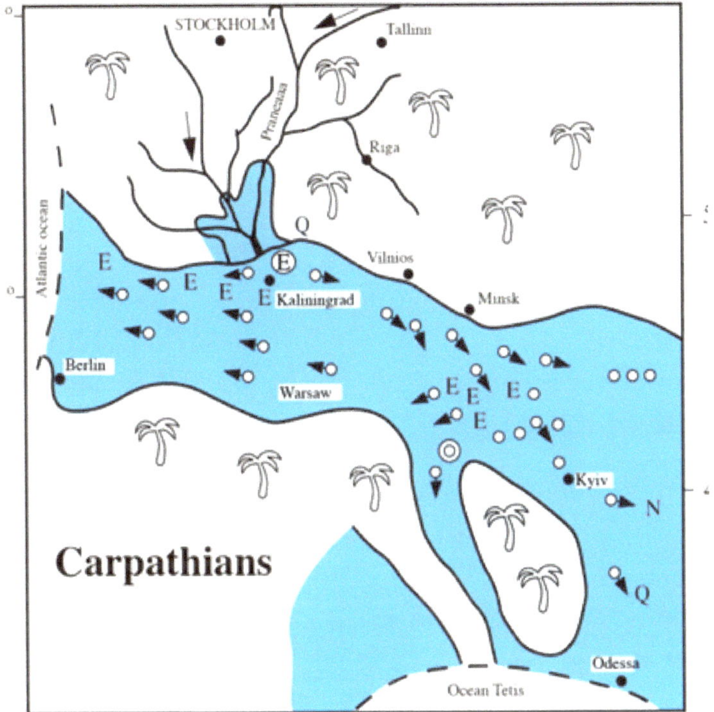

Fig. 9. The formation of amber deposits and their migration millions of years ago. Based on: Matsui, V. M & Naumenko, U. Z. (2017). "The Origin of Amber-Succinite: from Myths to Modern Theories". Bulletin of the NAS of Ukraine.

In 1893, P.A. Tutkovsky, in his book "South-Western Territory" [Tutkovsky 1893] expressed the idea of direct formation of amber within deposits of Eocene lignite, taking into account the brown coal fields at the Dnieper. The work referred to the origin of a lignite basin near the Dnieper. It has been established that the vegetation of this region consisted of evergreen forests growing in a humid climate, with a significant proportion of conifers,

mainly araucarias. These trees disappeared from the area at the end of the Middle Eocene. Forests drowned in peat bogs which, over millions of years, formed lignite deposits. In today's lignite deposits in this area there are fossilized resins – retinites.

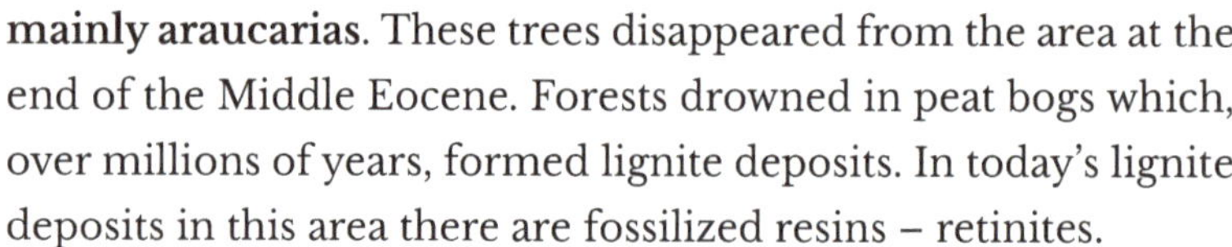

Amber mining in the lowland areas of the Kingdom of Poland was started in 1796 by Wolgram de Voza in Wołków. Similar amber harvesting took place in the Brda and Wda basins.

During the Eocene and the early Oligocene (33.9 to 23.03 million years ago) the territory of present-day Ukraine, Belarus, and Poland was under water. Lignite in this area eroded and amber outflowed into the coastal zone of sea basins. Seas containing glauconite and carbon mix converted the fossilized resin into proto-amber. Sediments enriched with oxygen and potassium interacted with the resin and stimulated a series of intermolecular transformations. It is known that, unlike minerals, fossil resins of inorganic origin do not have a stable elemental composition and do not have stable properties. Therefore – under the influence of external conditions - light, temperature – they change their properties. In addition to glauconite, it is possible that amber was also affected by calcite compounds. Formation of gypsum and sulfur could play an important role in the transformations.

Therefore, there are serious reservations about the hypothesis that amber from the north was brought by glaciers to the territory of present-day Ukraine. Proto-amber was shaped rather at the foot of the Carpathians, in the areas where the Eocene (Paratethys) sea was drying up and, before that, there had been a resinous subtropical and tropical forest.

Fig. 10. Areas of proto-amber formation in the Eocene - gray spots. Based on: Matsui, V. M. (2013). "Marine Stage of Fossilization of Resin Secretions of Conifers on the Way to Succinite Amber". Geology and Minerals of the World Ocean, (2 (32)), 101-108.

The peat bog stage in which the resin was hardened, lasted about 1 million years. The deposits were formed surrounded by peat vegetation, so volatile oils were also fixed in amber, in the atmosphere of which resin with unique properties was formed. A similar phenomenon of fossilization of cypresses is observed during the flooding of the Mississippi River in the USA. The formation of fossil resins in sand deposits is observed there, there is a formation of brown coal with pockets of burnt trees. In Ukraine, a similar phenomenon was observed on the territory of 100,000 km2. In over 120 coal open-pit mines, inclusions of fossil resins – retinites – were found. In paleo-peat bogs, the phenomena of fossilization and formation of proto-amber were documented. In 90-95% of cases they were formed by subtropical plants, such as cypresses, araucarias, *taxodium*, redwoods, legcarps, ginkgo trees, as well as conifers *Pinus Haploxylon, Diploxylon, Picea spruce*, and *Podocarpus*. From angiosperms, magnolias, palms, olives, madders, myrtle, holly, heather, Drimys (a tree currently growing in Argentina and Chile, whose bark and fruit are used to make the so-called Magellan cinnamon) have been documented in geological outcrops. The vegetation of the Eocene epoch found in peat open-pits was similar to the flora of modern southern Japan and southern North America [Matsui, Naumenko 2020]. This type of forest is also found in Greece.

Over millions of years, the land system has changed. Tiny valleys with biological matter and proto-amber were flooded with salt sea water with glauconite and sands playing the dominant role. This matter, however, for millions of years was covered with successive layers of sediments and rocks along with new types of amber formed in the Oligocene, Neogene, and Pleistocene periods. Amber moved along with the "blue earth", i.e., glauconite-quartz sands and mudstones. Amber deposits were found in the pre-valleys of the Dnieper (Ukraine), which confirms the south of the Carpathians as the place of the origin of amber. In the Odessa region, amber deposits lie at a depth of 2-10 m and in a deeper zone of 50-100 m.

The next stage of the formation of succinite was transported by river waters to various areas, where it was deposited together with glauconite sediment. This occurred both in fresh and sea waters. More than 780,000 years ago, the era of glaciation began. It afflicted the entire region of Central Europe, the ice reached as far as the Carpathians. During such a glaciation 780,000 years ago, huge layers of ice were formed, reaching maybe up to 1 kilometer in height, and during periods of warming huge glacial lakes occurred. The waters washed out all the ingredients,

including proto-amber. The entire amber-bearing region was in the middle of another glaciation, in which ice reached even the higher parts of the Carpathians. Warming caused another leaching of all matter, which together with water flowed to lower areas. 420,000 years ago, the glacier was already smaller, leaving lakes and frontal moraines consisting of tons of earth and everything that could be washed out in the foreground. The next glaciation occupied a smaller area, creating giant lakes in its foreground. Huge masses of land flowed down the valleys of rivers, including the Vistula and Pripyat. These constant changes in ice flow, periods of warming and glaciation, led to the formation of deposits in Central Europe. Most of the primordial amber was rinsed and went to river meanders, as well as to the Baltic coast.

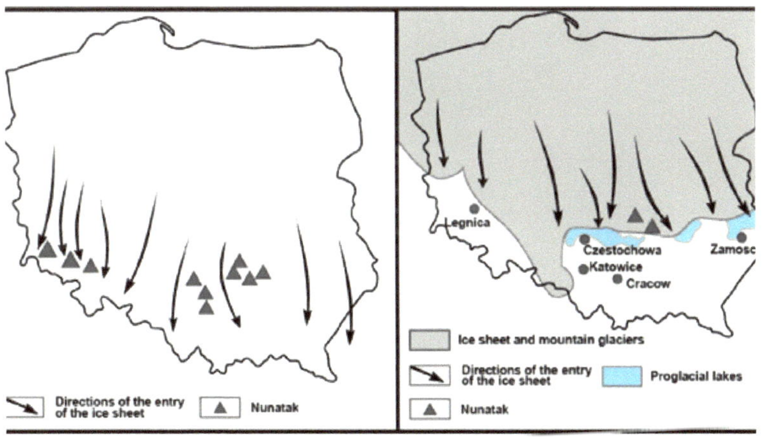

Fig. 11. Operation of glaciers in Poland in the glacial era. The ice absorbed the amber deposits from the south and carried them north. Ice age drawings based on: https://zywaplaneta.pl/.

Over millions of years, ambers different from succinite appeared in the layers of earth and coal – retinite, krantzit, rostornit, valchovite, cedarite. The listed fossilized resins do not have a constant elemental composition and stable properties.

They change depending on the temperature and depth at which they were accumulated. Under the influence of oxidation, the processes of formation of an oxidative shell on their surface take place. During the weathering of amber, the content of oxygen, carbon, and hydrogen increases, and the sulfur content decreases [Matsui, Naumenko 2020].

Photo 11. Valchovite – a unique fossilized resin. Cowshed near Moravska Třebová, Bohemia – Moravia. Photo by P. Barczak.

For millions of years, succinite was affected by glauconite, which formed compact rock-earth formations with it and – in the form of giant areas of the earth (rafts) – moved mainly north, pushed by waters produced by glaciers. It is worth mentioning a bit about the long-time companion of Baltic amber. The clay mineral glauconite has specific properties of ion exchange and absorption. Ion exchange consists mainly of cation exchange. Glauconite has magnetic properties that result from its internal structure. Glauconites are also characterized by the ability to absorb water, as they are organic substances. Glauconite contains pyrite, which was mentioned in scientific works as early as in the eighteenth century.

Let us recall that the specific electrical properties of amber are its important features, and they are similar to those of glauconite. Large deposits of glauconite and amber have been found in Poland in the region of Górka Lubartowska. However, they were not formed as a river delta, but as glacial dumps associated with one of the phases of the retreating ice sheet in Europe [Łozińska-Stępień, Rytel, Saliński 1986].

The ability to absorb elements from its environment probably caused that succinite, which for millions of years was staying in amber absorbed iodine. This was confirmed by research carried out by Warsaw University of Life Sciences and P. H. Royal LLC from Warsaw as a part of a project financed by the National Centre for Research and Development (NCBiR), from European Union funds. The main object of research were ethanol extracts obtained from amber. In each of the tested samples, iodine compounds were found in succinite. This confirms the fact that the examined part of amber has been in sea water and ion exchange has been taking place between amber and its environment. Maybe other batches of amber do not contain iodine, which may mean that they formed in fresh water. The study of iodine may explain the ways in which particular types of Baltic amber were formed.

Photo 12. Cypress Cupressus sempervirens, Athens Greece. Photo by P. Barczak.

The amber-bearing forest is still a mystery. It is not clear what trees formed it, although the climate in this zone has varied over millions of years from tropical to subtropical to moderate. The pointed trees, fragments of which have been preserved in amber, have become extinct or still exist – e.g., *Coniferous calocedrus* (a cedar from the cypress family), *Taxodium* (a kind of the marsh cypress) and *Quasisequoia couttsiae* from the cypress family (*Cupressaceae*). In addition, pollen of representatives of *Geinitziaceae* plants (*Cupressospermum saxonicum*), which is known only from fossils, was recognized in Baltic amber as well as pollen or pieces of plants existing today, for example pine trees (*Pinaceae*). They include *Abies* (firs), but also *Cathaya* (a pine currently found in China [Sadowski et al. 2017]. This extensive list does not end there. Fossil researchers point to trees currently growing in Australia and New Zealand that may be suspected of producing amber. And these are Agathis (*Araucariaceae*) trees and various species of the araucaria. However, no one has indicated the *eucalyptus* tree so far as a potential creator of Baltic amber. Perhaps because these trees do not produce typical resin. However, they grow intensively in Australia and New Zealand and the chemical compounds found in Baltic amber during chemical research indicate these trees as important in the creation of Baltic amber.

Photo 13. Growth buds in *eucalyptus*. Photo by P. Barczak.

1.1. THE AMBER FOREST

SUMMARY

Amber-bearing forests grew in Central Europe from 33.9 to 56 million years ago in the Eocene era. The age of amber is very approximate, as the date of the formation of the resin can only be assessed on the basis of the surrounding rocks or on the basis of the organic debris contained in this fossilized resin. It has not been established that Baltic amber occurs in only one primary rock, therefore the place of its origin is still in the sphere of scientific research. In the case of Baltic amber (succinite), its age is determined mainly by the sediments of the blue-green soil in which it is most often found, and which was formed in the Eocene. It is assumed that the Late Eocene lasted 22 million years. The Eocene is part of a more complex period, the Paleogene, which lasted from 66 to 23 million years ago. Over such a long period of time, the climate in areas where amber-bearing forests grew was different to nowadays, there were various climatic zones and part of the area was the sea.

At the turn of the Paleocene and the Lower Eocene, i.e., between 56.0 and 47.5 million years ago, the Northern Hemisphere experienced the greatest warming of the climate in the Cenozoic era. Lush tropical and subtropical forests grew. However, about 49 million years ago, there was a gradual cooling, possibly due to the development of the Azolla water fern, which absorbed large amounts of carbon dioxide, thereby cooling the climate. Vast areas were drying up and becoming steppes [Słodkowska, Kramarska, Kasiński 2013].

Fig. 12. The Cretaceous period – the last period of the Mesozoic era, lasting about 80 million years (from ~ 145.0 to 66.0 million years ago). Painted by K. Doszla.

On the basis of the studied samples, 215 species of plants living in the amber forest at that time have been distinguished, and the list of plants occurring there is likely to be significantly expanded. It was found that in this ancient primeval forest there were cypresses, pines, broad-leaved trees such as oak, beech, date palms, chestnut trees, magnolias, and cinnamon trees. The trees probably differed in appearance and resin composition from today's species. Many of these families and types of plants have ceased to exist. Most likely, there was no single tree from which this fossilized resin, today called succinite, was formed. It can also be presumed that it constitutes a compilation of geological processes that have taken place in the long history of the earth. Over the course of millions of years, some of the ancient trees have evolved into modern trees and plants. One of the first suspects of creating Baltic amber is *Agathis australis*, a tree that grows in Australia and New Zealand today. Its area is located thousands of kilometers from the Baltic Sea. However, after adding succinic acid to it, present-day resin obtained from the *Agathis australis* tree resembles succinite in its chemical composition. On the other hand, the experiment involving the removal of succinic acid from succinite has made the obtained resin similar to the resin of the modern *Agathis australis* tree – belonging to the *Araucariaceae* family [Anderson, 1995].

Another group of scientists, who observed the infrared spectra, concluded that the spectrum of succinite best matches the spectrum of the resin from the *Pseudolarix wehri* tree, a species that exists only in fossils. The resin obtained from this fossil showed similarity to Baltic amber. It also contained succinic acid. At present, trees of this type are represented, among others, by the species *Pseudolarix amabilis* (Chinese larch).

New scientific concepts and the search for the biological source of succinite are constantly being developed. Recently, the theory pointing to extinct trees of the *Sciadopityaceae* family has become popular, because needles and pollen have been found, similar to that of *Sciadopitys verticillata*, which grows mainly in Japan. In turn, the stress associated with cooling could cause increased resin secretion by trees such as *Glyptostrobus, Sequoia, Metasequoia* and *Sciadopitys* [Słodkowska, Kramarska, Kasiński 2013].

The fossil studies have proven the existence of about 300 types of amber-bearing trees (and new species are continually being found). Rivers transported resin and tree remnants to the shallow

Photo 14. Pseudolarix *amabilis*. Golden larch. Photo by P. Barczak.

sea that existed in Central Europe. River deltas could also occur in the area of today's Poland, Belarus, Ukraine, and Germany.

The amber forest which gave origin to Baltic amber was not the oldest amber forest in the history of the Earth. The oldest amber samples come from 200 million years ago. The amber deposits in Lebanon date back about 135 million years. Burmese amber dates back 100 million years and is also older than succinite. Chinese amber from Fushun is 53 million years old. Kuji amber from Japan was probably created 90-86 million years ago. However, succinite – amber from the Baltic Sea – is a unique resin, one of the most famous in the world. It was created under specific conditions. Its deposits were transported by glaciers and prehistoric rivers. During this "journey", the chemical compounds contained in amber were constantly polymerized, and each piece obtained specific properties of a biological nature. Baltic amber components reacted with rock components and with compounds contained in bacteria, absorbed essential oils contained in peats and coals. The resin also absorbed elements of the sea water environment [Matuszewska 2010].

Photo 15. Modern insects love *Pseudolarix amabilis*. Spiders known from the Carboniferous period are especially eager to make their webs there.

There are several concepts and hypotheses about the formation of the "amber forest". In 1880, H. Conwentz and O. Völkelin stated that succinite was created 50 million years ago by subtropical forest trees. On the basis of their observations of fossils and the then forests, it seemed to them that in the past redwoods and pines were the tallest, while oaks, maples, beeches, and sycamores grew below. In their opinion, those trees were mixed with laurel, magnolias, and myrtle. Moreover, the shores of a warm sea were covered with palm trees and ferns. According to this concept, the ancient pines belonged to the species P*inus baltica, Pinus cembrifolia, Pinus raruflora* and *Pinus sylvatica* [Conwentz 1890]. Contemporary geologists are inclined to suppose, however, that the chemical composition of Baltic amber results from the fossilization process of ancient trees deposited in peat and brown coal deposits, and of the specific conditions of the formation of Baltic amber forests, rather than just from the fact that trees produce resin. An important concept in this context is the idea of polluting the Eocene Sea with hydrogen sulfide, which is supposed to result in an intense release of resin in the forests growing in this area [Maidanovich 1988].

Generally, since the first scientific publications 200 years ago, no decisive answer has been given to the important issues related to the creation of Baltic amber, its genesis and origin. The known sediments near the Baltic Sea made it possible to evaluate the color, type, and healing properties of amber – but not much more. Only modern measuring devices have created a possibility of obtaining a more complete picture of the issue.

Fig.13. The Triassic – the oldest period of the Mesozoic era. It lasted from 251.9 to 201.3 millions of years ago. Painted by K. Doszla.

The main feature of succinite is its **material stability**. The elements of the tree – cellulose and lignin – transformed over millions of years into peat matter and then into coal deposits. The size of the tree crown, trunk diameter, sun exposure and temperature influenced the amount of released resin. The trees that generated Baltic amber were therefore of enormous size. The trees of the amber forest had increased resin channels, which is typical of ancient natural processes. This is evidenced by the finds of large pieces of fossil resins. Baltic amber resins were formed in a warm and humid climate. It could also be a subtropical climate, as it changed and evolved over millions of years. It was not only the type of substances flowing out of the trees that played a significant role determining the properties of succinite, but the geological conditions in which the substance was formed as well. Otherwise, today we would have amber deposits wherever brown and hard coal is. However, this is not the case. Even if there is amber, its quality is very low, and the chemical composition has no pharmacological or jewelry significance.

Photo. 16. Upper left and right – *Araucaria Juss.* Lower left – the Japanese umbrella pine (*Sciadopitys verticillata*). Photo by P. Barczak.

Baltic amber is the largest deposit of fossil plant resin in the world. Its formation is still controversial. However, it probably contains the largest collection of insects placed in resins. It has not been fully established what tree the amber-bearing resin comes from. Over the course of millions of years, trees have transformed. The continents were shifting around the globe. Some of the trees have traveled thousands of kilometers together with the continents. The trees growing in the Northern Hemisphere occurred in the Southern Hemisphere. Therefore, despite the fact that amber deposits are mostly found in the Baltic Sea, the trees that perhaps formed succinite, nowadays grow in the south of Australia.

Fig.14. The Carboniferous period – from 358.9 million to 298.9 million years ago. Painted by K. Doszla.

You can also try to recreate the amber forest based on fossils. On this basis, it was hypothesized that in the area of today's Central Europe, about 50 million years ago, three types of forests could be distinguished:

1. Broad-leaved or mixed forest, with a predominance of broad-leaved trees and representatives of ginkgo, conifers, and ferns, adapted to a cooler climate (**Mixed Mesophytic**),
2. Evergreen broad-leaved forest (**Notophyllous Evergreen**), with evergreen *Fagaceae, Lauraceae, Altingiaceae, Myrtaceae* and some conifers (*Pinus, Doliostrobus, Cephalotaxus*),
3. Evergreen/quasi paratropical forest, descended from tropical families with little variety of conifers (**Quasi-paratropic**) [Kvaček 2010].

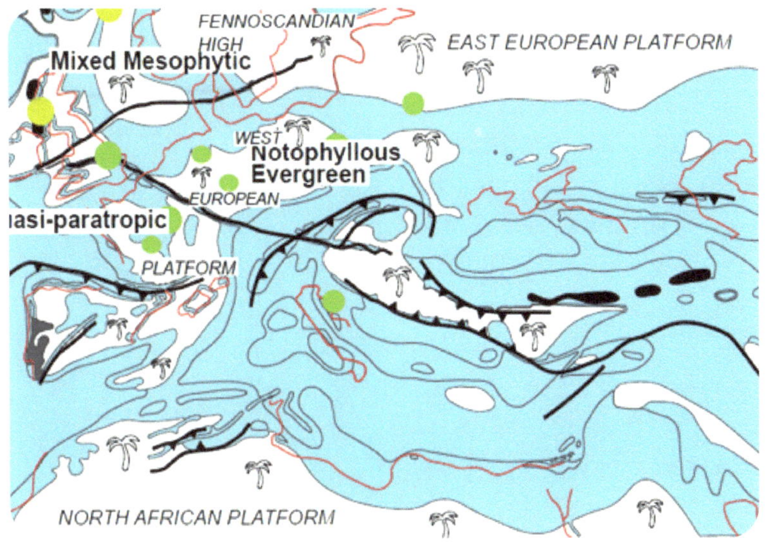

Fig. 15. Layout of land, the sea, and climatic zones in the Eocene. White color – land, blue color – seas, green points – found fossils. Based on: Kvaček, Z. (2010), "Forest flora and vegetation of the European early Palaeogene —a review. Bull Geosci 85:3–16.

Re. 1) The last group of trees – that from the cooler zone – includes trees that currently exist, e.g., gingko, metasequoia, *Mesocyparis*, *Pinaceae* (*Pseudolarix*). The fossils with preserved leaves are the evidence. **However, more than 90% of petrified tree specimens can be attributed to the currently existing species of trees and plants.** Traces of plane trees are found, and the excavated material is dominated by traces of unknown broad-leaved trees with an admixture of conifers.

The fossils found in the north of the planet also included *Glyptostrobus europaeus*, an extinct coniferous species of the Cypress family (*Cupressaceae*). There were also numerous *Pinaceae*, i.e., larches, spruces, pines, and birches. There were specimens of the trees *Larix altoborealis* (larch), Picea sverdrupii (spruce) and Picea nansenii (pine). Today some of those species are known from warm regions. The Grewia, a shrub with medicinal properties, grows in Africa today. The larch (*Pseudolarix wehrii*) has rarely been found in fossils from the north, but it is indicated by some scientists as the source of Baltic amber [Eberle, Greenwood 2012].

Fossils from the British Columbia province of Canada show the biodiversity of the flora of the early and middle Eocene (56 million-33.9 million years ago). At least 62 species of plants and trees have been identified that were related to each other. Several

Photo 17. *Ceratonia* siliqua, Kepatonia H Kepatea, Leguminosae, Greece. Photo by P. Barczak

species of *Ginkgo biloba* trees have been found. The Cypress family (*Cupressaceae*) played an important role.

Photo 18. Japanese cryptomeria, Japanese maple, *Cryptomeria japonica*, *Cupressaceae*. Cypress and its needles. Photo by P. Barczak.

There were also thermophilic Chinese tunics (also known as the Chinese fir (*Cunninghamia lanceolata*)), several types of sequoia including *Metasequoia occidentalis*, an evergreen sequoia, and species of sequoia that now are found only in fossil forms. Another family of conifers known to have fossils in this climatic zone of mixed mesophytic forest is the pine family (*Pinaceae*), including firs (*Abies milleri, Abies*), pines, the thermophilic larch (*Pseudolarix*), hemlocks (*Tsuga*). Yew trees were a large group. The ancient forest also consisted of broad-leaved species such as birch (*Betulaceae*), palm trees (*Arecaeceae*), boxwoods (*Buxaceae*), *Cercidiphyllaceae* (which are today an important family of trees growing in Japan), dogwoods (*Davidiaceae*), sea buckthorns (*Elaeagnaceae*), and beeches (*Fagaceae*), including chestnuts (*Castanea*) and oaks (*Quercus*). In the forest that existed about 50 million years ago, in the cooler zone, witch hazel (*Hamamelidaceae*), representatives of the walnut family (*Juglandaceae*, e.g., walnut), thermophilic laurel trees (*Lauraceae*, e.g., cinnamon), Platanaceae, willow trees (*Salicaceae*), and thermophilic soapstones (*Sapindaceae*, represented today by species such as: clones, lychees, cashews, rambutans) were also found. Most of this forest (North America), however, were conifers and broad-leaved trees were supplements [Dilhoff et al. 2013].

Photo 19. *Lauraceae*, *Laurus* nobilis, bay tree, the Mediterranean. Photo by P. Barczak.

Photo 20. Japanese pine and its young shoots. Photo by P. Barczak.

By studying the infrared spectra, several scientists have independently concluded that the spectrum of succinite emitted in the infrared most closely matches the spectrum of the fossilized tree Pseudolarix wehri, therefore this type of tree should also be considered as a potential amber resin donor. Today the Pseudolarix genus is represented by the species Pseudolarix *amabilis*, whose occurrence is restricted to mountainous areas in Southeast China. It currently tolerates moderate climatic conditions.

Re. 2) In the forests of the southern coast of the Eocene epicontinental sea in the foothills of the Carpathians (Central Europe), there were mainly evergreen broad-leaved forests (*Notophyllous* **Evergreen**). Their make-up has not been determined to this day, although the chemical compounds contained in amber may suggest their possible composition.

Photo 21. *Myrtaceae*, Psidium cattleianum, strawberry guava, Brazil
Photo by P. Barczak.

50 million years ago, the area of today's Czechia belonged to a zone where conifers of the cedar type grew. However, no succinite deposits are found in this area. Amber obtained in the Czech Republic has a different structure and composition than Baltic amber, whereas it contains borneol, cycloalkanes, and cadalene isomers. Interestingly, there are differences in the composition of amber in the Czech Republic, which may also indicate that they were shaped by other plant environments than those in which Baltic amber was created. Contrary to Baltic amber, succinic acid is not found in Czech amber. The trees that gave origin to Czech amber are *Glyptostrobus* (a cypress growing nowadays in Vietnam and Laos, threatened with extinction due to intensive logging) and *Quasisequoia* – a representative of the *Cupressaceae* family (cypresses). Czech ambers contain regular labdane structures [Havelcová et al. 2014].

Photo 22. Leaking petrified resin (duxite) in the middle of the trunk of a fossilized tree. Photo by P. Barczak.

In the warmer part of Europe, in the evergreen forest, *Lauraceae* tropical and subtropical plants dominated, often of a utility character, such as Ceylon cinnamon, camphor cinnamon, bay leaf, avocado. Malvaceae plants were also widely spread in this zone, among which we could find mallow, but also cocoa, hibiscus, baobab, and durian. Conifers were rarely found in this zone. Among the fossils, there is also the Tetraclinis cypress. Thujas (*Thuja articulata*), sandaraku trees, and yew trees, such as the *Japanese Cephalotaxus,* grew there too.

The fossilized resins also indicate trees growing in this zone. In the mine "Jauling" in Austria, there has been identified a hyacinth-red mineral named "Jaulingite". It was produced by trees from the Miocene period (23 to 5.3 million years ago). Resins of this type have been found in many Austrian towns. The resin contains a chemical – phyllocladane – a tricyclic diterpene, which is also found in trees growing today.

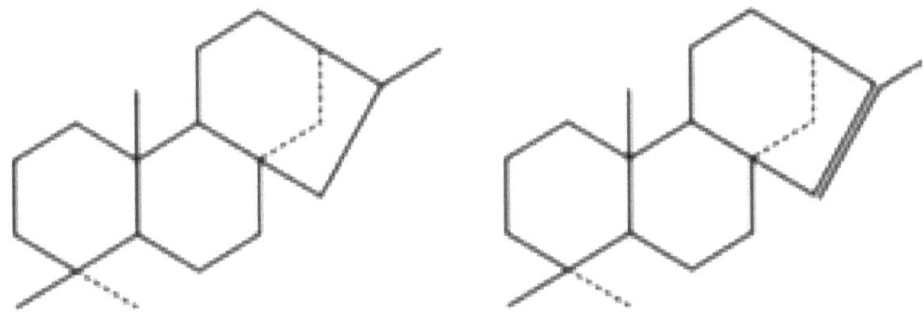

Fig. 16. Two examples of the carbon backbone of phyllocladane: α-phyllocladane (left) and isophyllocladane (right). Source: Vávra, N., "The Chemistry of Amber - Facts, Findings and Opinions", Annalen des Naturhistorischen Museums in Wien 111 A, 2008.

The essential oils of many plants contain this compound. Tricyclic diterpenes are known from coniferous resins. The most numerous among them are representative of the pine family, but compounds of this type are also found in heathers and in laurels (*Thymelaeaceae*). Nowadays, this kind of plant readily grows in Australia, the Mediterranean and tropical Africa. Steppes are covered by them. Another family that contains this compound is Euphorbiaceae. It is a family of trees and shrubs that thrive in tropical regions, although its representatives are found all over the world. Among the representatives of this family there are edible cassava, Brazilian rubber tree and oil plants, as well as those that are the source of Euphorbia Cerifera wax.

Duxite is also a fossilized resin from the Mesozoic period found in Czechia. Many hydrocarbons can be found in this resin: drimane, labdane, simonellite, retene, a C_{16}-bicyclic sesquiterpene and a C_{18}-tricyclic diterpene [Vávra 2008]. Drimane is a sesquiterpene. Sesquiterpenes are a group of hydrocarbons with the formula C15H24 and a molecule half the size of the monoterpenes. Such compounds are contained in essential oils, most commonly known from plants which are pollinated by wind and pollen. Sesquiterpenes have also been found in mushrooms.

Re. 3) In the third zone (Quasi-paratropic), in the area of today's Western and Central Europe, fossils of palm trees and conifers (*Pinus, Cupressaceae*) have been found. The area was abundant with numerous palm trees. There has been found a great variety of subtropical dicotyledons of the Annonaceae family, which are similar to magnolias, now growing mainly in the tropics, and cornelian trees (*Cornaceae*), which are also found in the tropics, but in the moderate climatic zone too. Characteristic of this region were trees of the families *Lauraceae, Rutaceae* (which include the genus Citrus) and *Vitaceae* (a type of the grapevine). The found conifers are redwoods. The *Myrtaceae*, which currently include common myrtle, *eucalyptus*, and clove tree, played the dominant role in the ecosystem [Kvaček 2010].

Photo 23. Cocoa (left), hibiscus (right). Photo by P. Barczak.

The area of the para-tropics currently covers parts of Africa. Lebanese amber is one of the oldest fossilized resins in this area. They came into being earlier than Baltic amber, because 125-135 million years ago. Amber found in Lebanon is very rare, but we find traces of plants in it. The chemical compounds contained in this fossil structure are present, among others, in the *Agathis* tree growing in New Zealand [Poinar, Milki 2001]. Lebanese amber was probably created by trees of the *Araucariaceae* family. Today trees of this type also exist in the world, mainly in Australia. Lebanese amber could also be formed by an extinct plant family *Cheirolepidiaceae*. They are known only from fossils, and unfortunately the fossils make it difficult to identify the chemical compounds that were present in this type of trees. Thus, it is almost impossible to establish how this amber was made, from what exactly trees. Plants known from Lebanon created also Jordanian amber in the Lower Cretaceous period (approx. 145-66 million years ago). Attempts are being made to determine some chemical compound, some mineral that occurs in fossilized trees and is also found in plants today. The trees growing in this area were a mixture of tropical and subtropical plants with floristic elements characteristic for the moderate zone: the *Lauraceae*, the *Myrtaceae*, the *Proteaceae* and the *Moraceae*.

However, Baltic amber in Ukraine was formed in different time and under different conditions. The amber forest from Rovno (Ukraine) seems to represent the flora of drier habitats compared to the Scandinavian or Lebanese amber forest [Perkovsky, Zosimovich, Vlaskin 2010].

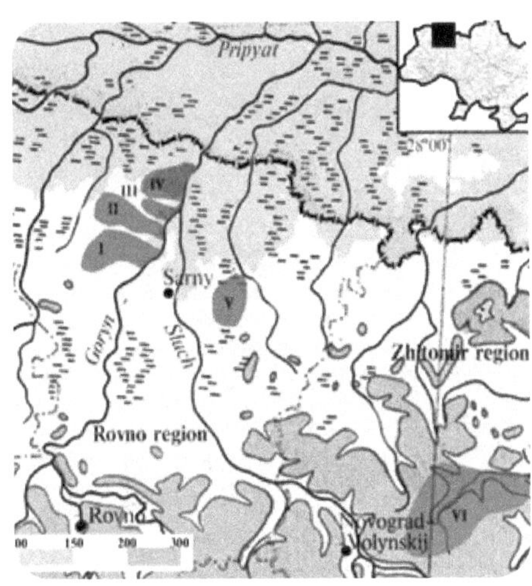

Fig. 17. Amber occurrence area in Ukraine, Rovno region. Source: Perkovsky, E.E., Zosimovich, V.Yu. & Vlaskin, A.P., 2003a. "A Rovno Amber Fauna: A Preliminary Report". Acta Zoologica Cracoviensia 46 (suppl. – Fossil Insects): 423–430.

In a part of modern France, which in the Eocene was also in the para-tropical zone, there were trees of the genus Hymenaea, which are now known from Africa. Today, copal resin is obtained from them. It should be noted, however, that particles and pollen of more than 750 species of plants have been found in Baltic amber, so describing all species is beyond the scope of this publication [Selden, Nudds 2004]. However, it is possible to outline the structure and character of the forest and the prevailing environmental conditions.

Photo 24. White Baltic amber (left) and a translucent type of amber - gedanite (right). Photo by P. Barczak

The amber forest is still an unsolved mystery. In rocks from 50 million years ago, we can find fossilized plants that show the nature of this mixed forest. The temperature in winter was around 20 degrees Celsius and there was a lot of rainfall. In the Eocene, the temperature on earth increased by approximately 6-8 degrees C in relation to the previous climatic periods. It was due to the natural emission of carbon dioxide into the atmosphere. The coasts of the Baltic Sea and the masses of sand with remnants of trees sometimes lying there inform us about this greenhouse catastrophe lasting millions of years, which destroyed and transformed the amber forest. In 1816, J. F. John wrote about the shores of the Baltic Sea in the following way: under these layers of sand there are layers of clay, which is tinted blue, and in it there are visible parts of trees saturated with amber. Sometimes whole trees of varying length and thickness are found that are larger than most known trees on earth. The wood is brittle and soft, mixed with blue earth, has a bituminous smell, sour taste, which is caused by the presence of sulfuric acid. Pyrite is also formed within. The wood is saturated with various salts, sometimes acorns of ancient trees are found. Palms were also found near amber deposits. Mithridates reported that the resin was flowing out of the cedars and that it formed the Baltic amber. Areca palm nuts were also found, and it was completely astonishing then that in the areas which were freezing cold then there had been tropics [John 1813].

Photo 25. The shrub is suspected of creating Baltic amber. Callistemon citrinus, *Myrtaceae*, Australia. Photo by P. Barczak.

On the Baltic coast there are stones like these in the pictures below. They look like giant palm nuts. In the past, the trees looked completely different, at least they were much larger.

Photo 26. A stone resembling a giant walnut from millions of years ago found on the Polish coast – a figment of imagination or a proof? Photo by P. Barczak.

However, did tree species and their resin secretion play a key role in the creation of Baltic amber? Well, it seems that it was not the most important element in the creation of this unique resin. After all, during the Eocene, in the territory of today's Austria and Czechia there were also forests which were resinous. Yet, there are no succinite deposits there. There are numerous amber stones, small, brittle fragments in rocks, which are of purely scientific and hobbyist value. It is possible, of course, that in the area where amber was formed, there were unique trees which were abundantly resinous. In our opinion, however, it was the type of geological processes that determined the formation of succinite. Essential oils – which in closed niches flooded with water, in conditions of a quite high temperature and pressure transformed residual organic matter into a new chemical structure, which we are happy to admire today, and which, for hundreds of years, has been serving man as a precious medicine – were of great importance in shaping the final chemical composition of succinite.

Photo 27. Today, the thermophilic trees that produced Baltic amber are often found in Australia. The silk oak *Grevillea robusta*, *Protaceae*, Australia. Photo by P. Barczak.

Oils are released at an increased temperature in the vicinity of water vapor. Such conditions may have prevailed in niches with woody matter covered with sediments of the Eocene Sea. It is also known that essential oils "resinify" after some time. It is a slow process of their transition from liquid to solid [Kohlmünzer 2007]. In the solid state, oils have gained strength over millions of years. Hence their diverse composition in the succinite. Essential oils changed the primal matter and created – together with tree resin – a chemical structure which is Baltic amber succinite.

Photo 28. Glessite from Indonesia, Miocene epoch, Sumatra. Photo by P. Barczak.

There are many trees suspected of producing Baltic amber millions of years ago. One of the first is *Agathis australis*. According to scientific concepts, trees of this species gave off enormous amounts of resins, which, like amber, have survived in the ground and are exploited to this day. Tons of fossilized resins have been excavated in the recent past and have been a significant source of export for Australian and New Zealand companies. Today, a similar situation exists in Borneo and Malaysia. Large amounts of fossil resins are exported.

Nowadays in those vast regions, three tree families grow and are resinous. These are *Araucariaceae* (araucarians), *Cupressaceae* (the cypress family) and *Pinaceae* (the pine family). However, it is worth noting that the dominant monoterpenes in pines – α-pinene and β-pinene – are not the dominant compounds in Baltic amber. Rather, it is dominated by terpenes known from broad-leaved plants. One of such plants, previously not included in scientific descriptions, is the *eucalyptus*.

Photo 29. The araucaria, New Caledonia (upper left) and the Monterey cypress (*Cupressus macrocarpa*), California USA (upper right). The Paraná pine (*Araucaria angustifolia*) (lower left). The Norfolk Island pine (*Araucaria heterophylla*), the Norfolk Island (lower right). Photo by P. Barczak.

1.1.1. ARAUCARIACEAE

Agathis australis could have contributed to coming into being of amber bearing resins. When scientists added to the *Agathis* resin a compound characteristic of Baltic amber, succinic acid, the infrared spectrum was very similar to that of succinite. Removal of succinic acid from succinite resulted in the fact that the obtained substance was similar to *Agathis australis* – a tree of the *araucaria* type.

Agathis (Araucariaceae) – is an important genus, because the existing *Agathis australis* secretes abundant diterpenoid resins that polymerize, which causes massive accumulation of raw material in the ground. As many as 13 species of *Agathis* are limited to the Southern Hemisphere of the planet. *Agathis* damara is valued for its leaking resin and wood. It is found in the southern part of the Indochina Peninsula, Philippines, northeast Australia, New Caledonia, and New Zealand. Fossils of this type of trees have been found in South America and China. Chemical research also confirmed that plants producing Baltic amber were related to ancient *Agathis* trees. *Agathis* resin consists mainly of diterpenes. It can be pointed out that in Australia and New Zealand, the local amber was formed mainly from *Agathis* trees in the times from the Cretaceous geological era to the Tertiary. Also, in the Northern Hemisphere, including Jordan, Lebanon, and Israel, *araucarias* created amber.

Agathis australis still produces enormous amounts of resin, which makes it different from the *araucaria*, although they are related plants. 450,000 tons of resin were exported from Australia, which may indicate the type of trees that made up Baltic amber. However, the deposits in New Zealand were formed around 100,000 years ago, as a result of climate change and glaciations on the planet [D'Costa, Boswijk, Ogden 2009].

Photo 30. Modern kauri resin from New Zealand. Photo by P. Barczak.

The compounds communol and communic acid, found in the resin of this tree polymerized easily. The similarity of *Agathis* resin to succinite is due to the fact that amber contains agathic acid, which is characteristic of *Agathis*. Polymeric succinite fractions contain 75-85% of labdanes, 15-21% of pimaranes and only 1.5-4% of abietanes. *Agathis australis* also contains such proportions. The infrared spectrum of succinite is almost identical to that of *Agathis australis* resin. These trees probably disappeared as a result of the cooling of the climate and today they are not found in the Northern Hemisphere. On the other hand, the bisabolene type of compounds (bisabolenes and dihydro-ar-curcumene) are widespread in *araucaria* trees which are found in New Zealand and Australia.

Photo 31. *Agathis australis. Araucariaceae*, a look at leaves. The botanical garden of the Polish Academy of Sciences, Powsin. Photo by P. Barczak.

Of course, there are also some differences because, as mentioned, succinite was formed in the environment of the steaming essential oils of various trees. For example, *Agathis* atropurpurea resin is dominated by the compound called limonene (89.8%–97.4%) with much smaller amounts of other monoterpenes. There are few diterpenes in the resin itself. Only one β-bisabolene sesquiterpene was found. Interestingly, as much as 36.4% of diterpenoids and 22.5% of sesquiterpenoids were found in the leaves, and a very small amount of limonene (1.5%). However, none of the important compounds in Baltic amber – i.e., m-cymene, para-cymene, bornyl – have been found [Garrison, Irvine, Setzer 2016]. In turn, in another species, *Agathis* atropurpurea, there are compounds known from amber in leaves of the tree: **1) α-pinene, 2) α-fenchene, 3) camphene, 4) β-pinene, and 5) limonene.** Camphene, edehydro-1,8-cineole have been found in the Australian *Agathis* resin. In some varieties there are only few compounds – e.g., in *Agathis australis*, only α-pinene and camphene. These types of trees could have had some influence on the formation of succinite, but most of the compounds found in them do not occur in Baltic amber [Frezza et al. 2020].

Today, *Agathis* produces kauri resin, kauri gum, kauri copal. Monoterpenes (C_{10}), sesquiterpenes (C_{15}) and diterpenes (C_{20}) have been found in the tree resins. Compounds such as pinene, thujene, and limonene are widespread in most species of *Araucariaceae*. unfortunately, volatile oils evaporate quickly, hence existing problems with their full identification. The chemicals found in the *Araucariaceae* are common in conifers. Bicyclic diterpenoids, composed mainly of labdane compounds, are common in all conifers. The tricyclic diterpenoids reported in *Araucariaceae* consist mainly of types of abietane and pimarane [Lu, Hautevelle, Michels 2013].

Photo 32. *Agathis australis, Araucariaceae,* a look at leaves. The botanical garden of the Polish Academy of Sciences, Powsin. Photo by P. Barczak.

Araucaria (the *Araucariaceae* tree family) has been known in New Zealand for millions of years since the late Cretaceous times (approx.145.0 to 66.0 million years ago). *Agathis*, Araucaria and *Wollemia* are trees of this family which have survived to this day. Although, according to current knowledge, araucarias known from Australia did not occur in the area of Baltic amber, but the resins and the method of their creation were similar. They may have been produced by extinct trees from the gymnosperm group that shared the features of the *Araucariaceae* and the

Pinaceae [Larson 1978]. Currently *Araucariaceae* are a component of the forests of the Southern Hemisphere. The trees are found in the tropics and subtropics. About 40 types of these trees have been found, including *Agathis* represented by 21 species, Araucaria represented by 19 species and *Wollemia* represented by one species (by *Wollemia nobilis*).

Photo 33. Australian Wollemi pine (*Wollemia nobilis*). The botanical garden of the Polish Academy of Sciences, Powsin.

All three genera are descendants from one ancestor. The first manifestations of these genera, in the form of fossil pollen found in research, come from the Middle Triassic (251 to 201.3 million years ago). In the geological era of the Jurassic, the Araucaria genus diversified and developed. In the Cretaceous era (145-66 million years ago), trees of the genera Wollemia and *Agathis* appeared at the end of the Cretaceous period, Araucaria began to leave the Northern Hemisphere and move to the Southern Hemisphere. The trees were very resinous then.

Araucariaceae are now found mainly in the middle and southern latitudes of Borneo, the Philippines, Chile, Argentina, southern Brazil, New Caledonia, New Zealand, the Norfolk Islands, Australia, and New Guinea. The insects have been always eager to attack trees, hence perhaps a large number of them in amber.

Generally, more than 50% of drugs come from plants, so it is no wonder that the araucaria that formed the amber forests also has pharmacological properties. Its resin contains significant amounts of terpenes and phenols, which protect the tree from insects. The Mesozoic forests that produced the resin consisted of *Araucariaceae* and the extinct *Cheirolepidiaceae* [Beimforde et al. 2017].

Photo 34. *Araucaria angustifolia*. The botanical garden of the Polish Academy of Sciences, Powsin. Photo by P. Barczak.

Araucaria bidwillii is known for antipyretic and anti-ulcer properties of its oleoresin. Its gastroprotective and wound-healing effects have also been found in the case of *Araucaria araucana* oleoresin. Of course, it also shows antifungal and antimicrobial properties because by creating resin, the tree protects itself against external threats [Devi et al. 2017].

Photo 35. *Araucaria columnaris*, New Caledonia (left side). *Araucaria angustifolia*, Brazil (right side). The botanical garden of the Polish Academy of Sciences, Powsin. Photo by. P. Barczak.

Individual species of *araucaria* have different pharmacological properties. *Araucaria* resins are cytotoxic and gastroprotective [Lee, Chenh 2001]. The terpenes and terpene derivatives of *araucaria* are gastroprotective in stomach diseases [Abdel-Sattara et al. 2009]. Brazilian *Araucaria angustifolia* contains mainly lignans but also biflavonoids. The extracts made of this resin have an antibacterial effect. Bark infusions are used in diseases of varicose veins [Koul et al. 2015], suggesting that *araucaria* terpenes found in amber may have a similar effect.g The following chemical compounds are found in *araucaria*:

Photo 36. Young shoots of araucaria (above), resin of araucaria (below). Photo P.Barczak.

Photo 37. Araucaria heterophylla. Cyprus. Photo P.Barczak.

Farnesene ($C_{15}H_{32}$) and bisabolene compounds. Farnesol (trans- and cis-) found in fresh plants is present in all *Araucaria* species. Bisabolene compounds, such as saturated bisabolene ($C_{15}H_{30}$), ar-curcumene ($C_{15}H_{22}$) and dihydro-ar-curcumene ($C_{15}H_{24}$) isomers, are observed in the aliphatic and aromatic fractions of *araucaria* trees, respectively. We know the compounds that occur in Baltic amber, and we can compare them with the *araucaria*. Ar-turmeric is often found in trace amounts for all species. There is also dihydro-ar-curcumen, which is found in amber too. Cadinol and cadinenes are ubiquitous in vascular plants. There occurs dihydro-cadinene, which is also in amber. Cadalane compounds represent one of the major sesquiterpenoids in *araucaria*. Abietane-type compounds, known also from succinite, are the major diterpenoids in *araucaria*. The *Araucariaceae* family is characterized by a remarkable predominance of saturated tetracyclic diterpenoids, including beyeran, filocladane, and kauran. This family is also characterized by a relatively high abundance of sesquiterpenoids (such as cadalenes, bisabolene, eudesmane and chamazulene), and diterpenoids (such as abietanes) [Lu, Hautevelle, Michels 2013].

Photo 38. *Araucaria heterophylla*, Norfolk Island pine, Norfolk Islands.
Photo by P. Barczak.

1.1.2. CUPRESSACEAE

The species from this family which is very resinous and suspected of producing amber is the Japanese umbrella pine (*Sciadopitys verticillata*). According to one of the scientific concepts, it was the main supplier of amber resins, although nowadays this tree does not excrete much resin [Seyfullah et al. 2018].

Photo 39. Japanese umbrella pine, *Sciadopitys verticillata*. A living fossil currently grows in the wild in a few locations, incl. in Japan.
Photo by P. Barczak.

The whole family *Sciadopityaceae* is connected with amber through chemical compounds observed in contemporary resin and Baltic amber. The only species from this family that currently exists on the planet is *Sciadopitys verticillata* – the Japanese umbrella pine that grows in a humid and warm climate, that is, in the area of the former Eocene Sea. In the Mesozoic, this species was widespread on the planet. Currently, it is known in its natural state from the Japanese islands of Honshu, Kyushu, and Shikoku [Philips, Rix 2002]. It grows in mountain, humid forests, on fertile and acidic soils. In fossils, it often appears in the area of Czechia. In the distant past, when the climate was warm on Earth, the *Sciadopityaceae* family conquered the northern regions, including Greenland, Spitzbergen, and the Baffin Islands [Bose, Manum 1990].

Photo 40. Cypresses in the snow. Athens, Greece, 03.2022. Photo by P. Barczak.

The cypress family (*Cupressaceae*) – including *Cupressus, Juniperus, Callitris* and *Tetraclinus* – currently includes over 130 species. They probably had a significant impact on the formation of Baltic amber. The names of trees of this family always appear when writing about the places where amber is formed; what is more, the numerous terpenes contained in Baltic amber indicate that it comes from *Cupressaceae*. Today, the cypress family is an important source of wood and resins. Trees and shrubs of this type grow all over the world, from arctic Norway to the central

Sahara. Only a few species can tolerate drought. It is a very viable family and belongs to the group of the longest-lived trees. Among the best-known representatives of this family, there are thuja (*Thuja*), juniper (*Juniperus*), cypress (*Cupressus*), and sequoia (*Sequoia*). Today, *Cupressaceae* do not excrete as much resin as in the Cretaceous period and they have been transformed since that time. It is known, however, that changes in the temperature of the external environment cause an increased release of resin, and temperature, as we know, was subject to changes over millions of years. This could be one of the reasons for the increased resin secretion by *Cupressaceae* in the Eocene.

Photo 41. Lawson cypress, Chamaecyparis lawsoniana. Photo by P. Barczak.

Cupressaceae contain a number of terpenes and terpenoids – sesquiterpenes and diterpenes. This substantiates the hypothesis that their resins influenced the formation of Baltic amber deposits. One species – *Cupressus sempervirens* – consists mainly of communic acid and related diterpenoids. Interestingly, the leaf resins of all species consisted mainly of a dozen or so of sesquiterpene hydrocarbons [Lanenheim 1995]. Communic acid is an important compound found in Baltic amber.

Photo 42. Nootka cypress, Chamaecyparis nootkatensis (left). The temple juniper, Juniperus rigida, *Cupressaceae*, Japan, China, Korea (right). Photo by P. Barczak.

Other known substances from Baltic amber are also present in the resin of *Cupressaceae*. Particularly numerous are diterpenoids in rosin, which is a component of the resins of conifers of the *Cupressaceae* family. Each of the trees and shrubs from this family has its own unique compounds. They are also present in Baltic amber, although it is difficult to clearly confirm that they work identically in amber extract or powder. It is worth mentioning, however, which terpenes are in *Cupressacea* and what activity they show. The Azores juniper (*Juniperus brevifolia*) shows properties of fighting specific kinds of tumor. Ferruginol – a diterpene contained in the cypress – exhibits antimicrobial, acaricidal and cardioactive properties [Ulubelen et al. 2002]. It has also antioxidant qualities [Ono et al. 1999]. The compound, sugiol, isolated from the Persian juniper (*Juniperus polycarpos*), manifests antimalarial activity. The compounds isolated from Sequoia sempervirens cones have a positive effect on inhibiting the development of colon cancer, lung cancer and breast cancer [Son et al. 2005]. The abietane compounds isolated from the tree Juniperus formosana – a thermophilic juniper growing in China

and Taiwan – show strong properties of fighting the Sars-Cov virus [Wen et al. 2007]. One of the tree species with interesting pharmacological properties is *Calocedrus decurrens*, also known as the incense cedar. It is characterized by low frost resistance. It needs full sunshine and is quite drought resistant. Two rare abietane diterpenoids found in *Calocedrus* exhibit significant cytotoxic activity [Hsieh et al. 2011]. Hinokitiol, a natural monoterpenoid isolated from *Calocedrus* formosana heartwood essential oil, has anti-cancer properties. Hinokitiol induces cell cycle arrest and influences the aging process. Hinokitiol may be a therapeutic candidate for the treatment of lung cancer [Liu et al. 2020].

Photo 43. An evergreen cypress. Botanical garden of the Polish Academy of Sciences, Powsin. Photo by P. Barczak.

The quasisequoia, another representative of the cypress family, is the tree which has remained only in the form of petrified sediments. Sequoias are evergreen trees. The trees reach enormous sizes and live up to 2,000 years. The wood is extremely rot-resistant. They prefer the Mediterranean climate, where winters are moist, and summers are dry. Fog coming from the neighboring seas and oceans play an important role. The quasisequoia has not been found to occur in the present-day natural world. However, *metasquoias* are known. Terpenes known from these trees have also been found in Baltic amber. Sugiol has been isolated from *Metasequoia glyptostroboides* found in Asia and North America. Interestingly, this compound – coming from amber – has been also found in balsamic substances used during the Roman Empire [Devièse et al. 2017]. Terpenes derived from M. glyptostroboides have properties known from Baltic amber, including antibacterial, antioxidant, anti-dermatophytic, and anti-fungal qualities. Sugiol is a diterpene, often found in conifers, but usually not in pines. It has energetic properties, affecting microparticles, also in mammalian cells [Bajpai et al. 2014]. Abietane diterpenes are used to prevent and treat type 2 diabetes. Sugiol – a diterpenoid – isolated from *Metasequoia glyptostroboides*, has also been evaluated regarding its anti-diabetic and anti-melanogenesis potential. As a result, sugiol in the concentration ranges 100-10,000 µg/ml and 20-500 µg/ml showed strong efficacy in inhibiting the α-glucosidase enzymes and tyrosinase in vitro [Ibid.].

Photo 44. Young amber from the Gdańsk region. P. Barczak.

Metasequoia glyptostroboides is a tree from the cypress family known as a living fossil. It is found mostly in Korea, China, and Japan, but also in North America and Europe. Needles of these trees are traditionally used in Chinese medicine for their antibacterial, antioxidant and antifungal properties. They are also use in inflammation, for pain relief and in the case of dermatological ailments. Active compounds were found in the needle extract. It inhibited the action of free radicals that damage neurons [Lee et al. 2016]. A similar effect has been found with succinite.

Taiwania cryptomerioides is the tree which is most abundant in Taiwan. It belongs to the evergreen trees and lives up to 2,000 years. It belongs to the cypress family. Taiwania does not like humid summers. Honokiol is a diterpenoid compound isolated from this tree. It has antioxidant, anti-cancer, and anti-inflammatory effects and causes the regrowth of neurites. It protects cellular mitochondria against oxidative stress. In mammals, honokiol has proven effective in controlling brain disorders [Chen et al. 2007].

Taxodium is a genus in the cypress family which grows in water, currently found in the south of North America. Trees of this type were one of the most important genera growing in the amber forest. In addition, there were such conifers as *Calocedrus* and *Quasisequoia couttsiae* from the Cypress family (*Cupressaceae*), *Cupressospermum saxonicum* (from the *Geinitziaceae* family), several genera and species of *Pinaceae* (*Abies, Cathaya, Nothotsuga protogaea, Pseudolarix* and several species of pines) [Schmidt et al. 2019]. *Taxodium* trees are especially valued for their wood, which is resistant to fungi and rotting processes. It has high resistance to biotic stress and some of its specimens can survive for thousands of years. The heartwood contains sesquiterpenes.

Fig. 18. Terpenoids suspected of forming Baltic amber, known from modern plants.

Pliocene extracts from Taxodioxylon gypsaceum wood contain sesqui- and diterpenoids that are found in the related species of the *Cupressaceae* today [Staccioli, Mellerio, Alberti 1993]. In fossilized resins of the *Taxodium* balticum tree there have been found also such terpenoids as ferruginol, sugiol, hinokiol and taran. Currently, they are components of resins and waxes found in conifers (*Cupressaceae*, *Podocarpaceae*, *Araucariaceae*) [Otto, Simoneit 2001]. The tree is extremely resistant to pest attacks. It reaches the height of 100-150 feet (30-46 meters). Taxodistines isolated from the fruit of this tree showed activity against tumor cells in mice. These compounds – diterpenes – have also proved effective against wood-decaying fungi [González 2015].

Photo 45. *Wollemia pine, Wollemia nobilis.* Resin leaks. Australia. Botanic garden of the Polish Academy of Sciences, Powsin, Poland. Photo by P. Barczak.

Glyptostrobus is another genus of the cypress family suspected of creating amber. Currently, it grows in its natural state in the area of Vietnam, formerly the tree was found in southern China. It is also referred to as the Chinese swamp cypress. Found in wetlands, it is most common along rivers and pond shores, in humid and subtropical regions of Asia. By growing in water, it produces knee-shaped roots that rise above the water, which allows oxygen to be delivered to the root system. At some point in the Eocene, this tree became dominant in certain areas. The trees growing in the tropics were replaced by subtropical vegetation, thermophilic conifers such as *Glyptostrobus, Sequoia* and *Metasequoia*, and extinct species of the Normapolles group. Chemical compounds from trees of this type have also been found in Chinese amber from Fushun coming from the Eocene period [Słodkowska, Kramarska, Kasiński 2013]. Today, *Glyptostrobus pensilis* can be found in Vietnam and Laos.

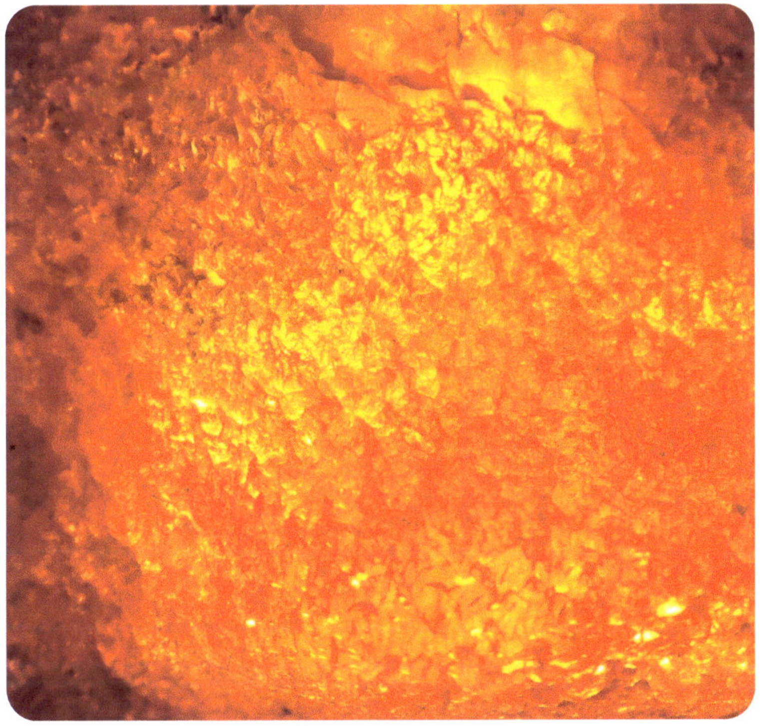

1.1.3. MYRTACEAE

A large number of members of the *Myrtaceae* family have been found in Europe and around the world. The fossilized geological outcrops suggest that this plant species has existed since 86 million years ago. The spread of this type of plant took place in the Eocene, where the forests initially resembled varied rainforests. *Myrtaceae* probably developed on the ancient continent of Gondwana, and plants of this species began to expand as early as in the Cretaceous period.

The Youganwo fossil excavations in the Guangdong province in southern China turned out to be interesting. Calycolpus, which has been found in the fossils, may be a sibling of the Mediterranean species *Myrtus*, which is an important plant of the *Myrtaceae* family. *Myrtus* differs from Calycolpus only regarding a few features of wood. The *Myrteae* tribe of the *Myrtaceae* family is known today from the guava, the common myrtle, and the allspice. It also grew in East Asia in the past. This is further proof that *Myrtaceae* spread over a large area and probably also in the area of potential places of Baltic amber origin.

Myrtales are a wider group, consisting of 140 genera, including over 5,800 species. They are found on virtually all continents, three species in Africa, and the species *Myrtus* in the Mediterranean. In the case of the preserved fossil and modern parts of these plants, the differences in the size of the leaves are insignificant. In the Eocene period, greenhouse conditions prevailed in the places

where Baltic amber occurs, so myrtles had good conditions for development here. However, the climate has changed and cooled over time. The vegetation adapted to the climatic changes. Today we find fossils of the *Myrtaceae* species also in present-day Germany, while the myrtle today has a different range of occurrence and it occurs, among others in countries of southern Europe. Fossilized leaves of Rhodomyrtophyllum sinuatum (*Myrtaceae*) are repeatedly found in the Middle Messel fossil formation near Darmstadt (Hesse, Germany) [Grein, Roth-Nebelsick, Wilde 2010]. Hence, the myrtle is also a pretender to indicate it as a species that contributed to formation of Baltic amber.

Plants of the *Myrtaceae* family have grown around the world since the Cretaceous times. As the Gondwana supercontinent split into smaller continents, families, and species of plants around the world divided. Some plants from the Eocene have consolidated their presence in, among others, Australia. That is why you can find there the floristic world of the Eocene, coming from the times of creation of Baltic amber. Most of Australian plants are endemic. The most famous species come from three families: *Myrtaceae*, *Proteaceae* and *Fabaceae*. This group includes trees that produce rubber: *Eucalyptus*, *Corymbia* and *Angophora*, *Melaleuca* and *Callistemon* trees growing on swamps and river banks, Darwinia and Leptospermum shrubs and, finally, the waxflower Chamelaucium uncinatum, which belongs to the myrtle family (*Myrtaceae*). Among the main Eocene families known from Australia, besides *Myrtaceae*, there should be mentioned *Proteaceae* (including such genera like *Banksia*, *Grevillea*, *Hakea* and *Telopea*), and *Fabaceae* (including Cassia and Acacia −e.g. golden mimosa).

Photo 46. Common myrtle, *Myrtus communis*, *Myrtaceae*. Photo by P. Barczak.

Melaleuca belongs to the myrtle family (*Myrtaceae*). This species could have contributed to the creation of Baltic amber. Melaleuca readily grows in swamps and along streams. The tree surprises with the diversity of secreted oils. Sometimes the content of methyl compounds (trans-methyl cinnamate/a-cubebene) can be up to 17%. The compound in the oil can be converted into menthol through alcohol extraction processes [Plößer, Lucas, Claus 2014].

Fig. 19. Trans-methyl cinnamate.

Menthol has been detected in amber extract tested by a team from Warsaw University of Life Sciences and PH Royal LLC during conducted research. Menthol content can suggest that Baltic amber was produced by trees which contain methyl. For millions of years, numerous chemical processes have been taking place in the active structure of Baltic amber, which is why menthol may indicate the plants that made up Baltic amber. Melaleuca also contains many other terpenes known from amber. So, there are α-pinene (even up to 21.5%), β-pinene (5.8%), para-cymene, 1,8-cineol, terpine-4-ol, γ-cadinene, and trans-kalamene.

The Melaleuca tree, also known as the tea-tree, is a species which quickly revives after fires. These are plants shaped in such a way that they develop when the competition has "burnt down". At the end of the Eocene era, the amber-bearing areas were steppe-like, many fires probably broke out during these transformations, and trees with features that allowed them to develop after fires had the best chance for development. This species is predisposed to grow in dry and barren areas. It currently grows in eastern Australia and New Zealand. It acclimates very well to lowland and alpine areas. In various regions of Australia and New Zealand, it can produce different chemical compositions. The tree is valued for its essential oils. An important feature of these trees is that after a fire, seeds are released from the plant and the tree can survive in extreme conditions.

Photo 47. *Akka Sellowa, Acca sellowiana, Myrtacea*. It occurs in South America. Not known from amber yet. Photo by P. Barczak.

Plants of this type could have had great development opportunities in the areas which potentially formed Baltic amber, i.e., in the areas at the foot of the Carpathians. With the end of the Eocene era, the climate changed to less friendly to tropical trees – it cooled down. The steppe formation was taking place, probably numerous fires broke out, and trees with features that allowed them to develop after fires had the best chance for development. Hence, many chemical compounds known from analytical research on Baltic amber are present in the resin oils of the above-mentioned species. Eocene *Myrtaceae* are found in Oligocene coal mines in Tasmania. Some of the forest complexes there were created by *Agathis* and *Melaleuca* trees [Rowell, Jordan, Barnes 2001]. They are also a very frequent element of fossils and pollen in amber from the area of Ukraine. Among the plants that grew at that time in the Rovno region in Ukraine, this includes tree groups such as palms, *Lauraceae, Myrtaceae, Proteaceae*, and *Moraceae*. Such tree families are associated with the southern and eastern parts of Asia, Australia, and South Africa [Sokoloff et al. 2018].

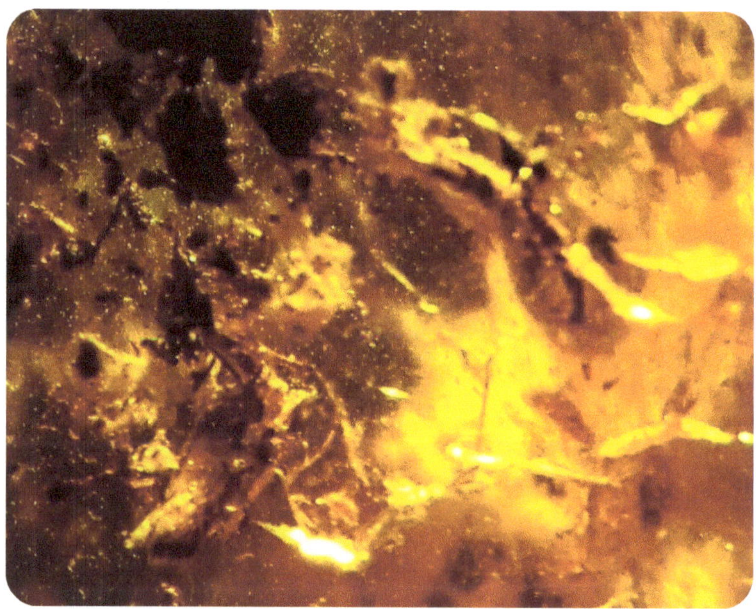

1.1.4. LAURACEAE

Another family suspected of producing succinite are the *Lauraceae* trees, so those which today are known for laurel, bay leaf, tasty avocado fruit, and even for camphor secreted by the camphor tree (Lat. C. camphora). Fossils from the region of Carpathian foothills coming from the Cretaceous period (approx. 145-66 million years ago), show that broad-leaved trees typical of the present tropical regions, including the species of *Lauraceae* trees, grew here. For example, Cryptocarya is currently a genus of trees growing in tropical-rainy regions, some species are frost resistant.

It is also not surprising that in the fossil pits of this region, traces of the original Cinnamomoides plants have been found from which plants such as magnolias, pepper species (known for the spice pepper), and, finally, the abovementioned *Lauraceae* family have originated.

In the area of the hypothetical formation of Baltic amber in the region of Krasnobród (southern Poland), fossilized leaves similar to modern *eucalyptus* (*Eucalyptolaurus depreii*) have been found. The entire region of the Carpathian foothills had a similar flora, so it was an island of plants on which a specific natural ecosystem was created [Halamski 2013]. (The Cretaceous period, about 145-66 million years ago).

Fig. 20. Halamski, A. T., "Latest Cretaceous Leaf Floras from Southern Poland and Western Ukraine", Acta Palaeontologica Polonica 58, 2013.

Among the plants that existed there, as evidenced by the fossils, were trees of the *Lauraceae* family, including probably *Cinnamomum camphora* – the camphor tree – i.e., the tree that produces camphor oil.

Photo 48. Cinnamomum cassia, *Lauraceae*, southern China.

Currently, this tree grows in southern China, Taiwan, southern Japan, Korea, and Vietnam. Its oils contain some chemicals known from Baltic amber: **cineol, α-pinene, ß-pinene, ß-myrcene, p-cymene, borneol, and camphene** [Chelliah 2008]. Some compounds, however, have not been detected, for example, eugenol, linalool, limonene, safrole, α-humulene, and nerolidol.

Another study concerned the composition of the fruit and bark of the cinnamon tree. It was found that the bark of the tree contains: D-camphor (51.3%), 1,8-cineole (4.3%), α-terpineol (3.8%), 3-methyl-2-butenoic acid, and oct-3-en-2-yl ester (3.1%). The leaves, in turn, contain: D-camphor (28.1%), linalool (22.9%), and 1,8-cineole (5.3%). Some compounds are unique in the bark of the tree, such as γ-terpinene, isoterpinolene, 1,3,8-p-mentatriene, terpinene-4-ol (known from succinite), α-terpineol, eugenol, β-cadinene, and α-cubebene. Some of these compounds have already been found in succinite, others may be found in it over time. There are also compounds such as p-menth-1-en-4-ol and p-menth-1-en-8-ol, which in reactions lasting for millions of years in a polymer such as Baltic amber – as well as through ethanol extraction – can turn into DL-menthol, and this compound was detected in the chemical analysis of Baltic amber samples.

Camphor tree resin has many pharmacological uses. One of the compounds found in it is borneol. It has a neuroprotective effect in ischemic stroke. It improves blood flow in the brain, blocks Ca+2 overload in neurons. It prevents excessive neuronal mortality. D-borneol is extracted from leaves and branches of *Cinnamomum camphora*. It has been known in China for 1,600 years. Borneol relaxes blood vessels, lowers blood pressure and the resistance of cerebral vessels, inhibits the development of a serious thrombosis disease. It also improves blood flow in the brain after cerebral ischemia [Yong et al. 2021].

1.1.5. GINKGOACEAE, FAGACEAE AND OTHER BROAD-LEAVED TREES

Palaeobotanists are well acquainted with fossils of the *Ginkgoaceae* family, which have been around the world for 300 million years. About 60 million years ago, all but two species became extinct. Today the tree species *Ginkgo biloba* is common. It is one of few species that survived the Ice Age. It is a natural relict and occurs usually in China. The plant has developed chemical compounds that protect it against pathogens and herbivores, although our own observations indicate that Ginkgo bark is a favorite delicacy for deer and goats living in the forests of Europe. *Ginkgo biloba*, the maidenhair tree, reaches 20-35 meters in height. The trees are deeply rooted and resistant to snow and frost, although they thrive best in warm regions of China. They are long-lived, with some lasting 4,000 years.

Photo 49. Young *Ginkgo biloba*. Runów near Warsaw, Poland.

The name Ginkgo comes from the Chinese word sankyo or yin-kuo, which means a hill apricot or a silver fruit. The leaves and their extracts have a number of potential therapeutic properties. In China, the tree is considered sacred for this reason. Leaf extract essentially consists of two types of chemicals, flavonoids, and terpenes. Ginkgo-flavone, a glycoside, is a key substance in the flavonoid fraction. It is assumed that the concentration of flavin glycosides in nutrients must be 24% or more, therefore increasing the concentration of this type of substance is the goal of many producers. Flavonoids are polar compounds forming electrical dipoles that can surround other electrical molecules. Together with amber, they can be an effective neuroprotective substance. When flavonoids combine with a sugar unit, they become glycosides. The polarity of ethanol increases with the addition of water. An ethanol concentration of 60% has the best effect on the water-ethanol polarity. Yellow and ripe leaves contain more flavonoids. Young and green leaves, on the other hand, contain more terpenoids than the yellow ones [Ding 1999]. The flavonoids known from Ginkgo include quercetin, kaempferol, and isorhamenetins. In the leaves, there are also trilactone diterpenes, including glinkgolides, sesquiterpenes (e.g., bilobalide). The main ingredients of the *Ginkgo biloba* (the maidenhair tree) are: flavonol and flavone glycosides, ginkgolides, catechin, diterpene lactones, ascorbic acid, iron-based superoxide, dismutase, sesquiterpenes, and P-hydroxybenzoic acid.

Countless studies have been carried out on the substances in Ginkgo. It has been established that Ginkgo extract can remove free radicals, protect endothelial cells of blood vessels, improve circulation in the brain. Hence, it has been used to treat cognitive dementia, problems with peripheral nerves, and tinnitus. There were different positive and negative results for dementia, which are discussed in more depth in the author's brochure, which can be found on Amazon: *'AMBER Elixir of Immortality: Health compounds with pharmacological action. Baltic amber, Succinite-scientific research on dementia and Alzheimer's disease.'*

Photo 50. Extract made from Baltic amber and lemon balm (left). *Ginkgo biloba* extract (right).

The flavonoids present in the Ginkgo leaf extract are flavones, flavonols, tannins, biflavonoids, and related glycosides of quercetin and kaempferol. The bioavailability of the flavonods is relatively low due to limited absorption and rapid elimination [Mullaicharam 2013]. Ginkgo substances eliminate free radicals, reduce nerve damage, Ginkgo has anti-inflammatory and anti-aging properties. Numerous terpenes, flavonoids and trilactones have been found in the leaves [Toghueo 2020]. Although no compounds identical to those in *Ginkgo biloba* have been found in Baltic amber, we consider – due to the commonality of this type of plants in the Eocene – its influence on the formation of succinite.

Fossils coming from the *Ginkgoaceae* family are often found in northern and southern parts of the planet, usual in China

and Japan. There have been found leaf traces in fossils on the Axel Heiberg Island in the Canadian Arctic [Anderson, LePage 1995]. However, no proof has been found that those trees were components of succinite polymer.

Fagaceae (*Quercus, Castanea sativa, Lithocarpus, Fagus*) are a family of plants consisting of over 900 species of trees. It is common all over the world. The main place of diversity of the Quercus genus is Mexico, with over 160 kinds of oaks, and China with 100 species. The species diversified over the course of millions of years, and some of these changes occurred during the Eocen. Therefore, in the preserved amber, one can find fossilized nuts and acorns. Amber is also a great place to store inflorescences of ancient trees, which enables their identification. The amber layers cover the age range from 48 to 23 million years ago. They often contain oak and chestnut inflorescences *(Lat. Quercus, Castaneoideae and Trigonobalanoideae)* [Sadowski, Schmidt, Denk 2020].

Photo 51. Various colors of Baltic amber. Photo by P. Barczak.

Castanopsis is a genus of plants of the beech family, nowadays it is found in warm regions of south Asia. The ancestors of these trees are found in amber. *Castanopsis* (e.g., *Castanea sativa*) and *Lithocarpus* (the evergreen oak) are the main components of forests in Southeast Asia. The entire *Lithocarpus* genus consists of 336 species. *Castanopsis* is a significant tree in the mountain and laurel forests of Southeast Asia. Some genetic lines of Eocene trees are completely extinct. Some, however, have survived, like *Fagus* or the beech, and are still dominant trees in the temperate zone of the Northern Hemisphere. We can take a similar situation in the case of the maple (Acer), the elm (Ulmus), and the walnut (Juglans). These trees already existed in the Eocene, and their fossils are found in Greenland, for example on the island of Axel Heiberg [Grimsson et al. 2015].

Bioactive compounds can be found in various parts of oaks, in their bark, roots, leaves, flowers, seeds and nuts. Of particular interest are oak galls formed by insects injecting chemical substances into the leaf matter, resulting in a leaf ball in which the insect feeds. Galls contain gallotannin and with it there is about 2-4% of gallic acid. Alcoholic extract of galls showed antitumor activity, especially it fought brain cancer cells [Kamarudin et al. 2021].

In the alcohol extract obtained from Baltic amber, the scientific team of the P.H. Royal LLC and Warsaw University of Life Sciences found the chemical compound plumbagin with the chemical formula $C_{11}H_8O_3$. This may be a clue in the search for more trees that formed Baltic amber.

Plumbagin is a structure associated with naphthalene and is a component of Baltic amber. The compound has a cytotoxic potential for cancer cells, it inhibits their development [Panda, Tripathi, Biswal 2020]. The search for the possibility of fighting glioblastoma cancer through the use of plumbagin continues [Liu et al. 2015]. Also, studies are being conducted indicating the effect of this chemical compound on prostate cancer [Powolny, Singh 2008]. Its usefulness in combating malaria has been

noted. The ingredient is found mostly, but not solely, in the roots of plants. It also turns out that plumbagin in plants has a number of functions of a biological nature. One of them is the fight against viruses and insects, because plumbagin is very effective in this action [Sumsakul et al. 2014]. Plumbagin is also a common component of the *Diospyros genus* of trees and shrubs. The species Diospyros lotus. comes from the temperate zone forests in parts of Asia and is also found in the north of Iran in the regions of the Caspian Sea coast. A tree called the date plum is found in Transcaucasia, in Asia Minor. It reaches a height of 5 to 10 meters. The genus consists of about 500 species. Diospyros plants were found in fossils from the Cretaceous period (from 145 to 66 million years ago), also in the Northern Hemisphere. Plants of this family were very numerous in the region of the Eocene Sea, as evidenced by numerous fossils, flowers, fruits of these plants. Most of them were found in Eastern Europe, Denmark, and Germany [Denk, Bouchal 2021]. It is worth mentioning in the context of the formation of Baltic amber that in the past the Caspian Sea was connected with the regions where Baltic amber was formed. It is possible, therefore, that this genus was also involved in creation of the resin from which succinite was formed.

The *Diospyros genus* plants are used as traditional medicines, e.g., as anti-fungal, in the treatment of hiccups, for internal hemorrhages, for bedwetting in children, as a female drug for insomnia, as an antihypertensive drug, in the treatment of shortness of breath, as a deworming agent, as a sedative, antipyretic, and bactericidal agent [Ganapaty et al. 2006]. The fruits of the tree are used in the treatment of diabetes.

Another kind of plant which contain plumbagin are those from the *Plumbaginaceae* family. It is a family of perennial herbs, shrubs, or small trees with about 30 genera and 725 species. The plants occur both on the southern and the Northern Hemisphere. They can exist in high concentration of salt, what – in the case of a drying Eocenian sea – could predispose them for producing amber-bearing sediments.

Plants such as *Plumbago*, *Plumbago* auriculata, *Limoniastrum* differentiated in coastal, inland, saline and gypsum ecosystems. Probably plants of this type existed in the Eocene and could grow on saline lakes producing sediments proto-amber has come from [Moharrek et al. 2019].

Plumbagin is also present amongst other structurally related naphthoquinones in the roots, leaves, bark, and wood of *Juglans regia* (English nut, Persian and Californian walnut), Juglans cinerea (musk nut and white walnut) and Juglans nigra (black walnut) [Checker et al. 2010]. Currently, there are more than 20 species of the genus Juglans in the world. These types of trees are now found in areas

Fig. 21. Plumbagin.

where proto-amber was formerly formed – the foothills of the Carpathians. Today they grow also in the Balkans and South-East Asia, at the Caspian and the Aral Sea [Chchagan et al. 2010]. A large expansion of this family of plants was noted. In the current conditions, it can master an area turning into a steppe even by inhabiting one hectare with over 2,000 trees. In addition, the tree releases substances that prevent the development of microorganisms. Juglon secreted into the soil clearly changes its physicochemical properties. Juglon, with the composition of $C_{10}H_6O_3$, contains plumbagin. Another mechanism of eliminating competition is the fact that trees develop a dense crown, which makes it difficult for other plants to germinate nearby. Individual Juglans live up to 500 years [Lenda, Skórka 2009]. The content of plumbagin and the developmental features of this type of tree could co-create Baltic amber.

Preparations obtained from the walnut are applied topically in the treatment of acne, inflammatory diseases, ringworm, fungal, bacterial, and viral infections.

1.1.6. CARNIVOROUS PLANTS AND AMBER

Plumbagin is also known from carnivorous plants *Drosera* and *Nepenthes*. *Drosera* is a genus of sundews that feed on insects, thanks to which they can feed in areas poor in nutrients. Perhaps these plants had been also placed in peat niches lying under the seabed, where fossilization of resins took place. They can function fairly well in nitrogen and phosphorus poor ecosystems. Currently, 125 species of *Drosera* are known, although they are becoming increasingly rare in the world. A good habitat for these plants is wet pine savannas of the southeast USA. Fires often occur in these regions, which again indicates the environment of the formation of proto-amber in the past. Fires contributed to destruction of amber forests in the area of the Carpathian Foothills. They cause the forest litter to burn out, which is beneficial for insectivorous plants. Scientific research shows that *Drosera* is predisposed to develop after wildfires. *Drosera* plants occurrence increased after such events. Their seeds endured fire better than their competitors. If plumbagin in amber formed in swamps and peat bogs occurred as a result of the decay of this type of plant, it means that the area where the succinite was hypothetically formed was transforming from a tropical forest into a steppe. The whole area was on fire from time to time.

The amber forest is still a subject of debate. Insects found in Baltic amber are suddenly found in Southeast Asia. It turns out that trees from New Zealand or Tasmania grew in the area of today's Poland and Ukraine. The search for the ultimate truth about the ancient forest is very difficult, because the whole puzzle was complicated by glaciers transporting amber in different directions, and this hinders paleontological work. Forests also have been constantly changing due to changes in weather and climatic conditions. Today, in Europe, there is no uniform forest growing, and each region is biologically diverse. This forest from 50 or maybe 25 million years ago will continue to arouse curiosity although, even today when we analyze the trees growing in it, ancient ecosystems surprise us with their beauty and amaze us with their unique character.

Photo 52. Young Baltic amber. Photo by P. Barczak.

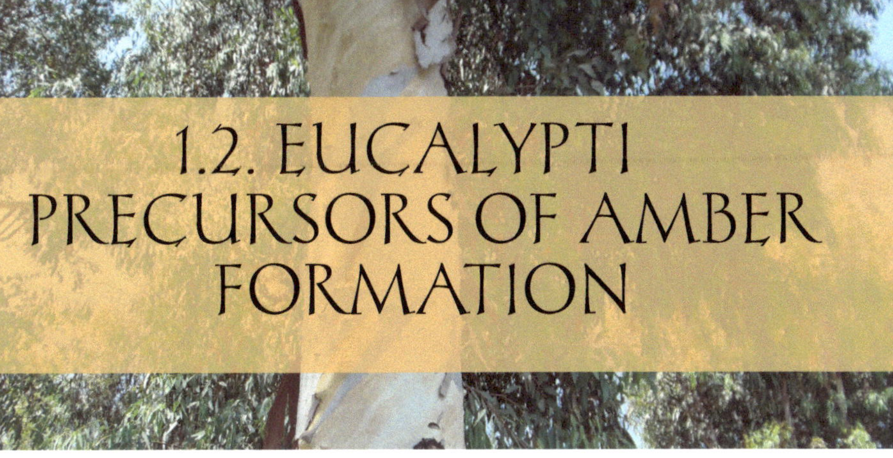

1.2. EUCALYPTI PRECURSORS OF AMBER FORMATION

Eucalyptus plants are currently growing in Australia, New Guinea, Malaysia, southern Philippines, New Zealand. They belong to the family Myrtaceae, which consists of seven types of trees. In the early Eocene, trees of this family were also found in South America, as evidenced by fossils. In the past, *Eucalyptus* trees probably also formed amber forests. They appeared in the territory of present-day Ukraine and southern Poland, and perhaps in the vast areas of the ancient Eocene era. *Eucalyptus* could have contributed to the creation of amber, although so far scientific discoveries only suggest this fact. This type of tree is common in Greece, Spain, and Portugal. The *Eucalyptus* genus includes over 700 species.

Photo 53. *Eucalyptus* leaves. Botanical garden, Athens, Greece. Photo by P. Barczak.

Eucalyptus trees have evolved with a special development strategy over the years. They can survive a forest fire better than other trees. Although the *eucalyptus* burns like most plants, it quickly regenerates thanks to the structure of the trunk in which, in the inner bark, the growth buds are deeply located. The trees of these species have adapted to the dry conditions prone to fire. Fires have become an ally of *eucalypti*; they help spread those trees. The competition is lost following the fires, which the *eucalyptus*, biologically adapted to development after a fire, makes use of. Burnout of the plant competition is one of the ways to master a new terrain and develop on the existing terrain. *Eucalyptus* trees can even explode, as the tree sap expands in a fire, and it can seep through the cracks in the bark. In addition, *eucalyptus* oil gives off flammable vapors in hot climates. It works like gasoline and the fumes can catch fire causing the tree to explode. The tree is therefore a barrel with flammable compounds that can explode during a fire, destroying the competition. However, it has a chance for development thanks to its unique structure. *Eucalyptus* trunk can sprout new limbs and regenerate the plant, unlike other types of trees that need to grow back from the roots.

Photo 54. Shedding bark constituting the cause of fires. *Eucalyptus* albens, *Myrtaceae, Athens*. Photo by P. Barczak.

The tree sheds its bark and dead leaves, which creates the perfect pile of kindling. When the oils in the tree heat up, the plant releases a flammable gas which ignites. On hot days, *eucalyptus* oil evaporates and creates a flammable mist over *eucalyptus* groves. This gas is extremely flammable and is the cause of many wildfires.

The high oil content in *eucalyptus* leaves makes them burn much better than others with less oil. Numerous fires could also have occurred in tropical rainforests turning into steppes. The changing climate and possible volcanic processes predisposed the area of future formation of proto-amber to the development of the *eucalyptus*.

Photo 55. *Eucalyptus albens* (*Myrtaceae*). Photo by P. Barczak.

The fossil records of the *eucalyptus* tree are still not well-researched. It is likely that the origin of the plant dates back to the Cretaceous-Paleogenic period. The *eucalyptus* is dominant among Australian trees, tropical forests, and mangroves. It is difficult to generalize all the characteristics of eucalypti, but they can survive fires also by maintaining their underground shoots. Their seeds are protected by a shell from high temperatures. *Eucalyptus* forests have evolved in an environment of frequent fires. The *eucalyptus* of today have several specific characteristics that allow it to fight off the plant competition through fires. It regularly sheds its bark, but not leaves. The leaves line up with their edges so as not to create a shadow. They need fire to reproduce. The seeds are surrounded by a hard husk that opens only when exposed to high temperatures. The *eucalyptus* germinates exceptionally well after a fire. *Eucalyptus* oil cumulated in the leaves is extremely flammable.

The oldest record of *eucalyptus* pollen in Australia comes from the Eocene period. Angiosperm trees evolved in conditions dominated by fires, and these were abundant in the Eocene in the regions of amber formation. During this period, there was significant volcanic activity in the area, which probably caused numerous fires in the vast area. The *eucalyptus* fossils discovered in the Laguna del Hunco deposit in South America were located just in the volcanic zone. In the Eocene, *eucalyptus* fruit were smaller than they are today, which may be a proof of their original adaptation or of a greater height of the then trees. Reports of the occurrence of eucalypti around the world have been prepared on the basis of fossils. There are reports from China, India, and South America [Hill et al. 2016].

The *eucalyptus* responds to external stresses and damage by secreting volatile organic compounds. *Eucalyptus benthamii* reacted within 24 hours after having been attacked by herbivores by emitting organic compounds, the secretion of which increased many times more, at 11 times, 59 times, and 69 times. First α-pinene occurred, the secretion of β-pinene increased several hundred times, the emission of para-cymene increased five times. At 72 hours after the event, the amount of limonene increased 10-fold, the content of linalool increased from 15.58 to 532, and the content of aromadendrene from 240 to over 16,500! Similar phenomena appeared in the case of all chemical compounds protecting the tree, i.e., 9-epi-(E)-*caryophyllene* (epi-(E)-*caryophyllene*), epiglobulol, globulol, and viridiflorol [Martins, Zarbin 2013].

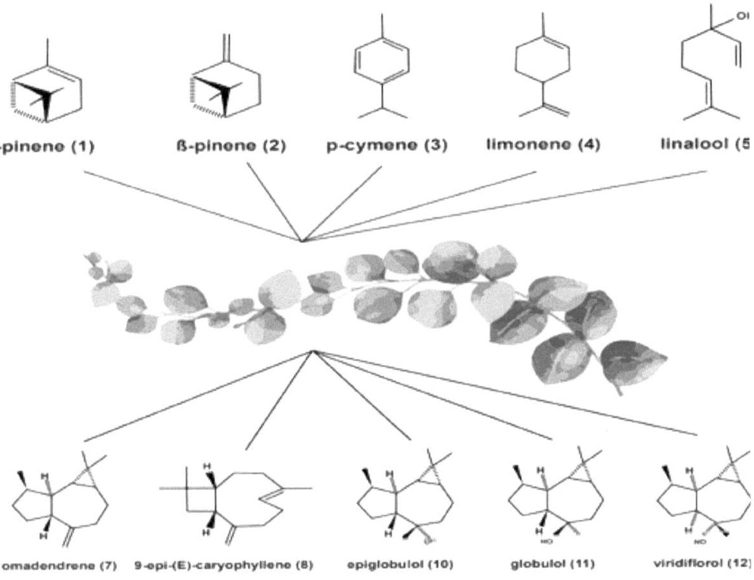

Fig. 22. Source: Camila, B. C. Martins, Paulo H. G. Zarbin, "Volatile Organic Compounds of Conspecific-Damaged *Eucalyptus benthamii* Influence Responses of Mated Females of Thaumastocoris peregrinus", J Chem Ecol (2013).

Volatile substances of various compositions are released during mechanical damage to the *eucalyptus* too. While conducting experiments on *Eucalyptus globulus*, its leaves and stems, the following vegetable oils were found to be released in response to mechanical damage:

- Monoterpene hydrocarbons: α-pinene, β-pinene, carene, and β-terpinene,
- Oxygenated monoterpenes: α-terpineol, α-terpenyl acetate, borneol, β-terpineol, cineole, thymol, and linalool,
- Sesquiterpene hydrocarbons, α-gurjunene, aromadendrene, alloaromadendrene, and eremophilene,
- Oxygenated sesquiterpenes: globulol, spathulenol, epiglobulol, α-eudesmol, γ-eudesmol, β-eudesmol, *caryophyllene*-oxide, and ledol,
- Hydrocarbons: pentadecanal, nonadecanone, tetradecanone, and heptadecanone,
- Aromatic carboxylic acid: benzoic acid [Tronosco et al. 2013].

It is worth recalling many of the compounds on the above list contributed to formation of Baltic amber. So far α-pinene, β-pinene, borneol, cineole (1,8-cineole), but also *caryophyllene*-oxide (a rare chemical characteristic for the *eucalyptus*), have been found in it. Volatile sesquiterpenes, which evaporate almost immediately from the leaves and the surface of the tree, have been preserved in Baltic amber too. Chemical compounds found in *eucalyptus* trees growing in present Australia and compounds in Baltic amber are partially convergent. Today we can still search for them in solidified resin.

There is evidence that eucalypti grew as early as the late Cretaceous period, over 100 million years ago, and that pollen of a species known as *Myrtaceidites eucalyptoidesm*, coming from the late Paleocene, has been found in the Lake Eyre area

of central Australia. This tree was found to have grown in the ancient continent of Gondwana 60 million years ago [Ladiges et al. 2010].

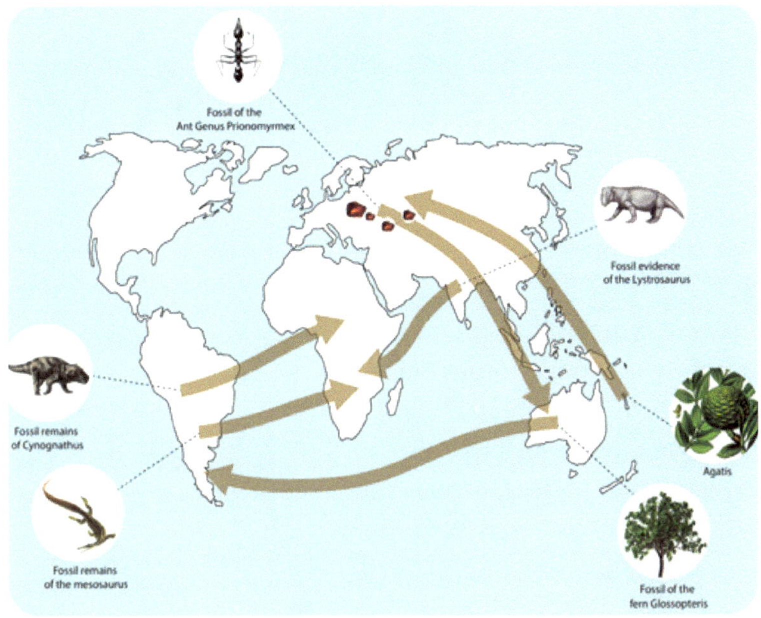

Fig. 23. Penetration of species in the times where the continents were connected and constituted one land – Gondwana.

The *eucalyptus* tree was most likely widespread as early as in the Paleogene (from 66 million years to 23 million years ago), although there are no numerous fossils. It is also possible that the ancestors of the *eucalyptus* tree had slightly different leaves than the trees present today. However, the conditions indicate that these species could occur and create amber deposits in Poland, Belarus, and Ukraine.

Photo 56. Young *eucalyptus* leaves. Photo by P. Barczak.

The development of forests millions of years ago is documented by found fossils. Tropical trees have been found in the fossils discovered in Serbia (southern Europe), i.e., in the areas which in the past were connected by a sea with northern Europe. Fossils of evergreen broad-leaved trees, such as *Laurus*, *Cinamomum (Daphnogene)*, *Persea* and *Ficus* have been found. There were also fossils of evergreen oaks and trees almost entirely resembling the *eucalyptus*, so far described by researchers as myrtle-affinity leaves. Conifers found in the fossils – among them *Quasisequoia* – constituted only 15% of the recorded species [Utescher et al. 2007].

Photo 57. Ficus plants could also create amber resins. Photo by P.Barczak.

The Miocene climatic optimum in the ancient parts of the Parathethys Sea (a currently nonexistent sea in Central and Southeastern Europe, reaching deep into Central Asia, formed about 160 million years ago) has been documented with a multitude of plant fossil finds from the period when present-day Croatia was connected by a common sea with the present-day territory of Poland and Ukraine. Among the fossils found in Serbia there were the *eucalyptus*, the ficus, the laurel, Persea (avocado) and Daphnogene (a fossil very similar to *eucalyptus* leaves). Currently, poplars, oaks, maples, pines, and elms grow in this area [Stanić 2018]. In the fossils from the island of Lesbos (Greece), the dominant species was Daphnogene polymorpha (21%), with leaves resembling *eucalyptus* leaves, and it was also a region connected with the Parathethys Sea and northern Europe [Kafetzidou 2020].

Photo 58. *Eucalyptus* leaves and a growing *eucalyptus*. Botanical garden, Athens, Greece. Photo by P. Barczak.

Eucalyptus leaves are especially interesting due to the amount of terpene oils produced by them. They contain flavones. These are compounds that can support the human body and protect the tree against pests. The *eucalyptus* produces a special leaf structure that allows certain chemical compounds to accumulate in them. A particularly large number of ethereal compounds – pinocembrin dimethylether and pinostrobin, depending on the type of the *eucalyptus* – is found in it. Trees adapt their production of compounds to local needs.

Photo 59. *Eucalyptus* seeds especially adapted to growing after fires. Photo by P. Barczak.

Today, New Zealand, with eucalypti growing there, is on the other side of the planet to Europe. In the past, the continents of Europe and Australia were connected with each other, and their worlds of plants and insects were common. This is evidenced by the preserved fossils and insects contained in amber.

Approximately 50 million years ago, Europe separated from the continent of Gondwana, parts of which were today's Australia, New Zealand, Antarctica, and South America. Theoretically, insects found in Australia should be different from those found in Europe. However, fossils and amber say something else. In the past, the *Cyclaxyridae* insects were widespread all over the world and survived at different times in different parts of the world. Today, those insects occur in New Zealand, but they have been also found in Burmese amber and, interestingly, in Baltic amber

from the area of Ukraine [Gimmel et al. 2019]. Eocene host plants have survived in New Zealand's geological history. The ancient forests of the Kauri and Nothofagus trees have survived to this day.

Lepidoptera fossils are another evidence of the occurrence of the *eucalyptus* in the amber-forming areas in the Eocene period. They confirm that the *eucalyptus* forest was responsible for the creation of amber. *Lepidoptera* is an insect that attacks the *eucalyptus*. A larva of this insect has been found in Baltic amber. Such finds in amber are very rare, probably due to the nocturnal lifestyle of these insects. *Eucalyptus* resin is secreted during the day and insects feed on the *eucalyptus* at night [Radchenko, Penkovsky 2020]. To this day, the Geometridae family is a problem for *eucalyptus* plantations.

A similar situation concerns ants and other insects found in Baltic amber. Today, they still like to seek food near *eucalyptus* trees. The *eucalyptus* wasp (*Hymenoptera*) sets galls on *eucalyptus* leaves, in which it lays eggs [Feyroz 2012]. A specimen of Chalcidoidea (*Hymenoptera*) [Simutnik, Perkovsky, Vasilenko 2020] was found in amber from the vicinity of Rovno (Baltic amber) and as mentioned, this insect attacks *eucalyptus* nowadays [Burks et al. 2015].

Unfortunately, there are still no perfect fossils proving the existence of *eucalyptus* in the area of amber formation. In the vicinity of Rovno in Ukraine, flowers of a species resembling the Myrtaceae family, i.e., the family *eucalyptus* belongs to, have been noticed in amber. In the Ukrainian amber deposits, which have been so far poorly researched, fossils of such plant families as palm trees, *Lauraceae*, Myrtaceae, *Proteaceae* and *Moraceae* are found. These types of plants currently grow in parts of Asia, Australia, and southern Africa [Sokoloff et al. 2018].

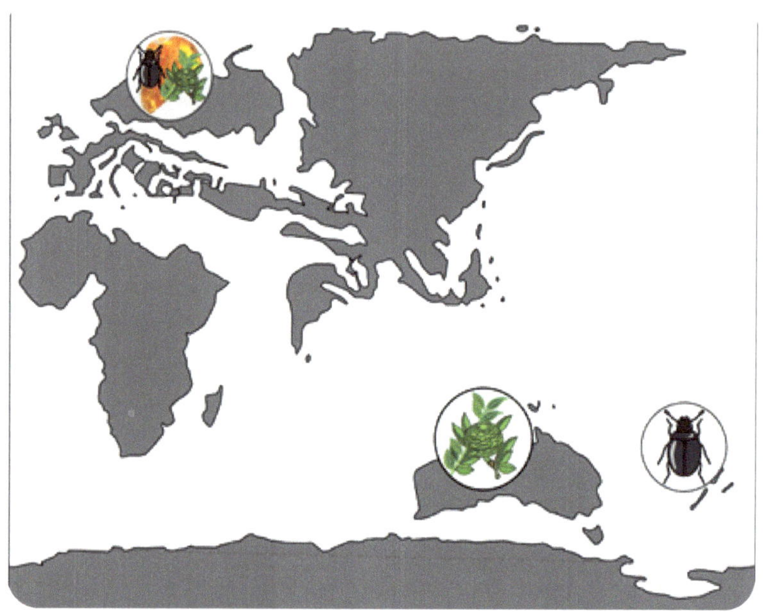

Figure 24. Today insects and pollen known from amber can be found in quite different parts of the planet.

The hypothesis on the co-creation of succinite by eucalypti may be consistent with the laboratory tests carried out by the team at the PH Royal LLC and Warsaw University of Life Sciences as a part of a project financed by the National Center for Research and Development in Warsaw from the EU funds. The sample, which was analyzed as a part of the work of the research team, consisted of a small batch of Baltic amber. It was bought at the Amber Museum in Kołobrzeg in Poland in 2020. It was milled amber, which constituted a certain statistical sample, as it was a collection of ambers from the Baltic Sea region (a part of the former Eocene Sea). The amber was extracted with 96% alcohol.

The results are presented in the table which shows the most important terpenes obtained through chemical analysis of the tested sample of milled amber. It is worth comparing it with the composition of *eucalyptus* oils presented here.

.dentification of samples:

No.	Description of the samples by the client	Description of the samples by the laboratory
1.	Milled amber 50 g P.H. Royal	SP 1439/20

Results of research:

No.	Sample Code	Description	Research Method	Result	Unit
1.	SP 1439/20	Camphene	Gas chromatography	5	mg/kg
2.		Cymene	Gas chromatography	212	mg/kg
3.		Borneol	Gas chromatography	42	mg/kg
4.		L-fenchone	Gas chromatography	12	mg/kg
5.		DL-menthol	Gas chromatography	3	mg/kg
6.		Eucalyptol	Gas chromatography	9	mg/kg

Tab. 1. Chemical compound found in amber analyzed with a gas chromatograph.

Analytical studies of Baltic amber allow us to compare the chemical composition of terpenes present in the currently occurring essential oils of trees with those that formed Baltic amber 50 million years ago. The analysis of amber samples with gas chromatography showed the existence of dominant compounds. A relatively large amount of cymene (212 mg/kg) was found, along with borneol (42 mg/kg). Both substances are present in Iranian eucalypti. Small amounts of L-fenchone (12 mg/kg) and camphene (5 mg/kg) were also found. These terpenes are also substances known from the Iranian *eucalyptus*. An important clue about the alleged origin of Baltic amber is the presence of eucalyptol (1,8-cineole), because this terpene is characteristic of eucalypti. It is almost their signature. For the first time eucalyptol was isolated from the essential oil from leaves of blue gum (*Eucalyptus globulus*) by F. S. Cloez in 1870. This compound has a multitude of applications in over-the-counter medicines intended for the treatment of inflammation of the throat and upper respiratory tract.

One more compound was found in analysis, DL-menthol; present in milled amber in the amount of 3 mg/kg. This compound may also be connected to *eucalyptus*.

Another study on amber extract using 96% ethyl alcohol, showed two dominant compounds: succinic acid in the amount of 141 mg/l and 1,8-cineole (eucalyptol) in the amount of 16 mg/l.

Identification of samples:

Description of the samples by the client	Description of the samples by the laboratory
BURSZTYN MARS 6 96% ethanol 27.11.2022	SP 1925/20

Results of research

No.	Sample Code	Described Parameter	Method	Result	Unit
1.		Plumbagin	HPLC MS/MS	<10	mg/l
2.	SP 1925/20	Succinic acid	HPLC MS/MS	141	mg/l
3.		1,8-Cineole	GC-MS	16	mg/l

Tab. 2. Chemical compounds found in ethanol extract obtained from Baltic amber.

This study also shows that Baltic amber was formed, among others, from compounds contained in eucalypti. The compound found through the gas chromatography analysis was 1,8-cineole (eucalyptol) known from eucalypti.

Eucalyptus (the family *Myrtaceae*) is a genus of myrtles. It includes about 700 species found in tropical and subtropical regions of the world. They are also being introduced in the Mediterranean, in Spain, Portugal and Algeria. In species of the *eucalyptus* growing in Iran (Lat. *Eucalyptus torquata* and *Eucalyptus oleosa*) numerous compounds known from Baltic amber are found, namely:

α-pinene (15.25-15.74%), camphene (0.86-0.88%), β-pinene (0.71-2.22%), p-cymene (0.38-2.46%), 1.8-cineol-eucalyptol- (28.57-31.96%), α-fenchone (1.14%), endo-fenchol (0.79-0.83%), camphor (0.49%), and borneol (endo and L-) (1.72-1.84%) [Ebadollahi et al. 2017].

These types of compounds are present in Baltic amber. They are mentioned in numerous scientific articles as important and dominant in this fossilized resin.

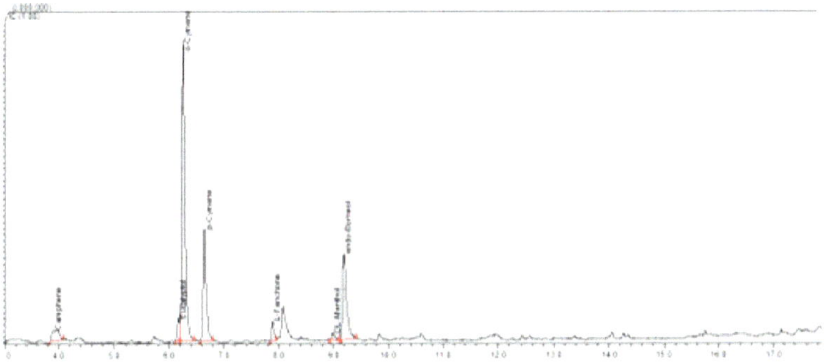

Figure 25. Graphical representation of a gas spectrometer test on ground amber.

Eucalyptol is one of the most important terpenes in *eucalyptus*. Its content in essential oil is as high as 94.03% ± 0.23%. Other studies indicate that there is 71.7% of 1,8-cineole (eucalyptol) and 9.14% of α-pinene% [Vivekanandhan et al. 2020].

In *eucalyptus* oils we find such compounds known from amber as: α-pinene, o-o-cymene, β-pinene, camphor, and terpentine-4-ol. The last compound – terpinene-4-ol – has been found in the stems and leaves of *eucalyptus* (E. globulus) growing in Thailand, Portugal, and Spain. The analysis of oils contained in eucalypti indicates that *Eucalyptus* globules – introduced in the Mediterranean region and currently growing in Spain, Portugal, and Tunisia – is more closely related to amber than *Eucalyptus* radiata currently found in Australia, because the latter contains mainly limonene (68.51%), and not the compounds known from amber, such as p-cymene or eucalyptol. Let us remember, however, about plant adaptations to the environment over the ages.

Photo 60. *Eucalyptus* and its unique bark. Photo by P. Barczak.

Today's *eucalyptus* oils lack DL-menthol, which was found in the study by Warsaw University of Life Sciences and the PH Royal. Menthol, on the other hand, was found in a plant species of the *Burseraceae* family, namely the *Commiphora mukul* tree. It is a resin-producing tree commonly called gugal, that is used in Vedic medicine. The tree is found in India and Pakistan. It also grows in Iran [Varamini et al. 2016]. It contains menthol in the amount of 1.77%. Menthol is also found in the oil of another plant of the *Burseraceae* family – *Boswellia sacra*, a tree native to Somalia and the southern part of the Arabian Peninsula. In aqueous extract from the resin there has been found 1,8-cineole (4.16%) and menthol (10.81%) [Asad, Meshal 2015]. Therefore, we assume that such species could also produce Baltic amber. However, the hypothesis requires evidence and further research.

1.2.1. EUCALYPTUS OIL AND AMBER COMPOUNDS

It was noticed that the compounds contained in *eucalyptus* oil are similar to those found in amber extract and Baltic amber. The difficulty with precise identification is that the compounds in amber have been formed for 50 million years and changed depending on the time and place of occurrence of a particular tree. The table below shows the differences in the content of compounds in *eucalyptus* growing in different countries. The genus *Eucalyptus* includes 700 species of evergreen trees.

Photo 61. *Eucalyptus* excreting resin under the influence of external heat.

COMPONENTS	CONTENTS (%)					
	CHINA [13]	INDIA [14]	AUSTRALIA [15]	SPAIN [13]	PORTUGAL [16]	MOROCCO [13]
ISOVALERIC ALDEHYDE	-	-	-	0.1	0.1	-
α-PINENE	6.0	11.0-29.9	2.3-14.8	17.4	18.5	7.6
CAMPHENE	traces	0.3-0.9	-	0.1	traces	0.5
β-PINENE	0.8	0.1-10.9	-	0.5	0.4	0.1
SABINENE	-	0.6-1.9	-	-	-	-
MYRCENE	0.5	-	-	0.3	0.3	-
α-PHELLANDRENE	0.2	-	-	0.2	0.1	0.3
1,8-CINEOLE	**80.5**	**28.4-53.2**	**39.3-52.7**	**61.8**	**62.5**	**74.2**
(Z)-β-OCIMENE	0.1	-	-	0.1	-	-
(E)-β-OCIMENE	0.1	-	-	0.1	traces	-
γ-TERPINENE	0.7	0.7-2.5	-	0.6	0.4	0.2
p-CYMENE	2.2	0.4-7.0	0.2-1.1	2.5	0.5	0.6
TERPINOLENE	0.1	0.1-0.2	-	0.1	0.2	-
ISOAMYL ISOBUTYRATE	traces	-	-	0.1	-	-
LINALOOL	0.1	-	-	0.1	traces	0.1
TERPINENE-4-OL	0.1	0.5-1.3	-	0.5	0.4	0.2
α-TERPINEOL	0.1	2.2-3.6	0.1-6.2	1.0	traces	1.8
NERAL	-	1.1-2.7	-	-	-	-

Tab. 3. Chemical composition of eucalypti in various countries. Source: Józef Góra, Anna Lis, "The Most Valuable Essential Oils". Part I. Monographs of the Lodz University of Technology, Łódź 2019.

A characteristic compound present in *eucalyptus* in considerable amounts is 1,8-cineole, which has been also found in Baltic amber. Analyzed oil sometimes contains more than 80% of it. In the study conducted by the research team of Warsaw University of Life Sciences and the PH Royal LLC company on amber extract, a terpene 1,8-cineole (also known as eucalyptol) was one of the two key compounds. In ethanol extract made of succinite, mainly succinic acid and just 1,8-cineole were detected. However, none of these compounds were found in pine (*Pinus*) oil and Siberian pine (*Pinus siberica*) oil. It is also absent from juniper (*Juniperus communis*) oil and fir oil, from the genus *Abies*.

Eucalyptus oil is considered to be one of the most important oils used in pharmacy. Produced with the method of steam extraction, it is applied in a number of diseases and is considered to be one of the most effective natural antiseptics. Steam distillation extracts 0.7-1.2% of the oil, and in modern distilleries the yield increases to the maximum level of 1.0-1.5%. As a result of the extraction from Baltic amber, a higher content of 1,8-cineole was obtained than in the known distillates. In Baltic amber it was found in the amount of 16 mg/l, i.e., over 1.6%, so more than in distillates obtained with the modern oil extraction equipment.

Photo 62. Baltic amber coming from Rovno in Ukraine. Photo by P. Barczak.

Eucalyptus oil is known for its numerous properties supporting human health. It has a diastolic effect, releases expectorant reflexes, and facilitates breathing. It is used in chronic and acute inflammations of the mouth, throat, and larynx. It has a positive effect on the respiratory and circulatory systems, stimulates the heart, reduces muscle swelling, soothes inflammation, adds energy, strengthens thought processes and supports remembering [Góra, Lis 2019].

The 1,8-cineole is currently being discovered by the world of science. Today it has many uses in protecting the brain and nerve cells. It is on the list of compounds closely related to the effects on Alzheimer's disease.

Eucalyptol plays an important role in restoring cell viability, reducing the potential of the mitochondrial membrane, suppressing the level of pro-inflammatory cytokines, and lowering the reactive oxygen species in the body [Khan et al. 2014]. A similar effect was noticed in the case of the joint action of α-pinene and eucalyptol. The action of these monoterpenes strengthened human cells and inhibited their inflammation [Porres-Martinez et al. 2016]. Both compounds, α-pinene and eucalyptol, are present in *eucalyptus* leaf oils.

A monoterpene 1,8-cymene also plays a role in reducing the symptoms of asthma and pneumonia, both of which are inflammatory conditions in the body. Research is ongoing into treating sinusitis in which 1,8-cineole (eucalyptol) plays a significant role. It is recommended to use the compound in the case of the induced effects of colds and inflammation of the upper respiratory tract. A monoterpene 1,8-cineole positively influenced the inflammatory processes in the lungs of smokers [Juergens 2014]. This compound also removes free radicals absorbed in cigarette smoke. It reduces lung damage and inhibits chronic respiratory disease [Juergens, Worth, Juergens 2020].

Borneol, another compound known from the *eucalyptus* oil, has also been found in amber. This compound inhibits cytotoxicity in cell systems, reduces the harmful effects of excessive oxidative stress. **Borneol is also known to relieve symptoms of anxiety, fatigue, and insomnia.** This monoterpene destabilizes peptide aggregates that are formed as a result of disorders in the human brain [Habtemariam 2018]. Finally, borneol plays an important role in cellular flow. It blocks the influx of calcium cations to cells, as a result of which cells – including nerve cells – located in the human brain are damaged. Therefore, a positive effect of borneol on the functioning of the human brain and protection against the effects of pathologies resulting from Alzheimer's disease has been found [Hur et al. 2013].

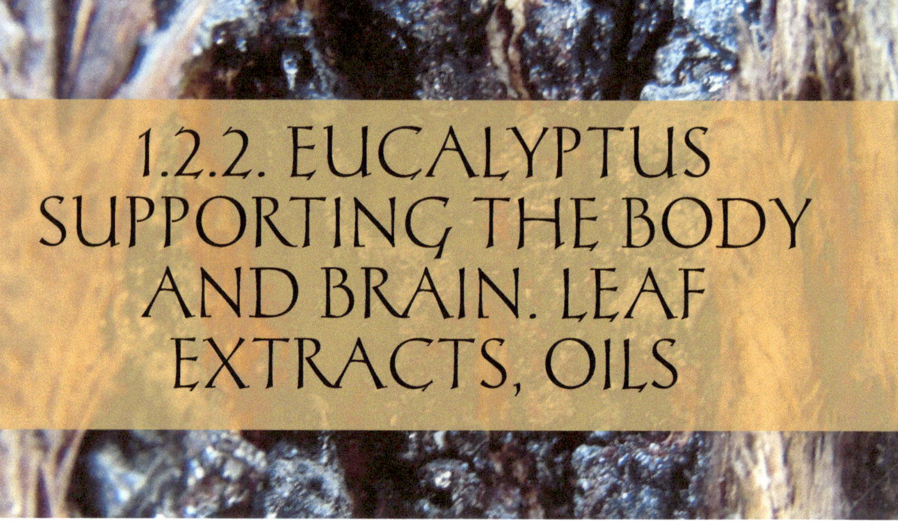

1.2.2. EUCALYPTUS SUPPORTING THE BODY AND BRAIN. LEAF EXTRACTS, OILS

Currently growing *Eucalyptus globulus* and its essential oils are in great demand on the market. They are used as anesthetics, antiseptics, astringents, diaphoretics, disinfectants, expectorants, antipyretic and hemostatic agents, sedatives, stimulants, and deworming agents. They are often used as folk remedies against arthritis, asthma, bronchitis, burns, cancer, and diabetes, diarrhea, diphtheria, dysentery, encephalitis, and enteritis. Oils of this type contribute to lowering fever, they work against influenza, larynx inflammation, leprosy, malaria, mastitis, pharyngitis, tuberculosis, rhinitis, ulcers, sore throat, and cramps. They support the trachea, fight worms, and heal wounds. They are eagerly used in the cosmetic industry [Bachir, Benali 2012].

Photo 63. Leaves of a mature *eucalyptus*. Photo by P. Barczak.

Eucalyptus oils contain monoterpenes including five atoms of carbon in their structure, marked by C5 by chemists. They have analgesic, antibacterial, antifungal, anti-inflammatory, antiseptic, antiviral, antioxidant, and even acaricidal, insecticidal, and nematocidal properties [Batish et al. 2006].

Photo 64. *Eucalyptus* leaves, Eucalyptus sp. Photo by P. Barczak

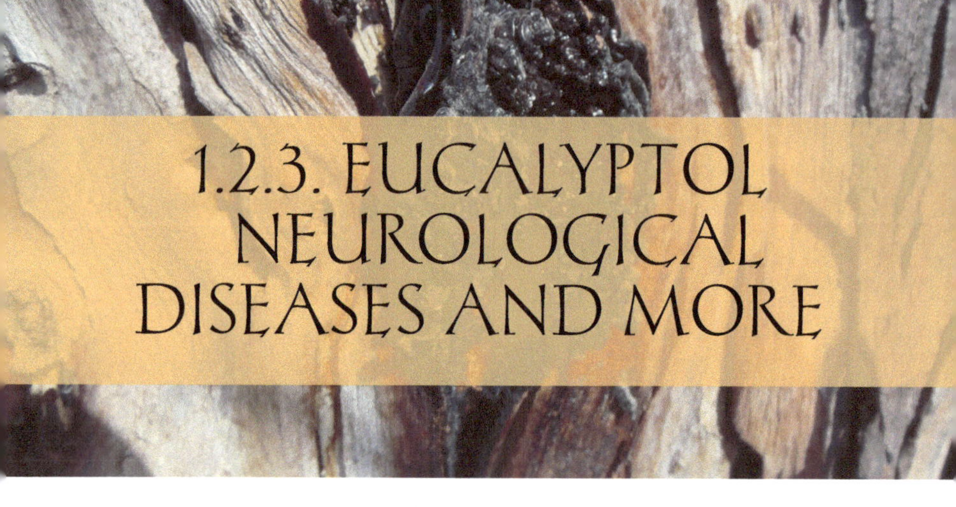

Currently the *eucalyptus* tree is grown in Mediterranean and subtropical areas and is an important firewood material. The leaves are used in brews for respiratory diseases, asthma, and bronchitis. Their antimicrobial, antifungal and antidiabetic properties have also been found. Ethanol extracts have a particularly interesting effect on oxidative stress in the brain. The brain and its cells are susceptible to the action of reactive oxygen species. Such a condition is especially harmful for nerve cells, neurons. These factors influence, among others, the development of Alzheimer's and Parkinson's diseases. Since there is still no cure for these diseases and the number of cases is increasing, dietary supplements slowing down the development of this type of diseases are increasingly used.

In the ethanol extract of the *Eucalyptus globulus* leaves, there have been found significant amounts of:

1. 3,4-dihydroxybenzoic acid (protocatechuic acid),
2. p-coumaric acid,
3. gallic acid,
4. quinic acid,
5. apigenin,
6. quercetin,
7. chlorogenic acid,
8. quercetin-3-O-rutoside (rutin).

The compounds which had the largest share in the extracts were flavonoids i.e. quercetin, luteolin, kaempferol, isorhamnetin, and florentina. It is worth highlighting that quercetin glycosides have antioxidant, anti-cancer, and protective properties in the circulatory system. The compounds found in *eucalyptus* have also a vasodilating effect. Damaged cells that have been applied with ethanol extract from *eucalyptus* leaves at a concentration of 5 µg mL showed an improvement in cell protection by 151.4%. The compound worked especially strongly in the protection of the liver [González-Burgos et al. 2018].

Amber and *Eucalyptus globulus* contain eucalyptol (1,8-cineole), which – as already mentioned – has anti-inflammatory properties. It is used as a pain reliever and to treat the upper respiratory tract. It has been found to have neuroprotective effects, reduce nerve inflammation, and alleviate oxidative stress in the brain.

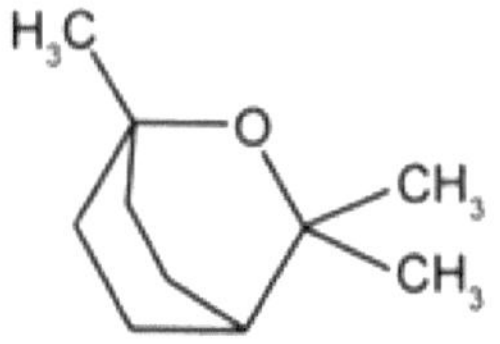

Figure 26. A molecule of Eucalyptol.

Research on rodents has shown that eucalyptol inhibits the development of oxidative stress in the brain, inhibits nerve inflammation, and blocks reactions to stress [Xu, Guo, Sun 2021].

Another disease which Baltic amber and *eucalyptus* may help to inhibit is epilepsy. Many works from the past indicated the positive influence of amber on epilepsy-related disorders. Compounds contained in *eucalyptus* also have this type of effect. Epilepsy is a disease that affects over 50 million people worldwide. It is caused by abnormal electrical responses from a damaged brain. Epilepsy leads to neuronal death, mitochondrial dysfunction in cells. The hippocampal circuits in the brain are destroyed. In inflammation, the brain is prone to increased oxygen requirements. Free oxygen causes damage to nerve cells. Free radicals are formed. The use of 1,8-cineole has turned out to inhibit this type of reaction. It is therefore one of the chemicals that contributes to relieving the symptoms of epilepsy [Rad, Anarkooli, Abdanipour 2022]. However, complete pharmacological studies have not been carried out yet.

Photo 65. Leaves of *Eucalyptus globulus*, Athens, Greece. Photo by P. Barczak.

Compounds contained in *eucalyptus* also support the fight against mental disorders. An extract from *Eucalyptus globulus* reduces pain – possibly due to its anti-inflammatory effects. Compounds in *eucalyptus*, such as alkaloids, flavonoids, saponins, terpenoids, have specific alleviating effects [Owemidu et al. 2020].

In the oil obtained from the fruits of *Eucalyptus globulus*, there have been found sesquiterpenes – mainly:

1. aromadendrene (31.17%),
2. 1,8-cineole (14.55%),
3. globulol (10.69%),
4. ledene (7.13%).

In turn, the following were found in the leaves: 1,8-cineole (86.51%), α-pinene (4.74%), γ-terpinene (2.57%), and α-phellandrene (1.40%). *Eucalyptus citriodora* leaves contained 90.07% of citronellal, a compound with a lemon scent, in one of the samples [Batish I in. 2006].

1.2.4. SEARCHING FOR COMMON COMPOUNDS OF BALTIC AMBER AND EUCALYPTUS

It is worth noting that many chemicals have a chemical structure like aromadendrene, which has been found in *eucalyptus*. They are organic chemicals, terpene compounds. However, it should be remembered that amber resin is an extremely delicate matter, subject to constant changes in the external environment. Museologists all over the world face this problem. Even in hermetic display cases, amber is subject to aging processes. It is particularly affected by light and temperature. Baltic amber, for example, contains a succinate ester which, under the influence of thermal and light factors, can be hydrolyzed into communol and succinic acid. It makes its structure change [Pastorelli, Shashoua, Richter 2013]. It is enough to examine a different piece of amber and we observe different chemical changes. The compound isolated from Lycopodium lucidulum (Huperzia lucidula) shows that abietic acid known from Baltic amber can be transformed into diabetic acid, a succinite monomer [Fujita 1974].

Figure 27. A succinite monomer formed as a result of transformations.

The terpenes in *eucalyptus* are also known from other plants, which makes identification even more difficult. For example, viridiflorol has been discovered in *the Melaleuca viridiflora* tree found in Australia. Viridiflorol has also been found in Baltic amber.

Perhaps such premises have caused that eucalypti so far have not been taken into account when considering the Eocene trees producing succinite. The convergence of the actions of Baltic amber and *eucalyptus* oils may, however, indicate that the two structures are of similar origin.

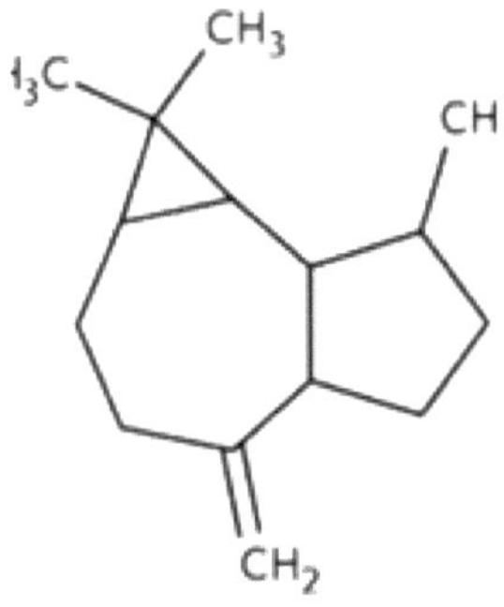

Fig 28. A molecule of Aromadendrene.

Viridiflorol, with the formula $C_{15}H_{26}O$, is present in Baltic amber in the amount of 0.93%. In turn, *Eucalyptus* oreades, which grows in the mountainous areas of Australia, has been found to contain 20% of this sesquiterpene. The same chemical formula $C_{15}H_{26}O$ characterizes β-Eudesmol, which has been detected in the amount of 40% in *Eucalyptus* brevistylis leaves.

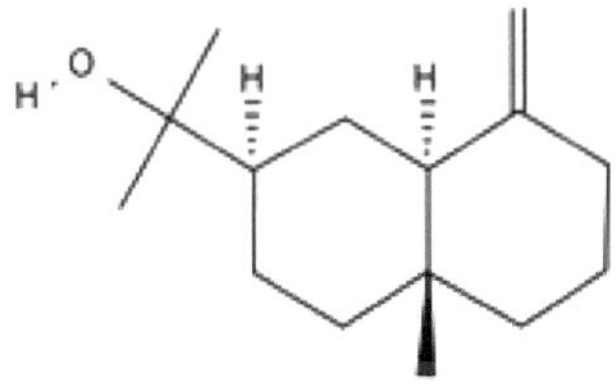

Figure 29. β-Eudesmol.

This type of *eucalyptus* grows in Australia. It is a tree that lives over 400 years and is now threatened with extinction. Although β-Eudesmol has not yet been detected in Baltic amber, it is only a matter of time. Compounds with such structures are found in Baltic amber in small amounts. There is also another sesquiterpene known from eucalypti. It is Spathulenol (from the same group of sesquiterpenes – Aromadendrenes) with the formula $C_{15}H_{24}O$.

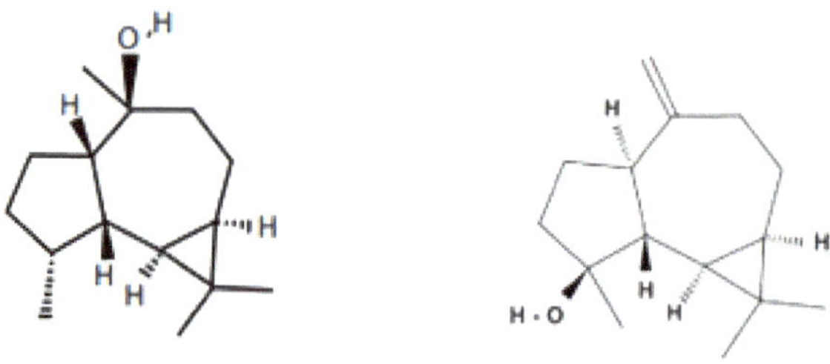

Fig 30. Viridiflorol $C_{15}H_{26}O$ (left) and Spathulenol $C_{15}H_{24}O$ (right).

In amber there has been found 0.72% of the compound called trans-beta-Santalol. It also has the formula $C_{15}H_{24}O$, but with a different structure. Millions of years of transformations of the polymer resin resulted in a change of structures, meaning that it is perhaps a hidden memento of the past, allowing the identification of plants that made up Baltic amber.

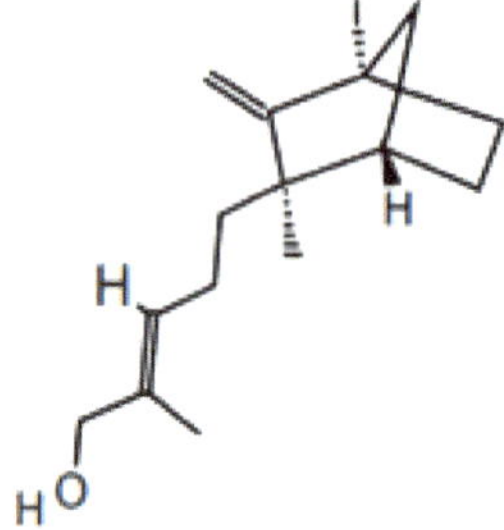

Fig. 31. Trans-beta Santalol $C_{15}H_{24}O$.

Nowadays *Eucalyptus* oreades grows in mountainous regions in Australia. It is extremely resistant to fires. In this *eucalyptus* there have been also found P - c y m e n e monoterpenes (as much as 41%) and trans-p-menth-2-en-1-ol ($C_{10}H_{18}O$) in the amount of 22%. Isoborneol and borneol, which are contained in Baltic amber, have a similar amount of carbon, hydrogen, and oxygen atoms. The *eucalyptus* contains 18% of trans-Piperitenol with the chemical formula $C_{10}H_{18}O$. All compounds are related to each other by a common chemical formula, although trans-Piperitenol has not been found in Baltic amber yet.

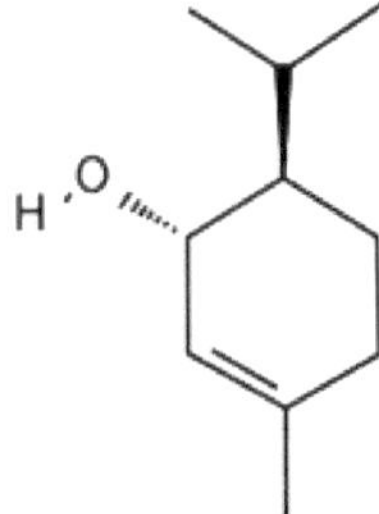

Fig. 32. Trans-Piperitenol.

Also, trans-p-menth-2-en-1-ol with the chemical formula $C_{10}H_{18}O$ has not been found in amber, although, as we have already mentioned, there are present borneol, isoborneol and 2-fenchanol, which have the same chemical formula ($C_{10}H_{18}O$) and are key compounds in the succinite.

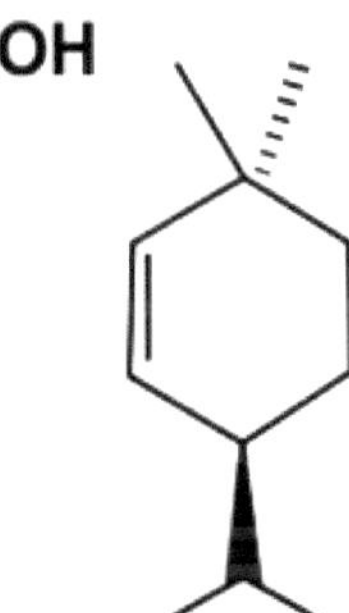

Fig. 33. Trans-p-menth-2-en-1-ol.

Other sesquiterpenes not found in Baltic amber yet are also interesting. Eucalypti contain the compound called *α-caryophyllene*. In *Eucalyptus* stellulata it is found in the amount of 10%. On the other hand, *Caryophyllene* oxide is found in *Eucalyptus* muelleriana in the amount of 11%. This tree is endemic to Australia. It grows in wet forests in coastal plains.

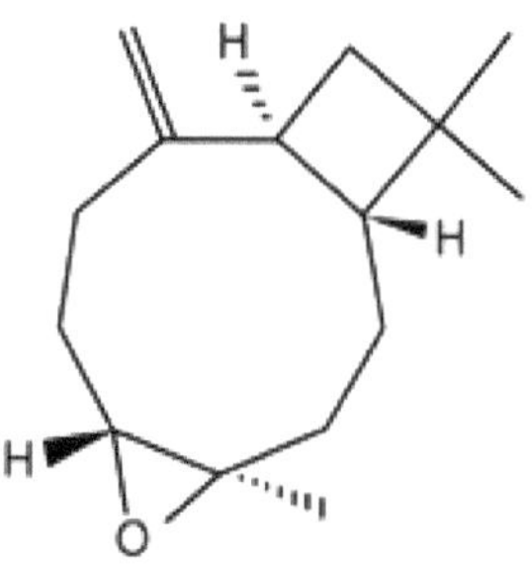

Fig. 34. *Caryophyllene* oxide.

In turn, as much as 30% of this sesquiterpene has been found in *Eucalyptus* apiculata.

Trans-*Caryophyllene* ($C_{15}H_{24}$) has been found in *Eucalyptus* dendromorpha in the amount of 25%. This species of eucalyptus occurs mainly in mountainous areas. *Caryophyllene* has a structure that is very rare in nature. It is found in the essential oils of clove, black cumin, hops, basil, oregano, black pepper, lavender, Cannabis sativa and in Copaiba oil. The presence of *Caryophyllene* oxide in Baltic amber is indicated by infrared spectroscopy [Wagner-Wysiecka 2018].

The infrared spectrum of succinite is very characteristic among the known fossil resins. This is due to the presence of succinic acid and its esters, i.e., compounds resulting from the conversion of acid and alcohol. Therefore, succinite has a constant curve, displayed on measuring devices, which is almost identical for all Baltic amber.

"Baltic arm" – a specific graphic pattern emitted by measuring devices – shows (while emitting in infrared) the chemical compounds contained in the succinite. The chart area always has the same layout. As a result, β-*Caryophyllene*, a sesquiterpene, has been found in Baltic amber [Ibid.], as well as compounds known from eucalypti, including *Caryophyllene* oxide.

In *Eucalyptus* dendromorpha we can also find a sesquiterpene called Calarene (13%), with the chemical formula $C_{15}H_{24}$, and Bicyclogermacrene ($C_{15}H_{24}$), in the amount of 48%, can be found in *Eucalyptus* suberea. The latter tree grows in Australia in moors and mountainous areas. Still another sesquiterpene is found in *Eucalyptus approximans*. It is α-Gurjunene, whose oil contains 23% of this compound. Its formula is also $C_{15}H_{24}$. This compound has not been found in Baltic amber either.

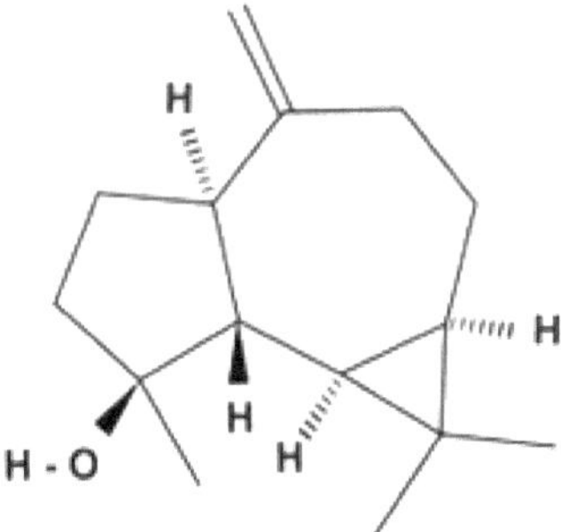

Fig. 35. Gurjunene (Dipterocarpus turbinatus).

Another compound which is found in eucalypti and has a great potential for supporting the body is a monoterpene called Cryptone ($C_9H_{14}O$). This compound has not been found in Baltic amber so far. Small amounts of this compound have been found in the *Eucalyptus* subarea, and in *Eucalyptus* globulus, growing in the Mediterranean along the coast of Montenegro and Spain, there have been found Cryptone (0.00-17.80%), 1.8-cineole (4.10-50.30%, depending on maturity and place of the origin and collection), as well as α-Pinene (0.05-17.85%), p-Cymene (27.22%) and Spathulenol (0.12-17.00%). On the other hand, oils from fruit, buds and branches contain α-thujene (sometimes 0.00% but also 11.95%), 1,8-cineole (15.31%, 36.95% and 56.96%, respectively) and Aromadendrene (23.33%, 16.57% and 8.24%, respectively) [Chalchat et al. 1995].

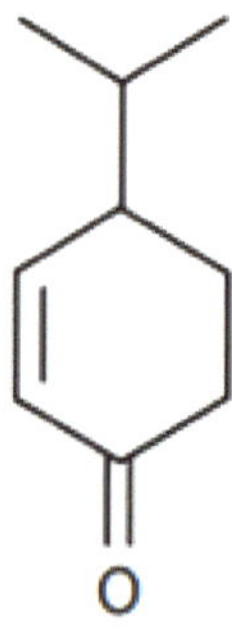

Fig. 36. Cryptone $C_9H_{14}O$.

The above-mentioned terpenes are known from Baltic amber, especially p-Cymene and α-Pinene. On the list of compounds known from Baltic amber there is also Eucalyptol, which indicates trees that make up Baltic amber which have remained completely omitted and unnoticed by the world of science so far.

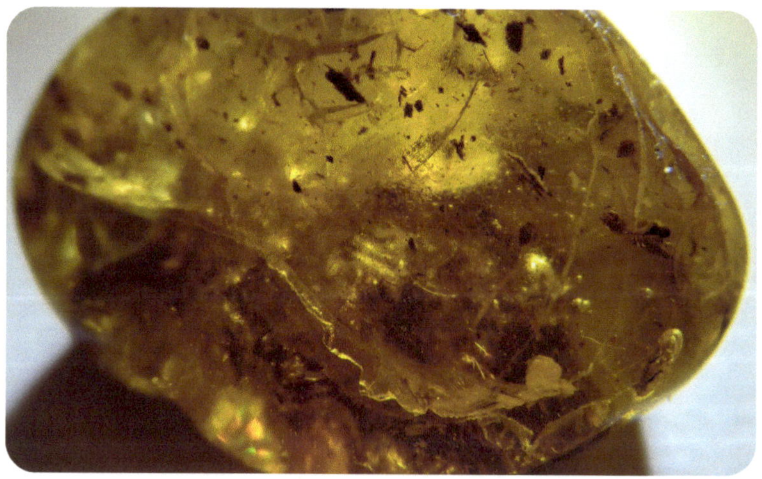

1.3. BALTIC AMBER FLORA AND FAUNA AND THE NATURAL ENVIRONMENT OF AUSTRALIA, NEW ZEALAND, AND NEW CALEDONIA

Amber deposits have been discovered relatively recently in Australia. They come from the Late Triassic period – 230 to 40 million years ago – so from the time when the continents were all connected with each other and first they constituted the supercontinent Pangea and then Gondwana. At the beginning of the Eocene, the area of today's Australia was covered with tropical vegetation. This was due to the large amount of carbon dioxide in the atmosphere. A similar climate prevailed then in the territory of what now is Europe, which was just emerging from sea waters. Over time, drifting apart of the continents caused that the sea currents around Antarctica changed and the climate started to cool slowly. *Araucariaceae* and *Nothofagidites* – an extinct genus of broad-leaved trees resembling beeches – developed intensively, as well as a number of species of *Casuarinaceae* – a family with specific needles, found in Eocene fossils in Germany and Czechia [Wilde, Frankenhäuser, Lenz 2021]. In contrast, *Myrtaceae pollen*, including pollen from eucalypti and *Melaleuca trees*, have been found in late Eocene fossils in Australia.

Photo 66. *Casuarina cunninghamiana (Casuarinaceae)*. Photo by P. Barczak.

Baltic amber was probably produced by some *Proteaceae*. They have been also found in open pit coal mines in Western Australia. Plants of this family now also exist in the tropical forests of Australia. As early as in 1864, Göppert found fragments of this type of plants in Baltic amber. Ultimately, however, the hypothesis that amber was produced by these species has not been confirmed to this day [Sadowski et al. 2019]. *Lauraceae*, on the other hand, have been found in European Tertiary amber [Tarasevich, Alekseev 2017]. Other fossils found in Australia include conifers – such as *Agathis* (*Araucariaceae*) and *Libocedrus* (*Cupressaceae*), representing families known from Baltic amber – as well as *Acmopyle* from the *Podocarpaceae* family and Dacrycarpus: an endemic tree native to New Zealand, not "yet" found in Baltic amber. On the other hand, *Sciadopitys*, the Japanese umbrella pine, whose fossil fragments are found in Europe, North America, China, and Japan, but also in Australia, was widely spread on the Australian continent. In the past, the natural world was more integrated than it is today, and plants present in Australia or New Caledonia may once have formed forest complexes in Europe. This hypothesis is confirmed by the amber deposits discovered in the Australian Fingal Valley Coal Measures mine in Tasmania in 2015. The amber found there is in small 1.5-millimeter pieces containing fragments of bark, plants, and organic debris. Sediments of this type are integrally connected with the European fossils known from the Dolomites in Italy, which contain spores of the same fungi. The resin that makes up amber was probably secreted also by the *Cheirolepidiaceae* conifers, among which some species preferred dry and saline areas.

Cretaceous amber has been found in New Zealand. The ambers here are slightly larger, and they contain fragments of trees already known to us from the region where Baltic amber was created, such as *Cupressaceae*, *Podocarpaceae* and *Araucariaceae*. Ginkgo, seed ferns and angiosperms also grew locally quite

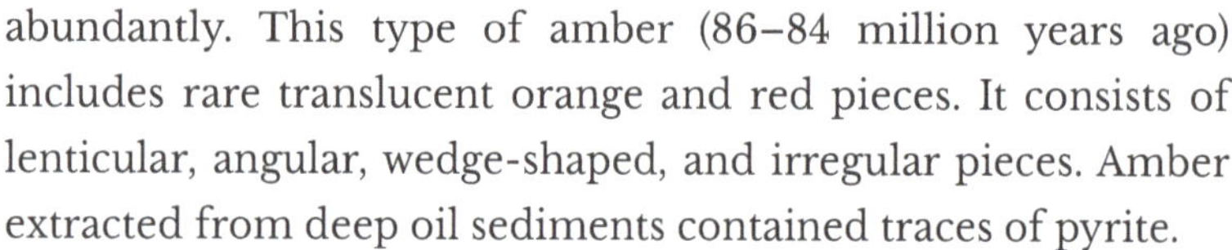

abundantly. This type of amber (86–84 million years ago) includes rare translucent orange and red pieces. It consists of lenticular, angular, wedge-shaped, and irregular pieces. Amber extracted from deep oil sediments contained traces of pyrite.

Mid-Paleocene amber (66 to 62 million years ago) obtained with core drills in Gippsland in southern Victoria (Australia) from a depth of 1967.9 m also consists of translucent orange and red pieces of fossil resin and the spore-pollen zone of Lygistepollenites balmeii [Stilwell et al. 2020].

Photo 67. Baltic amber – succinite. Photo by P. Barczak.

Early Eocene amber has been discovered in the Paleogene Macquarie Harbor rock formation in western Tasmania. These are small pieces of 1-2 millimeter, rarely 20 millimeters. Most likely, this amber originated from the resins of *Araucariaceae* and / or *Cupressaceae* trees, including Agathis or Araucaria. Eocene ambers were also discovered at the Anglesea Coal Measures lignite mine near the town of Anglesea, Victoria, Australia in 2014. They are inserts for coal, sometimes combined with it. Amber is most often light green or dark yellow and translucent with less common orange and red and rare white and brown amber. As in other places in Australia, it was created by plants of the families *Cupressaceae* and/or *Araucariaceae*, but also *Dipterocarpaceae* – that

is, the family of trees currently growing in the tropics. Certain insects (*Meunierohelea*) have been found in Australian amber, but also in Baltic amber from Bitterfeld and Rovno, and in Indian amber in Cambay. Some insects of this type are now found in the Northern Hemisphere, but in Australia they have remained only in fossils [Ibid.].

New Caledonia can also tell us about plants growing in the Northern Hemisphere during the Eocene. The archipelago lies in the southwest Pacific, 1,500 kilometers east of Australia. Due to this, it is a land of endemics. Plants inhabited the island in the Eocene period, i.e., the supposed time of creation of Baltic amber. The island emerged from under the water in the Eocene. However, it is not clear why some plant species are common in Australia and not in New Caledonia. This applies, inter alia, to eucalypti and acacia trees. It is hypothesized that 37 million years ago Eocene plants inhabited New Caledonia and hence such a large number of "living fossils" on the island. They interest us as a potential source of trees suspected of creating Baltic amber.

Fig 37. View of Pangea in the Triassic. "Palaeogeography of South-East Asia and Its Connection with Distribution of Early Jurassic Lithiotis-type Bivalve Buildups According to Triassic/Jurassic Mass Extinction Event", Krobicki M. & Golonka J., Geology, 2009, Vol. 35, No. 3/1.

New Caledonian Forest system consists of the tropics with evergreen forests and, in some regions, savannas. Due to the characteristics of the soil, local endemism occurs in New Caledonia. Sometimes it is one mountain, other times a group of mountains, with forests of exceptional natural value. They are called laurel forests. They are related to the forests of Australia, New Zealand, Tasmania, but also south Africa, Madagascar, southern Japan, Taiwan, the coasts of southern China, and the warm coasts of North America.

Fig 38. Plants of New Caledonia – a formula for Baltic amber.

One of the tree genera in New Caledonia are Metrosideros (*Myrtaceae*) [Pillon 2012]. All species of New Caledonian Metrosideros are 37.1 million years old, and the genus emerged 42.2 million years ago. These trees are also known from New Zealand and Hawaii. Fossil pollen, known from fossils and found in amber, dates back to the Miocene and Oligocene periods.

Nothofagus (Nothofagaceae) is a type of beech-like trees dating back 30 million years – is another endemic originating from the prehistoric continent of Gondwana. It is known from Chile and Argentina because South America used to be connected with Australia. A genus of the southern beech family, frost-resistant to -5 degrees Celsius, likes the sun. It is the equivalent of the beech from the Northern Hemisphere.

Regarding *Araucaria (Araucariaceae)*, it is unclear whether this genus of trees is native to the younger island of Norfolk. Those trees are now quite popular in the north of the planet, where they are planted in warm parts of Europe, especially in greenhouses and botanical gardens. Some species naturally grow in Chile, New Caledonia, New Guinea, and eastern Australia. This genus has been known since the Cretaceous, so it influenced the creation of Baltic amber.

Dacrydium is a genus of the *Podocarpaceae* family. It occurs in subtropical climate and is known from New Caledonia and the Fiji Islands. It can form hybrids with other species, hence its small area of identification. *Pycnandra (Sapotaceae)* is recognized as a unique endemic genus of New Caledonia. The origin of the genus was 35.8 million years ago. The shrub has developed a system of accumulation of nickel, that is abundant in the soils of New Caledonia. This element is extremely harmful to plants, but Pycnadra has developed mechanisms to protect its tissues.

Tree ferns have a trunk that makes them look like trees. Currently, they grow in New Caledonia, but also in the rainforests of Australia and New Zealand. Ferns were one of the first plants to dominate continents millions of years ago, probably their remains formed layers of peat and influenced the formation of succinite.

The trees of New Caledonia show the ancient ecosystems that ruled the planet millions of years ago. It should be assumed that their migration was possible also in the past when the continents were integrated. Therefore, more and more often pollen of plants which now occur in New Caledonia and Australia is found in Baltic amber, despite the fact that today the planet has a completely different climate and land configuration.

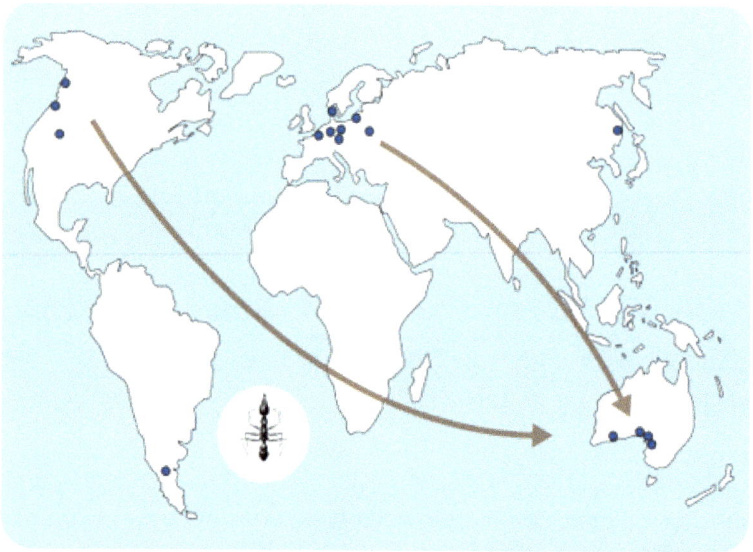

Fig. 39 Radchenko, A.G., & Perkovsky, E.E., (2020). "New Finds of the Fossil Ant Genus *Prionomyrmex Mayr* (*Hymenoptera*, Formicidae, *Myrmeciinae*) in Late Eocene European Amber". Paleontological Journal, 54 (6), 617-626.

The survival of certain species of plants and insects in Australia and finding their progenitors in Eocenian fossil amber in Europe is shown by research on ants of the subfamily *Myrmeciinae*. For a long time, they have been found in Baltic amber, in amber from the Rovno region, in German fossils from Rott in Westphalia, and also in the USA. Today, *Myrmeciinae* ants live in Australia, New Caledonia, and Tasmania. They still climb trees, drink nectar, and hunt small invertebrates [Radchenko, Perkovsky 2020].

1.4. CYPRUS - POSSIBLE REMAINS OF THE EOCENE SEA ENVIRONMENT

The Eocene Sea, existing about 50 million years ago, and stretching from the area of present Jutland to the present Black Sea, was probably a warm sea. During the Eocene (55.8 ± 0.2 - 33.9 ± 0.1 million years ago), global warming occurred, forests developed even in the present polar regions. The Eocene thermal maximum lasted for about 200,000 years, the global temperature increased by 6-8 degrees C. Currently, since the pre-industrial period (1850-1900) compared to the years 2011-2020, the temperature on earth has increased by 1 degrees Celsius. drying up of lakes and seas, melting of glaciers and numerous climate changes. The temperature jumps in the area of amber formation were therefore much greater and took place several times in the distant history of the planet about 53.7 million years ago, 53.6 million years ago, 53.3 million years, 52.8 million years ago [Szwedo , 2015]. Under such conditions, the amber forest changed and it contained over 750 species of plants. The growing trees probably had a diversified character, occurring in regions at different latitudes. Perhaps those in the north were different from those in the south. It is impossible to determine exactly today. The northern regions of the former Eocene Sea have been heavily leveled by glaciers. However, the trees of the southern areas may have survived. Perhaps the current warm climate prefers plants from the Eocene. The oldest elements of the flora are known from the regions of Australia, New Zealand,

Tasmania and New Caledonia. Some species are assumed to date from the time of the continent of Pangea, when the lands were interconnected. A subtropical, greenhouse place is the coast of Cyprus in the Mediterranean Sea. We can observe here trees known from scientific literature, suspected of creating Baltic amber, as well as trees on which the world of science has not yet made a firm statement. These trees are known for their chemical compounds found in Baltic amber. The terpnoids of these trees are found in the fossilized Baltic resin. Moreover, the insects found in Baltic amber are also known from the Mediterranean region. For example, termites have been found in the Baltic resin, and their present-day relatives of cockroaches (*Blattodea*) and praying mantises (*Mantodea*) are known from warm climatic zones [Szwedo, Kania, 2015]. *Mantodea* is known from the region of Turkey and Cyprus [Ehrmann, 2011]. Similarly, the *Blattodea* insect has established itself in Cyprus [Ghesini, Marini, 2015]. It is worth taking a special look at the Cypriot trees growing on the coast. The Eocene Sea was a place where numerous bays created a sphere of development for plants. Currently, in the Limassol region, *araucaria* (*Araucaria*), *eucalyptus* (*Eucalyptus*) and *Casuarina equisetifolia* are thriving. They are mostly introduced. However, it is worth mentioning them, because they thrive in this region, drawing on the resources of subcutaneous saline water and the moisture of the nearby sea. *Ceratonia* (carob) and *Olea* (olive) are also found on the coast.

Photo 68. Leaking *eucalyptus* resin due to high temperature (left). Photo 69.Falling off *eucalyptus* bark (right). Photo by P.Barczak.

The trees in this subtropical zone produce chemicals known from Baltic amber. The *Casuarina equisetifolia* tree in the essential oil contains 13.1% eucaliptol, which is found in a significant amount in Baltic amber. This compound is also known from *eucalyptus* and ficus leaves (8.2%) [Ogunwande et al., 2011]. They now cover the coast of Cyprus.

Photo 70. *Casuarina cones* (left photo).
Photo 71. Leaking resin from *Casuarina*. Cyprus. Photo by P. Barczak.

We write more about *eucalyptus* in Chapter 1.2. However, it is worth mentioning the reactions of these trees to the high temperatures observed in the summer of 2022, as well as the leakage of resin and sap as a result of mechanical damage to these trees. When exposed to intense sunlight, *eucalyptus* leaves emit a resin in the form of red-black exudates that flows down the fallen bark. Additionally, black tar-like resin structures form in the damaged areas. Therefore, it seems that these trees could be an important element creating amber deposits, so far neglected by science.

Photo 72. Resinous *eucalyptus* in places where the bark is cracked. Photo by P.Barczak.

Ficuses also grow in Cyprus. They do not emit too much resin, although the terpenoids known from their oils are found in Baltic amber. Ficus Benjamina is the tropical plant most widely distributed in North Australia and South Asia. Its extracts are effective in the folk treatment of cancer, tumors, leprosy, skin diseases, inflammations and malaria. The tree is able to differentiate secreted compounds depending on the environment and time of day or night.

Photo 73. Ficuses in Cyprus in Limassol. Photo by P. Barczak.

It mainly contains α-pinene (13.9%) known from Baltic amber as well as germacrene-D4-ol (9.7%) and cis-α-bisabolene (8.4%). The number of compounds is extensive. Let us mention only the most important, known from Baltic amber: α-pinene (13.9%), 4-terpineol (2.5%) ρ Mentha-1,5-dien-8-ol (2.1%), 1,8-cineole (2.1%), dehydro-ρ-cymene (2.0%), limonene (2.0%), p-cymene (1.2%), camphor (1.2%), camphene (1.1%), β-caryophyllene (0.9%), myrcene (0.8%), α-phellandrene (0.7%), δ-cadinene (0.2%), ρ-cymen-8-ol (0.2%). This tree could form former peat bogs formed in the course of climate change.

At present, scientists more and more often point to the family *Moraceae*, from which ficuses originate as the primary trees of the European continent [Pederneiras et al. 2018], although there is an ongoing discussion about the fossilized precursors of the tree.

In the region of Cyprus, the *Araucaria*s do particularly well, which are found on the coast rather than in the local mountains. They are treated as exceptionally interesting ornamental plants. The tree releases large amounts of resin as a result of changing environmental conditions. Each damage is caused by a waterfall of secreted resin. In this respect, it perfectly meets the conditions of the Baltic amber source. The resin contains over 39 terpenoids. Germakren-D (9%) as well as hibaen (30%) and phylocladene (20%) diterpenes were found in the *Araucaria* angustifolia tree. These are the main ingredients of the essential oil. Chemical compounds of this type are found in residual amounts in Baltic amber, which suggests the role of this type of trees in the formation of fossilized resin, the more so that forests consisting of *araucaria* occurred in the Cretaceous period in the present part of Europe. Moreover, trees are able to adjust the amount of secreted oils to the surrounding natural conditions. *Araucaria bidwillii*, *Araucaria columnaris* and *Araucaria cunninghamie* all of these trees in one study were low in mono- and sesquiterpenes and high in diterpenes. In the first of the above-mentioned trees, the hibaen compound was the main component (76%) of the resins, the next *araucaria* species contained only 9%. hibaenu, in the latter case the most important component was 16-kauren (53%). *Araucaria heterophylla*, on the other hand, consisted of 52% of the monoterpenoid α-pinene and 32% of phylocladene [Brophy et al. 2000]. On the other hand, in the obtained oil from *Araucaria heterophylla* growing in Egypt: monoterpenes (66.53%), followed by sesquiterpenes (30.85%). Also found α-pinene (44.88%) [Elshamy et al. 2020].

Photo 74. Young shoots grow up from *araucaria*. Cyprus, Limassol. Photo by P. Barczak.

Cupressus sempervirens cypress is also popular in coastal areas in Cyprus. The terpenes found in it are similar to those found in Baltic amber. Gas chromatographic analysis showed the presence of α-pinene-$C_{10}H_{16}$- (38.47%) and δ-3-karene-$C_{10}H_{16}$- (25.14%). In trace amounts, we also find other substances known from Baltic amber: borneol $C_{10}H_{18}O$- (1.37%), terpinene-4-ol $C_{10}H_{18}O$- (1.55%), camphene $C_{10}H_{16}$- (0.29%), p-cymene $C_{10}H_{14}$ - (0.86%) [Akermi et al. 2022].

Photo 75. Cones of cypresses. Cyprus, Limassol. Photo by P. Barczak.

The mastic tree (*Pistacia lentiscus*), another "suspected" of Baltic amber formation, occurs in the region of Cyprus and is a popular species throughout the Mediterranean. The shrub / tree tolerates dry, salty coastlines. Its resin is today a well-known delicacy, an additive to beverages. In the past, it was a well-known pharmacological drug, although today it is also used in ethnopharmacy. Its terpenoids are similar to those known from Baltic amber. The tree-shrub could also be a precursor to the formation of proto-amber.

Photo 76. Fruits of *Pistacia lentiscus*. Cyprus, Limassol. Photo by P. Barczak.

In different countries, different terpenes are found in the obtained Pistachio oils, although they are always similar to those known from Baltic amber. In Corsica, for example, the principal compounds are tepinen-4-ol and α-pinene [Castola et al. 2000]. In Greece, α-pinene accounts for 58.9-70% of the essential oil derived from mastic gum, in Israel it is characterized by a high content of α-terpineol [Fleisher and Fleisher, 1992], in Egypt the main ingredient was car-3-en, constituting 65% of the composition essential oil [Barazani et al. 2003]. It is worth noting, however, that the tree tends to prefer valleys and sunny places with some water.

The *Pinus* family in the region of Cyprus and Turkey is represented by five species of pines. These are *Pinus nigra* (black pine), *Pinus brutia* (Turkish pine), *Pinus sylvestris* (Scots pine), *Pinus halepensis* (Aleppo pine) and *Pinus pinea* (stone pine). In the pine trees attacked by pests, compounds known from Baltic amber are found, i.e. α-pinene (20.2%), camphene (4.2%), β-pinene (9.4%), γ-terpinene (10.1%)), β-carophyllen (10.8%) although a very large group of terpenes and terpenoids found in pines are not known from fossilized resin [Öz et al. 2015]. Contrary to appearances, despite the considerable resinization of this type of trees, the composition of the chemical compounds contained in the oleoresin may indicate a minimal impact on the formation of Baltic amber.

Photo 77. *Pinus* brutia tolerates the salty waters of the coast of Cyprus well. Photo by P. Barczak.

In Cyprus, you can also find cedars (*Cedrus brevifolia*) belonging to the pine family. However, they do not grow on the coast but in the mountains, from 600 m above sea level. However, it seems that the cedar forest complexes growing in the mountainous areas did not affect the formation of amber. Cedar wood does not contain the main compounds known from Baltic amber. No terpenoids known from this tree were found in Baltic amber: sesquiterpenes α- and β-himachalene, and they together account for approximately 53% of volatile substances in the analysis of cedar wood. There is also no manool, atlantone, β-himachalene oxide and longiborneol [Fleisher, Fleisher, 2000] known from cedar. No significant amounts of compounds known from the Lebanese cedar resin were found in Baltic amber, which contains dehydroabietic acid including 7-oxo-dehydroabietic acid with pimaric [13 (R) -podocarpa-8 (14), 18-diene-15-oic acid] and sandarakopimaric [13 (S) -podocarpa-8 (14), 18-diene-15-oic acid] [Brody et al. 2002].

Another tree known from Cyprus is the rock oak *Quercus coccifera*. It is rarely seen on the coast. The terpenoid betulina (Eng. Betulin) known from oaks [Sousa et al. 2021] is also not found in succinite, although this relationship does occur in younger ambers, described as Baltic amber.

It can therefore be concluded that it was rather trees growing on the shores of the Eocene Sea that created amber deposits over millions of years, and not those existing in the mountains or inland. It seems logical, for millions of years, numerous fires broke out on the continents, destroying the harvest of trees and resins, and these are known to be flammable. Thus, the conclusion is that coastal places, where there was no fire, produced resins, which in the sea, covered with layers of sand, formed succinite. The question is, however, how did the vast areas of proto-amber in Ukraine and Poland arise? If the period of cooling began in the Eocene period, periods of widening and alternating diminishing of flood waters may have followed. The shores of small sea basins, where low-salinity tolerant trees grew, were absorbed by sea waters over the course of millions of years, and they kept increasing or shrinking, absorbing new emerging coasts, islands, bays and the trees growing on them. During these changes, the resin settled in the sand deposits, because although it is quite light, it always sinks to the bottom, and does not float like a piece of wood. Dynamic climate change has devastated coasts and continents. Destroyed coastal regions formed proto-amber deposits. Former seas turned into peat bogs with significant amounts of proto-amber. The next millions of years of geological processes changed the marsh and peat structures. This is what the very beginning of today's Baltic amber could have looked like, which is today a valuable fossilized resin called succinite.

Not everyone knows that Baltic amber burns, giving off a pleasant resinous smell. After sticking a very hot needle into the amber, it can be easily removed from the amber. On the other hand, it will not be easy to remove it from the young copal resin. When real amber treated with high temperatures, a characteristic resinous aroma will appear.

CHAPTER
II

2. FORMATION OF THE "LIVING STONE"

SUMMARY

Succinite is not the oldest fossilized resin, as much older resins already existed in the Triassic period – 250-200 million years ago. Fossil deposits have been found with fossil resins from as far back as the Devonian period - 400 million years ago. Fossil resins have also been found in Upper Jurassic sediments from 200 million years ago.

Succinite, or Baltic amber, comes from the resin of trees that grew in the area we now know to be Eastern and Southern Europe. For millions of years, slumps of petrified resin were transformed into amber. They have been stored in silt, peat, or inside tree trunks flooded by seawater. Most of the amber can be found in the Eocene layers of greenish-blue earth, which is a form of glauconite rock. This sedimentary rock now serves as a mineral fertilizer containing potassium, magnesium, and iron. It is used to purify drinking water and is a so-called ion exchanger. Passing a solution through an ion exchanger layer enriches it with an additional electric ion.

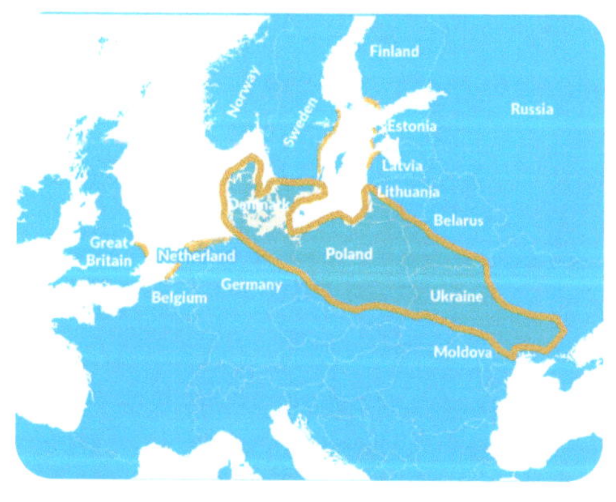

Fig.40. The Eocene Sea area. Source: Based on Bogdasarov, M.A., "Amber from Anthropogenic Deposits", Belarus, 2001.

There are many theories on the creation of Baltic amber, but none of them has been formally accepted by the world of science. There is no one recognized model for the formation of succinite. The problem is that resins cannot be fully recognized in terms of their chemical composition, as a polymer Baltic amber has been subject to constant chemical changes. This chemical structure is still active. In a way, it is a "living structure" that constantly interacts with the outside world. It is possible to classify ambers according to geological eras based on the rocks above and below the amber layers. They can also be described according to their chemical composition, which is becoming the most popular method. Various types of fossilization of resins have been discovered, including fossilization via the sea, via land-swamp, and one resulting from combining elements of rocks and resins. Sometimes methods of creating the fossilized resin were combined, for example, plants drowned in a swamp, were covered with layers of clay, and then the whole area was flooded by seawater. This may have been the most common course. The area where the resins were produced in this way was over 100 000 km2. One hundred and twenty geological occurrences of various resins, mainly retinites, have been found in the Carpathian foothills [Naumenko, Matsui 2020].

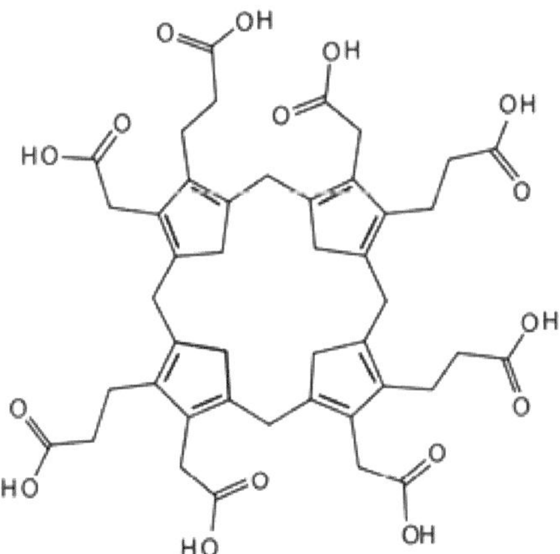

Fig. 41. An example of a polymer structure

Ambers can be divided into five classes according to their main chemical components. Class I, which has three subclasses, namely Ia, Ib, and Ic and includes most of the amber in the world, is characterized by a polylabdanoid structure. It is produced by trees of the *Araucariaceae (Agathis), Leguminosae (Hymenaea, Copaifera) and Cupressaceae* families. Class Ia and Id differ from Ib and Ic in terms of the succinic acid, as both Ib and Ic do not contain succinic acid.

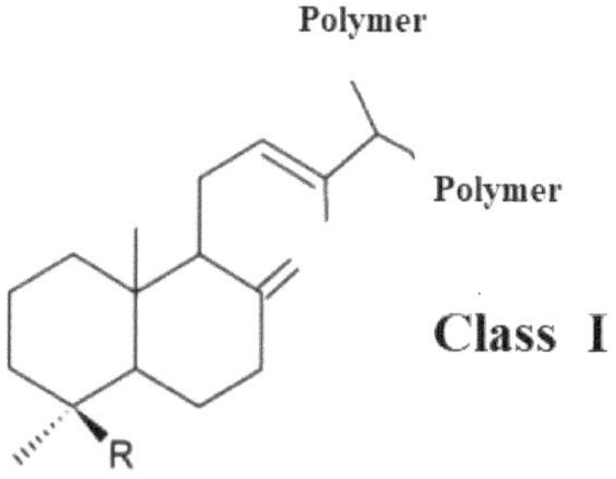

Fig. 42. Class I. A polyabdanoid structure.

The place of the letter R can be taken by: 1) communic acid (COOH), 2) communol (CH_2OH), or 3) CH_3. Class Ib is constituted mainly by New Zealand and Australian lignite bonded resins.

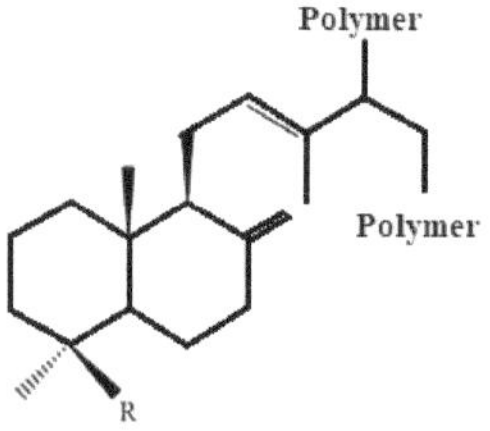

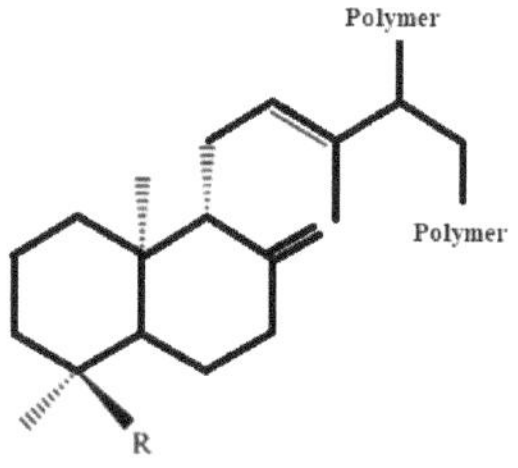

Fig. 43. Class Ia / Ib.   Class Ic / Id.

In class Ic, instead of communic acid, it contains ozic acid found in lotuses ($C_{20}H_{30}O_2$); instead of communal, there is ozol. This group includes amber from Mexico and the Dominican Republic. Mexican ambers were produced by *Leguminosae trees* (especially *Hymenaea*).

Class II ambers are polycadinenes, and are much less common than succinites. Their main plant producers are *Dipterocarpaceae* trees.

Fig. 44. Class II – polycadinenes.   Class III – polystyrenes.

Class III ambers are resins of fossilized polystyrene. Class III includes siegburgite, which is a compilation of resin and silica most likely formed from *Liquidambar* trees – including, among others, the oriental sweetgum growing in Cyprus and Turkey.

Photo 78 and 79. Leaves of the *Liquidambar* tree and "amber" siegburgite, Westphalia, Germany. Collection: Piotr Barczak. Photo by P. Barczak.

Class IV ambers are made of a non-polymeric substance based on a cedrane structure linked to terpenes and terpenoids. Phyllocladanes are one of the Class IV subgroups made of hydrocarbons, found in lignite in Australia, Austria, China, Czechia, Germany, Hungary, Italy, Nigeria and Russia. Class IV specimens contain a mineral called hartite, which was found in the Bílina opencast mine in the Czech Republic. Another specimen contained ixolite, which can be considered a mixture of hydrocarbons, the main component of which is phyllocladane, derived from the *Sciadopitys* or *Cryptomeria* trees [Vávra 2009].

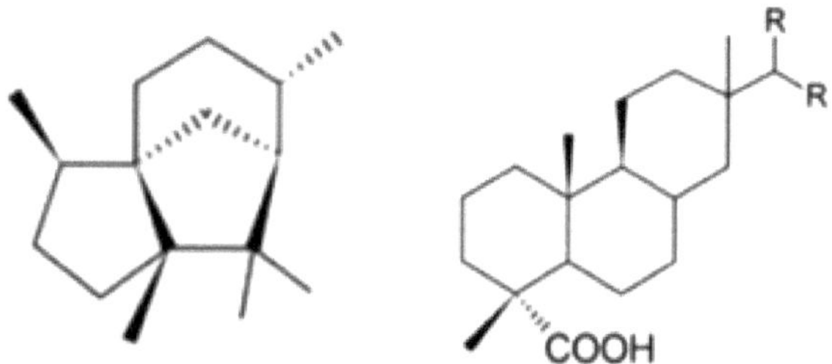

Fig. 45. Class IV - a non-polymeric substance. Class V - diterpenoids: abietane / pimarane.

Summarizing the ambers from around the world, Baltic amber has been established as class Ia, Burmese amber and amber from Fushun,China are class Ib, and Dominican amber is class Ic [Kong, Meng, McKenna 2022].

Another issue is the type of trees that grew where proto-amber was formed. In 1889 H. Conwentz and O. Völkelin theoretically studied the existing trees of subtropical forests and compared them with fossils. They imagined that the redwoods and pines were the highest and oaks, maples, beeches, and sycamores were lower. They were mixed with laurel, magnolias, and myrtle. In addition, there were palm trees and ferns. They believed that the pines were *Pinus Baltica, Pinus cembrifolia, Pinus parviflora, and Pinus sylvatica* [Conwentz 1890]. According to another concept, formulated based on research on coal pits, the presence of the

following subtropical trees was found in the areas of amber formation: cypresses, araucaria, *taxodiums*, sequoias, *Ginkgo*, the conifers *Pinus haploxylon* and *Pinus diploxylon*, *Picea*, and *Podocarpus* (plum pine). Magnolias, palms, olives, myrtles, hollies, and heathers were also found [Naumenko, Matsui 2020].

It has also been noticed that the chemical composition of amber depends on the fossilization process and different conditions in the formation of Baltic amber forests. The majority was formed in the Carpathian foothills, Volyn, where sea floods occurred. An essential concept in this context is the idea of polluting the Eocene Sea with hydrogen sulfide, which resulted in an intense release of resin in the forests growing in this area [Maidanovich 1988]. This concept is consistent if we take into account the sulfur deposits formed in the vicinity of Tarnobrzeg, Poland, i.e., in the region flooded by the sea in the distant past.

Generally, since the first scientific publications 200 years ago, no complete answers to important questions related to the creation of Baltic amber, its genesis, and its origin have been given. In the case of succinites found in the Baltic Sea, it was possible to evaluate their color, type, and healing properties in the beginning. Only modern measuring devices have created the possibility of a more complete picture of the issue. The main feature of succinite is its material stability. The tree elements – cellulose and lignin – turned into peat matter and then into coal seams. Fires, which allegedly affected the release of the resin, intensified the release of this substance. The amount of resin released has always been influenced by the size of the tree's crown and the diameter of the trunk, as well as by sunlight and temperature. Baltic amber resins were formed in a warm and humid climate, although a subtropical climate (similar to the climate of Greece) could also work. However, it was changed because of millions of years of evolution. The trees of the amber forest had increased resin channels, which is typical of ancient natural processes. This is evidenced by large pieces of fossil resin found. The type of substances flowing out of the trees was of major importance for the properties of succinite as well as the geological conditions under which the substance formed. For if this were not the case, we would have amber deposits wherever there is brown and hard coal. However, we do not have them. Even if there is amber in coal, its quality is very low, and the chemical composition has no pharmacological significance.

The process of resin fossilization is a fascinating issue. Amber is created through the process of terpenoid fossilization, and they influence the polymer structure being formed, which is why each piece of amber is a work of art formed from resin and essential oils over millions of years. Pimaric and abietic type resin acids have remained unchanged in amber for millions of years. Various species of trees made resins, but only a few have been fossilized. A strange transformation turned volatile compounds into permanent, chemically stable substances preserved in amber.

Another process took place in peat. After a peat bog had been created, the resin was compacted, and its mass increased. Despite these changes, the basic structures have consolidated and remained unchanged. These types of phenomena also lasted for millions of years.

Succinite, amber of the Baltic Sea, is a natural polymer in which numerous vegetable oils, amino acids and terpenes have been preserved. It contains ingredients of resins and oils derived from trees and plants from the distant past. These structures were additionally affected by the processes of transformation of coal, oil, and natural gas taking place in the ground. The hardening of the substance may have occurred under the influence of pressure generated by the environment. The fairly shallow Paratethys Sea could cause a pressure of three

to five bar. We know this because we can observe this type of pressure at a depth of 20m in salt water. The pressure probably hardened the succinite formed. Perhaps the pressure was higher because today, the amber pressing processes use a pressure of 3000 bar and a temperature of 250 degrees Celsius [http://www.baltic-amber.eu/metody-obrobki.html].

The succinite production process could be as follows. Resin in the form of a sticky substance found in fallen trees, leaves, and branches formed a proto-amber layer. Bacteria could have influenced the shaping of the raw material. In the depths of the earth, mixed with wood, hardening processes took place. The areas that were flooded with seawater hardened due to the pressure. As a result of being covered with the oxidation crust, the nuggets obtained greater hardness. An example of this is gedanite, which exhibits such a layer. It is amber which does not contain succinic acid, so it was probably formed under different geological conditions than succinite.

Photo 80. Gedanite from Bitterfeld. Photo by P. Barczak

In the depths of the earth where succinite was formed, the formation of amber polymeric structures (related chemical molecules constantly looking for appropriate compounds) took place. "Living matter" interacted with its biological-geological environment over millions of years. Over time, it turned into lignite, bitumen, crude oil, and natural gas. Under anaerobic conditions, the resin gained entirely new properties. Synthetic polymers are now produced from hydrocarbons such as petroleum and natural gas. Some of the fossil resins contained in brown coal, formed in different geological conditions, have not turned into succinite. These resins did not have an oxidation crust, so they were unstable and had a different chemical structure than succinite. An example of this is amber from Czechia and Moravia. Succinite has a durable, hard structure, but Czech and Moravian ambers are only of collector's importance.

Photo 81. Duxite in a petrified tree, Břežánky, Czech Republic. Photo by P. Barczak.

Succinite always occurs in so-called blue earth glauconite along with siderite, and pyrite. This is because amber has specific properties of permeating liquids and gasses as well as swelling in water. Therefore, the continuous influence of the external geological and environmental environment (glauconite) on amber changed its internal molecular systems. Over time, these primary deposits of rocks, minerals, and amber were mixed and displaced. As the climate changed, glaciers began to work, and over time, floods and rising rivers during warming periods made their marks too.

In the Black Sea region, succinite is found on the Dnieper River, near the towns of Berislav and Kakhovka, in the Kremenchuk region, as well as on the banks of the Khorol and Samara rivers [Bogdasarov 2005]. Amber was transported to the Baltic Sea and numerous regions in Poland by post-glacial waters, which deposited amber in lakes and rivers. A huge Vistula sedimentary delta was formed near Gdańsk (Poland). Sea currents also carried amber along the southern shores of the Baltic Sea, and glaciers and emerging waters transported amber throughout the Polish lowlands. This is how, among others, amber deposits in the vicinity of Ostrołęka (near Warsaw) came into being.

There are some pieces of hard evidence for such a hypothesis. One of the key ones is geological studies conducted by the Polish Geological Institute in Warsaw and presented, inter alia, at the International Geological Fair in 2014. Drilling carried out on the Baltic Sea coast showed that amber was formed in the Eocene layers in various places on the Baltic coast near Gdańsk. Pieces of Eocenian amber have been found while analyzing the layers of earth and rocks to a depth of approximately 170m, solely in the layers of clay and sands from the Eocene period.

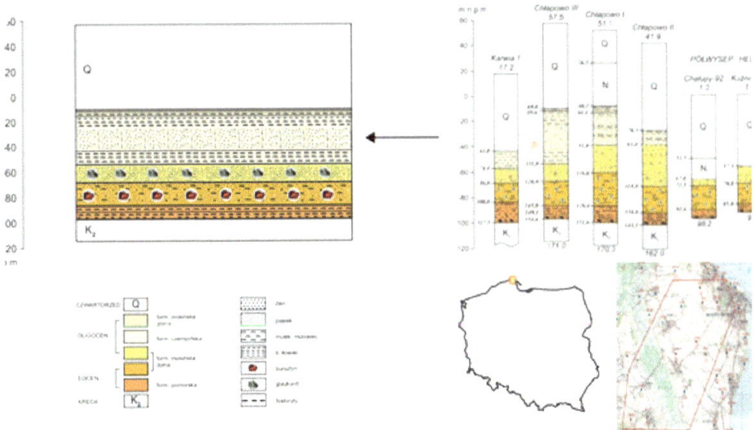

Fig. 46. Distribution of amber in particular layers of the earth in the Baltic Sea coast region. The amber color marks the layer from the geological period of the Eocene. The brownish, yellow, and amber zones show the existing layers of glauconite. Drilling area is on the right. Based on: https://www.pgi.gov.pl/en/dokumenty-przegladarka/aktualnosci-2014/2562-geologia2014-bursztyn/file.html

However, there is a problem with the identification of ambers found. They were found at shallow depths. Currently, there are no mines with depths greater than 100m in Poland. As such, the Baltic amber found does not have a documented age of origin. Pieces of raw material found could be from the Eocene period, but also from the Cretaceous or the Miocene.

During this period, the climate was moderate, with an average temperature of 25 degrees Celsius, without any division into climatic zones. The warmest period of the Eocene was EECO time (Early Eocene Climatic Optimum). It was a period of lush vegetation growth and the development of numerous plant species. Approximately 45 million years ago, climatic conditions changed, and consistent cooling began across the planet. The forests began to resemble the present-day warmth-temperate, and subtropical forests of China and Japan. The Oligocene period (about 33 million to 23 million years ago) featured the total collapse of existing forests. Peat bogs and swamp forests formed in the valleys. The swamp cypress (*Taxodium distichum*) was an important species, which now grows naturally in the muddy areas of Florida (USA). Other vital species were *Nyssa sylvatica* (the black tupelo, a tree species of the *Nyssaceae* family growing in wetlands in the east of the US) and trees from the *Glyptostrobus* genus of the *Cupressaceae* family, now known from China and Vietnam and growing on moist soil along rivers and streams [Słodkowska, Kasiński 2016].

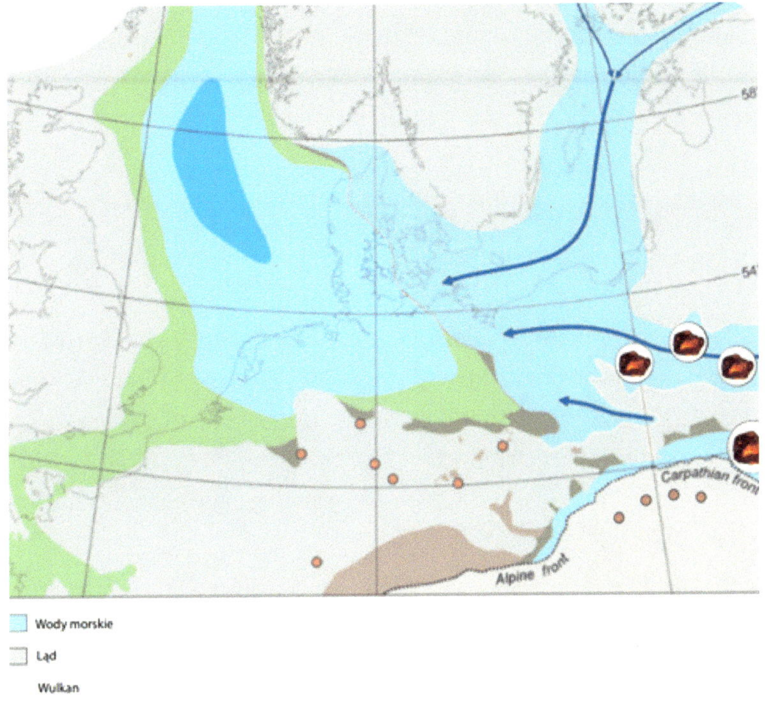

Fig. 47. A hypothetical outline of the North Sea (left) and rivers flowing into it in the Middle Miocene. Based on: Gibbard, PL, & Lewin, J. (2016). "Filling the North Sea Basin: Cenozoic Sediment Sources and River Styles (André Dumont Medallist Lecture 2014)". Geologica Belgica.

Over millions of years from the Eocene to the Miocene, 23.03–5.33 million years ago, there were systemic changes in the seas and land of modern-day Europe. These changes were associated with the degeneration of forests, the formation of resin deposits, and their transport to the North and South. There was more and more ice on the planet and the seas grew shallower. The Sea of Paratethys began to dry up about 13.7 million years ago. In the Carpathian foothills region, deposits of rock salt and gypsum formed, i.e., minerals containing sulfur. Finally, sulfur deposits were formed in the hypothetical region where the main deposits of Baltic proto-amber were created. The "boiling of resins" in the drying sea of sulfur hardened them and transformed them into Baltic amber, unlike Czech amber, which has not undergone this type of process.

After rubbing against cotton or natural fabric, amber attracts pieces of paper and dry grass. The electrons from which the substances are built pass from the fabric to amber. The generated local amber electric field attracts pieces of paper with the opposite electric charge.

The first person to notice the existence of an electrical phenomenon was the Greek scientist Thales of Miletus (620– 540 BC). Thales noticed the attractive power of amber rubbed with animal fur. He ascribed a soul to objects. The soul was static electricity and magnetism.

The "soul" was static electricity and magnetism. And: Pieces of the paper attracted by Baltic amber rubbed with paper.

2.1. THE INFLUENCE OF GLAUCONITE ON BALTIC AMBER

Glauconite is a distinctive green sedimentary rock, classified as a component of sandstone. It was described as an independent raw material in 1928 by Keferstein, who called it glauconite from the Greek word "glaucus," which means blue-green. The binder of glauconites is silica, calcite, and, to a lesser extent, zeolites. As a rule, this rock is made of clay and silica-aluminum coagulum. Its crystallinity increases with the maturity of the rock, then it becomes richer in potassium and iron, and its color moves closer to green. Glauconite contains swelling layers, which is also characteristic of amber, as succinite also swells. During maturation, the amount of potassium cations in it increases. Glauconite has unique ion exchange and absorption properties in comparison to metal cations. It is also characterized by exchanging its active ions with ions in the surrounding solution. Uncompensated negative ionic charges resulting from internal exchange cause constant chemical imbalances and reactions with environmental ions. This type of condition most often occurs in the outer layers of minerals.

Photo 82. A glauconite layer in a cliff by the Baltic Sea. Photo by P. Barczak.

The mineral is often found in Cretaceous and Tertiary sediments. It was then formed near coasts. The unripe glauconites are white, while the dark green ones are well crystallized. The exchange capacity of glauconite depends on the pH of the solution in which it is located.

The largest glauconite deposits are in Podolia and Volhynia (Ukraine) and the vicinity of Lubartów (Poland). The local types of sand and clay contain varying amounts of glauconite, sometimes 2% and sometimes 70% [Franus 2010]. When combined with amber resin, which is a polymer, silica-containing glauconite and calcite can form a polymer hybrid that becomes a succinite used for jewelry purposes. Hybridization may take place through reactions in chemical bonds [Nanko 2009]. If exposed to a UV lamp, calcite, gypsum, and amber, all fossilized specimens react similarly, "shining" under the influence of UV rays. Glauconite behaves slightly differently.

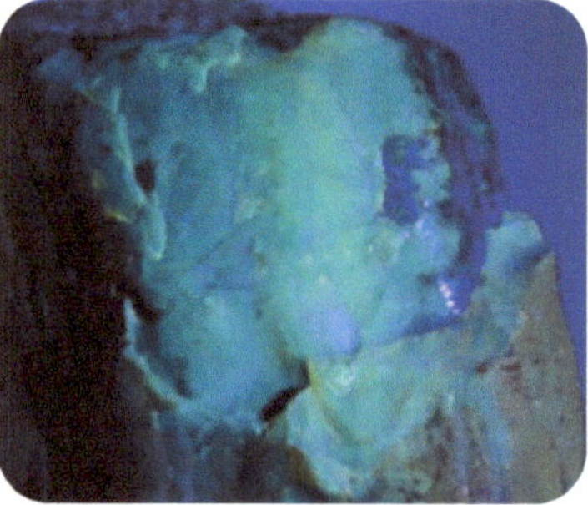

Photo 83. "Glowing" at night in the light of a UV lamp: gypsum (left, top), succinite (right, top) and calcites (bottom) against a backdrop of other minerals. Photo by P. Barczak.

Only mineral particles shine. Perhaps just these "glowing" particles react with amber if both substances stay together for millions of years, hidden under deposits of more modern clays, sand, and rocks.

Photo 84. The effect of UV light on glauconite and microparticles reflecting UV rays. Photo by P. Barczak.

It is not entirely certain which minerals "shine," as such investigations are beyond the scope of this publication. However, it is known that glauconite is rich in Fe (iron). It also contains K_2O and mica. The content of potassium – K – determines its color, though the smectite silicate also has an influence. Sometimes the composition of aluminum and iron elements can also determine the "glow." It is therefore possible that UV irradiation changes the value of the iron cations from Fe^{+2} to Fe^{+3}, which causes the glow. It is also possible that coexisting iron cations in the structure of the silicates cause the mentioned effect [Sánchez-Navas 2008].

2.2. THE INFLUENCE OF GYPSUM, ANHYDRITE, AND SULFUR ON THE UNIQUE PROPERTIES OF AMBER

The succinic acid probably got into the succinite through the diffusion of hydrogen sulfide with the fossilized resin. As a result, the oxygen contained in it formed strong internal macromolecules. Consequently, the succinite also acquired a certain hardness. The observations of amber oxidation confirm such a hypothesis. With higher oxygen content, microcracks appear on the surface, and amber becomes dull. In such layers of fossilized resin, the sulfur content decreases, and the carbon and hydrogen content increases [Savkevich 1983]. Similar hardening processes are used in modern industrial processes, where the hardening of resins occurs through contact with the hardener, i.e., sulfur. It is worth recalling here that the elemental composition of succinite is as follows: C 61 – 81%, H 8.5 – 11%, O – complement to 100%, S – 0.5%. Up to 7% sulfur is found in Ukrainian Miocene amber from the Carpathian Foredeep and in krantzite [Kosmowska-Ceranowicz 2017].

Photo 85. Cracks in the Baltic amber. Photo by P. Barczak.

A regularity has been noticed related to the formation of sulfur deposits in the areas where proto-amber was formed. It is possible that in the Carpathian Foredeep region, which was formed by the former Paratethys Sea, gypsum ($CaSO_4 \cdot 2H_2O$) and anhydrites ($CaSO_4$) were transformed into sulfur deposits. It always happened in the presence of bitumens, which formed a kind of gas niche from the deeper layers. Under such circumstances, gypsum and anhydrite could transform into sulfur, and Baltic amber was thus transformed into a strong gemstone [Osmólski 1963]. We have observed a similar influence exerted by sulfur compounds in German amber deposits in Goitzsche. The lakes left after the lignite mine are fed with sulfides [Trettin et al. 2007]. The amber-bearing regions are also known for the presence of gypsum and anhydrite [Stollberb 2013].

Photo 86. Native sulfur, Tarnobrzeg, Poland. Photo by P. Barczak.

If amber was created in the Forecarpathian in Poland and Ukraine, then we should find places where sulfur can be found in this region. Such places can be found in the areas where Baltic amber can potentially be formed. One such area is the Carpathian Foredeep, i.e., Tarnobrzeg, Machowa, Piaseczno, but also the region of western Ukraine near Lviv. The sulfur deposits found there were created with the participation of bacteria. It is assumed that living organisms contributed to the formation of this mineral in the vicinity of Tarnobrzeg. The bacteria *Desulfovibrio desulfuricans* have been found in hydrogen sulfide deposit samples from the Forecarpathian region. As part of the experiments, these bacteria were grown. and then, with their help, an experimental conversion of gypsum into sulfur was carried out. In the world's scientific literature, these bacteria have been attributed with a role in converting gypsum into hydrogen sulfide in the presence of natural gas and crude oil. This reaction can be as follows:

$$CaSO_4 + (C+H_4) \rightarrow H_2S + CaCO_3 + H_2O$$

However, there is no consensus among the scientific community on how gypsum is converted into sulfur. It is assumed that the *Thiobacillus thioparus* or *Thiobacillus thiooxidans* bacteria transform the gypsum, or the transformations take place chemically without the participation of bacteria.

Bacteria of different specializations could cooperate in the process of sulfur formation: older ones – reducing, and younger ones – oxidizing. Older reducing bacteria have been found directly on the fracture planes of post-gypsum limestones or in small caverns [Ryka 1986].

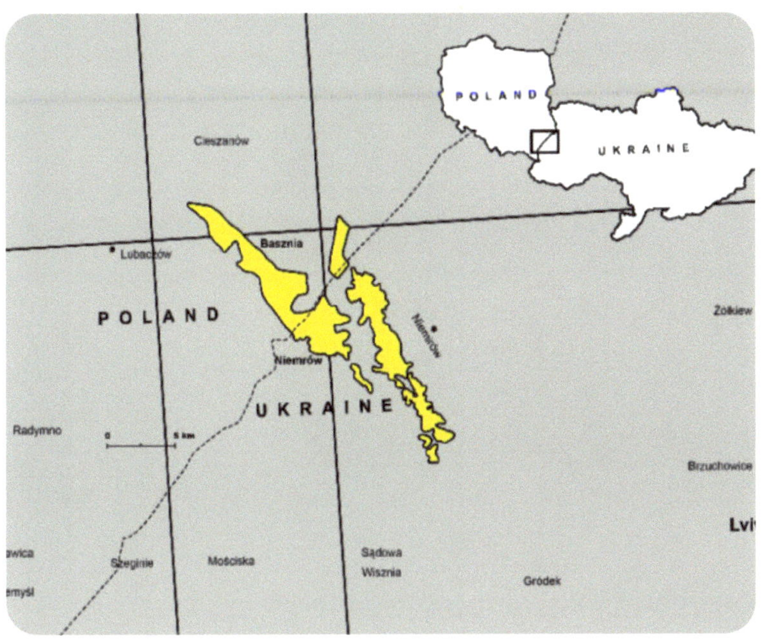

Fig. 48. A simplified map of sulfur deposits in southern Poland and Ukraine. Based on: Gąsiewicz, A., Jasionowski, M., & Poberzhskyy, A. (2012). "Wpływ eksploatacji siarki na cechy geochemiczne środowiska powierzchniowego złóż siarki z pogranicza polsko-ukraińskiego". Biuletyn Państwowego Instytutu Geologicznego.

As early as the 19th century, there were discussions about sulfur formed in Ukraine and Poland as a result of organic matter's influence on gypsum in the "atmosphere" of liquid bitumens. Over time, laboratory studies have confirmed the bacterial conversion of sulfates to sulfur [Gąsiewicz 2006]. So there is hard evidence of the sulfur formation in combination with bitumens in proto-amber formation.

Photo 87. Selected Baltic amber from the Rovno region in Ukraine. Photo by P. Barczak.

Similar sulfur deposits have been found in Ukraine and the vicinity of Lublin in Poland, i.e., not far from currently exploited amber deposits. The deposits were formed in the Miocene epoch (from 23.03 million to 5.333 million years ago), i.e., in the period when most resins were probably flooded by waters of the Paratethys Sea, which, moreover, was slowly disappearing. Sands and loams were deposited into sinkholes, and extensive peat bogs were created [Gąsiewicz, Jasionowski, Poberzhskyy 2012].

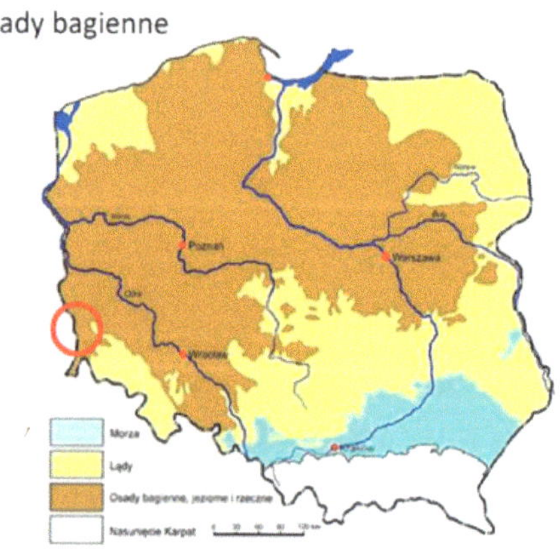

Fig. 49. Koźma, J. (2016), "Anthropogenic landscape changes related to the former lignite mining on the example of the Polish part of the area of Łuk Mużakowa". Górnictwo Odkrywkowe, 57.

Amber contains some sulfur, which is its specificity. Some scientific concepts assume that succinic acid – a unique substance found in significant amounts in Baltic amber – was created due to a volcanic catastrophe. Volcanic ash could have significantly changed the ancient world of plants and entire ecosystems. The area where Baltic amber was formed was indeed subject to volcanic processes in the past. During the Eocene epoch, volcanic activity began in the Carpathian region, as evidenced by numerous inserts of volcanic tuffs, which resulted in the development of a biochemical cycle in the sediments [Gucwa, Poprawa 1996].

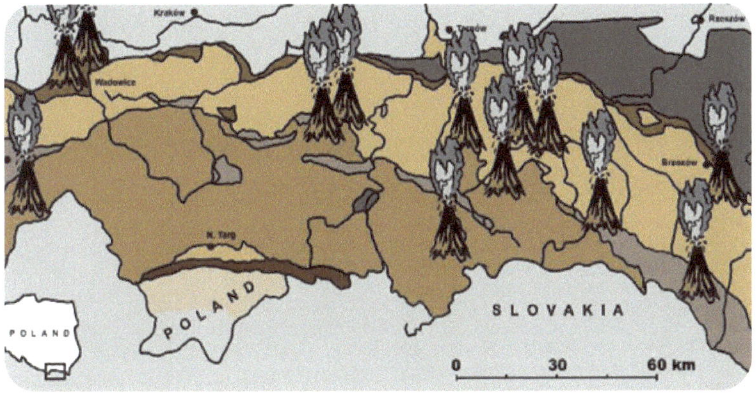

Fig. 50. A simplified map of the volcanoes in the Carpathian Foredeep area, Poland.

The volcanoes were active for a long time in the area where Baltic amber formations occurred (the Carpathian Foothills in Poland and Ukraine). In the Slovak Carpathians, traces of volcanic eruptions have been detected in the Vysne Zbojne village, and the volcano's creation was dated back to the Middle Eocene epoch. Volcanic sediments have been found in the town of Danilovka in Transcarpathia (Ukraine) and are classified as coming from the Upper Eocene epoch. Volcanic ashes have also been found in the village of Gernyes in Hungary. Descriptions of rocks and tuffs show that in this region, the Eocene period was characterized by strong volcanism. Centers of Carpathian volcanism existed at different times. The volcanic ash minerals have transformed into glauconite [Koszarski, Wieser 1960].

Photo 88. A petrified tree, Miocene, Slovakia. Photo by P. Barczak.

There are also known occurrences of volcanic ash in Ukraine a short distance from the documented deposits of "Baltic amber" in the vicinity of Klesiv, which supports the theory that volcanoes may have influenced the formation of succinic acid. It can also be assumed that volcanic processes led to the burnout of forests. Certain kinds of trees and other plants developed in areas burnt by volcanic ash, on steppes whose emergence was related to climate change. One of them could be the *eucalyptus*, a plant predisposed to the colonization of warm areas, which survives particularly well after burning out competing trees.

In the Miocene period, amber in the Carpathian Foredeep was formed in the vicinity of sulfate rocks, i.e., gypsum and anhydrites. Anhydrites in this area occur deeper than gypsum, as gypsum can be found almost on the surface. The anhydrites mineralized as a result of the evaporation of brines from the vanishing Paratethys Sea, especially at very high salinity [Kasprzyk 2005]. The whole Carpathian Foothills area in the region analyzed is covered with gypsum and anhydrite deposits. In such conditions of the perishing Paratethys Sea, the world of trees and other plants were also dying, and the resins were hardening. Huge deposits of plant origin came into being. Today's succinite has been shaped from this matter. **Sulfates and sulfur shaped the structure of Baltic amber**.

Fig. 51. The area of formation of sulfates-gypsum and anhydrites. Based on: Kasprzyk, A. (2005), "Genetic Models of Baden Anhydrites in the Carpathian Foredeep in the Area of Poland", Przegląd Geologiczny, 53(1), 47-54. Panow, G. M., & Płotnikow, A. M. (1996), "Baden Evaporates of the Ukrainian Fore-Carpathian: Lithophages and Thickness". Przegląd Geologiczny, 44 (10), 1024-1028.

Similar gypsum and anhydrite formations took place in what is now Ukraine. Here, too, almost the entire area is covered with gypsum and anhydrite. Amber formed in such a geological structure by absorbing sulfur compounds from disintegrating mineral structures. Today, limestone nanoplankton and conglomerates consisting of gypsum and anhydrite are found in sediments there. Is it possible then that some of the amber found in the exploited shallow areas of the Gdańsk region come from the Miocene and not the Eocene?

2.3. BITUMENS, LIQUID COAL AND CRUDE OIL, AND SUCCINITE

There is a hypothesis that amber formation occurs due to the influence of bitumens on proto-resins [Matsui 2013]. It is assumed that in areas where lignite erosion occurs and where bitumen – a solid substance, sometimes of high viscosity, which is a mixture of numerous hydrocarbons – has been found, bitumen influenced the amber structures. Several chemical compounds in amber suggest this may be true. There are examples of such processes taking place, such as the degradation of lignite deposits at the Mir mine in the Czech Republic, which resulted in the oxidation of aromatic structures and fission of chemical compounds contained in amber in the presence of bitumen. This is how succinic and malonic acids have been formed in Czech amber [Doskočil et al. 2014]. The oxidation took place at a temperature of 30-50 degrees Celsius. Therefore, it is possible that succinite was formed in places where we can find asphalt, crude oil, and natural gas today. Thus, bituminous lignite was one of the components in the formation of succinite, which is in line with previous observations. In previous centuries, studies noted that bitumen spilled in areas where ambers were found. It was even assumed that amber is made of bitumen.

The bituminous environment, along with glauconite, can influence the structure of succinite. Bitumen causes gelation of peats with natural resins from fallen trees. A law exists on the transformation of minerals, which states: for the phenomenon of complete conversion of gypsum and anhydrite to native sulfur to occur, a continuous supply of bitumen is necessary [Osmólski 1963].

Covering the sediments with seawater created closed niches beneath the seafloor in which the processes of succinite polymer structure recombination took place. Layers of mature amber were washed away from peat bogs, and different types of amber were deposited in different places. For example, the amber known as retinite can be found in the Dnieper basin. The amber known as gedanite, with a minimum succinic acid content and more brittle nature than succinite, can be found in the Gdańsk region and in Bitterfeld in Germany. Similarly, stantienite – a black amber which does not contain succinic acid – can sometimes be found in Ukraine as well as near Bytów in Western Pomerania in Poland, and in Sambia near Gdańsk. Amber stones that have undergone a transformation below the seabed in an alkaline environment of bitumens have ethnopharmacological and jewelry values [Matsui 2013]. The other can be parts of hobbyist collections.

Photo 89. Black amber – stantienite – found near Gdańsk. Photo by P. Barczak.

It is worth noting that bitumens are released in many places of the Carpathian Flysch (Poland), where gas and oil deposits exist [Karnkowski 1962]. Western Ukraine is similar, as bitumens are present all over the region suspected of creating proto-amber

[Więcław et al. 2008]. It cannot be ruled out that Baltic amber was also formed in this region according to the "bituminous" concept. But the glaciers have washed away almost all amber deposits and transferred them through the river valleys towards Gdańsk and the Baltic coast.

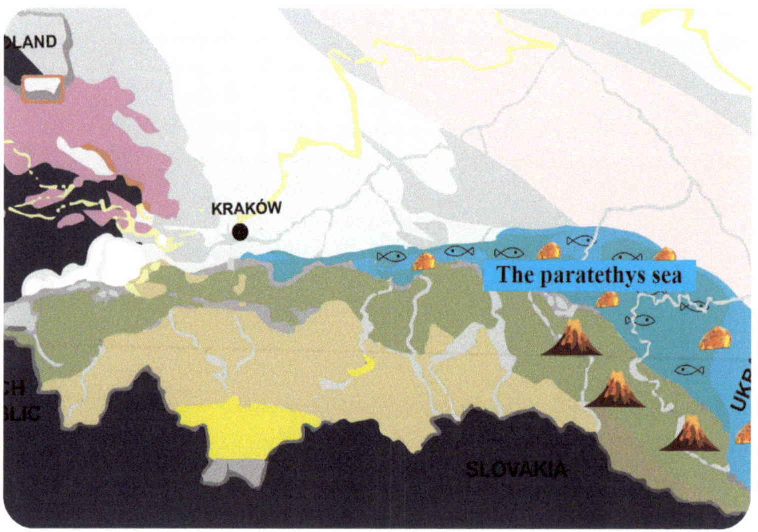

Fig. 52. The area of the Carpathian Flysch with deposits of crude oil and natural gas.

Huge coal deposits created due to geological evolution influenced the emerging structures of amber. Chemical compounds found in bitumens have been found in Baltic amber i.e., naphthalene compounds. For example, $C_{13}H_{18}$ – naphthalene, 1,2,3,4, -tetrahydro-1,6,8-tri methyl – has been found in trace amounts (0,97%) in Baltic amber. Moreover, $C_{14}H_2O$ or naphthalene, (1,2,3,4, - tetrahydro-5,6,7,8-tetramethyl) is also present in trace amounts of 0.72% in succinite. Another structure found in amber is heptane (bicyclo [2.2.1] heptan-2-ol, 1,3,3-trimethyl-, acetals, (1s-exo) -) in the amount of 3.75% ($C_{15}H_{26}O$). Heptane is involved in the processes of oil formation. These types of compounds are also formed in the processes of coal liquefaction. In these processes, naphthalene is the main volatile compound formed under the influence of temperature, as the resulting naphthalene vapors are flammable [Buku et al. 2006]. Interestingly, amber's benzene compounds do not always come from oil or coal. The compound 2,6,6-trimethylcyclohex-1-enylmethane-sulfonyl, i.e., benzene, was found in the plant *Embilica Officinalis*, specifically in the ethanol extract of this tree's fruit [Hamsarekha et al. 2017]. The question regarding the influence of bitumens on the formation of Baltic amber remains unexplored in the laboratory, which does not mean that such an influence did not exist.

Photo 90. Petrified bitumen. Photo by P. Barczak.

Also, laboratory experiments show that sulfur compounds influenced the formation of Baltic amber. A mutual interaction of water and organic matter produces hydrogen sulfide (H_2S), sometimes in the vicinity of asphalts, which react with surrounding matter. Hydrogen sulfide is often formed in swamps as a result of the metabolism of rotting plants in the absence of oxygen. It is anaerobic fermentation with the participation of sulfate-reducing microorganisms. Hydrogen sulfide is also present in volcanic gasses and natural gas deposits, which also indicates how the structure of Baltic amber is formed and preserved.

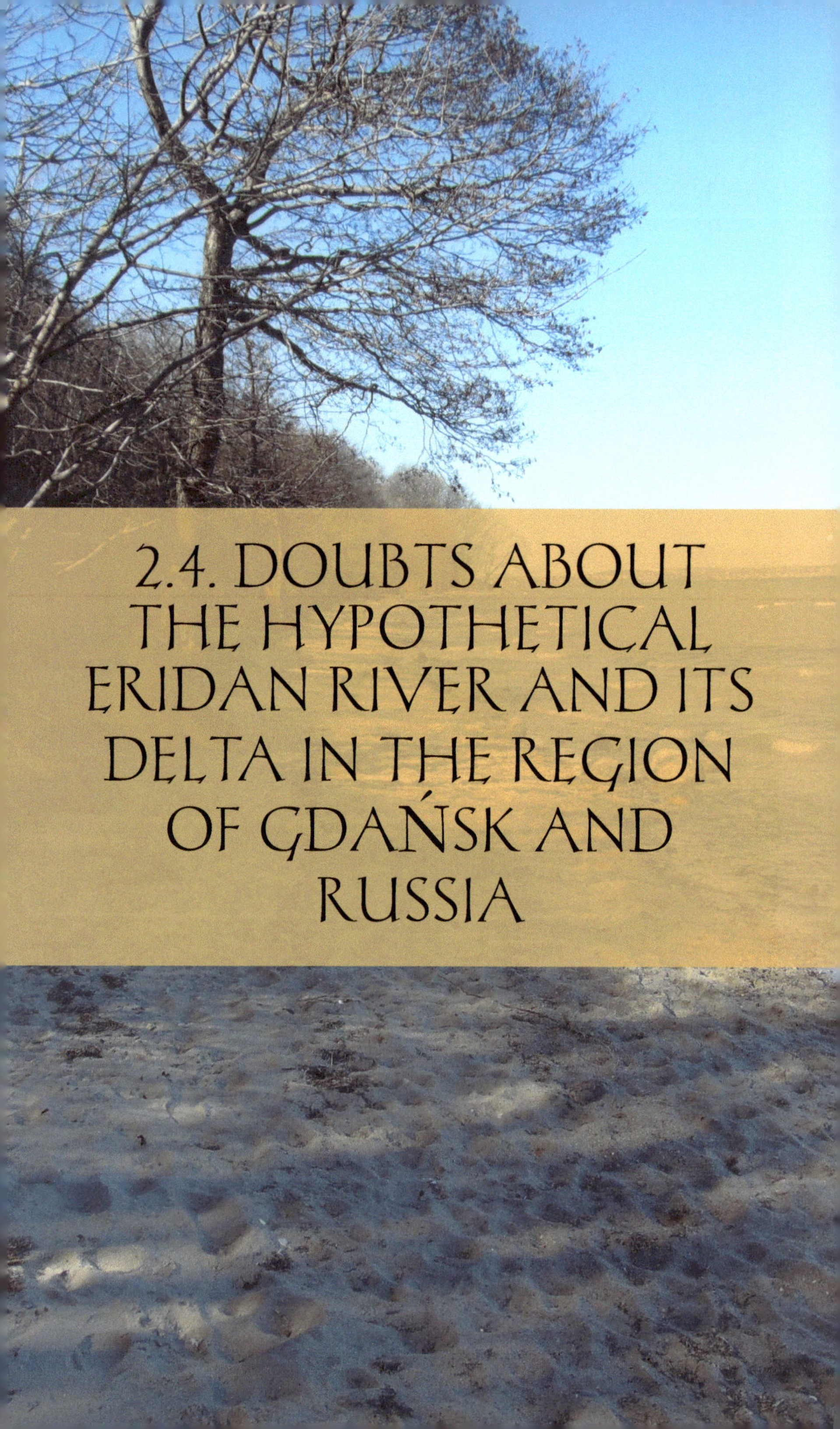

2.4. DOUBTS ABOUT THE HYPOTHETICAL ERIDAN RIVER AND ITS DELTA IN THE REGION OF GDAŃSK AND RUSSIA

Some scientific materials hypothesize that Baltic amber was formed by a great prehistoric river called Eridan, flowing from the area of present-day Sweden towards Poland. It was supposed to be the main supplier of amber to the Gdańsk region in Poland and to the present-day Kaliningrad region in Russia, but there seems to be no geological evidence documenting the existence of this type of watercourse. Also, the search for sulfur deposits in Pomerania, because sulfur could hypothetically preserve amber deposits, has not brought any results. Structures containing sulfur, specifically sulfide deposits, have been found the area of the German brown coal deposit Goitzsche [Trettin et al. 2007], but no sulfur deposits have been found in the area of the Vistula delta or in Sambia (Palvininkai -Yantarnoye).

Concepts about amber and the alleged delta of a great river, which had its mouth near Gdańsk and Kaliningrad, refer to press articles quoted in the "Pravda" newspaper printed during the USSR [Katinas 1987].

Fig. 53. Vision of the Permian period – a part of the Paleozoic era, lasting from about 298.9 ± 0.15 to 252.17 ± 0.06 million years ago. Painted by K. Doszla.

Admittedly, experimental drilling has confirmed the existence of amber-bearing layers from the Eocene era in the area of Chłapowo near Gdańsk in Poland, but has not confirmed the existence of an ancient river delta, for example, from the Eocene epoch or the Permian period. It was only found that amber could be carried away by surface runoff and deposited in the coastal areas. The nearby brown coal, in which the succinite could be formed, is similar to deposited glacial rafts, not primary deposits formed in that place. The deposits are similar to those found in the Carpathians' foothills and Polish lowlands [Wagner 2007]. They have been most likely brought in by glaciers.

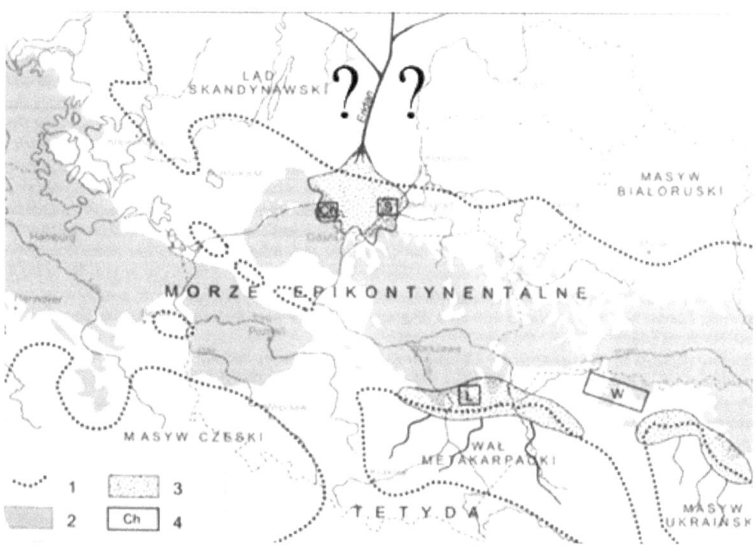

Fig. 54. The Epicontinental Sea with the hypothetical Eridan River marked, presented in the Amber Museum in Malbork (Teutonic Castle). Source: Kramarska, R., Kasiński, J., & Sivkov, V. (2008), "Paleogene Amber in situ in Poland and in Neighboring Countries - Geology, Exploitation, Perspectives". Górnictwo Odkrywkowe, 49 (2, no. 2-3), 97-110.

Another uncertain hypothesis is that 3.8 billion years ago, the Fennoscandia continental platform came into being (which has been scientifically proven) and from this region, trees with fossilized resin shifted to Gdańsk (no scientific evidence). The question remains as to whether it was covered with amber-bearing forests when amber was formed, as there is no evidence regarding this point. Furthermore , if there is no scientific evidence supporting such a supposition, did a mythical river carry amber to the Eocene sea reservoir in the Baltic Sea region river 50 million years ago [Kosmowska-Ceranowicz 2017]? The hypothesis, including the concept of creating proto-amber, has not been confirmed by scientific evidence. The Eridan hypothesis is also excluded by evidence due to the lack of gypsum or anhydrite deposits in the Gdańsk area. There are also no such deposits in Palvininkai (Yantarnoye) – current-day Russia. Though there are large deposits of glauconite in these places,no gypsum or anhydrite can be found nor any sulfur compounds. On the other hand, large deposits of these minerals have indeed been found in the Ukraine, and Podkarpacie in Poland. Deposits of gypsum and anhydrite have also been found in the area where German Bitterfeld amber deposits are located. At Puck Bay, in Chłapowo near Gdańsk, drilling has been carried out, and deposits of anhydrite and gypsum have been found, but they exist below the depth of 660m, in the layers from the Permian period, that is, from 298.9 to 252.17 million years ago [Domagała 1982] – not from the Eocene.

ERA ERA	OKRES PERIOD	EPOKA EPOCH	WIEK (MLN LAT) MILLIONS OF YEARS AGO
ENOZOIK ENOZOIK	CZWARTORZĘD/QUATERNARY	HOLOCEN/HOLOCENE	0.0117
		PLEJSTOCEN/PLEISTOCENE	2.58
	NEOGEN/NEOGENE	PLIOCEN/PLIOCENE	5.33
		MIOCEN/MIOCENE	23.03
		OLIGOCEN/OLIGOCENE	33.9
	PALEOGEN/PALEOGENE	EOCEN/EOCENE	56.0
		PALEOCEN/PALEOCENE	66.0

Tab. 4. Source: "Succinite, Amber". Polish Geological Institute National Research Institute, Warsaw, 2018. https://www.pgi.gov.pl/dokumenty-pig-pib-all/foldery-instytutowe/foldery-surowcowe-2018/6215-folder-bursztyn/file.html

It is important to note that Baltic amber that is sold is not always of Eocenian origin. Rather, it is a mixture of ambers from different time periods. There are Eocenian deposits (56-33.9 million years), as well as Pleistocenian (2.58 -0.0117 million years) and Holocenian, i.e., modern ones. Such a mixture of differently colored ambers makes identification even more difficult.

The nature of a small amber deposit in the refractory clay mine Stanisław in Lower Silesia (Poland) also confirms that amber in Poland was dispersed as a result of the "work of glaciers" and not due to a specific river. The statements of geologists show that in small caverns where lignite amber accumulated, it was also surrounded by marcasite, a mineral from the sulfide group.

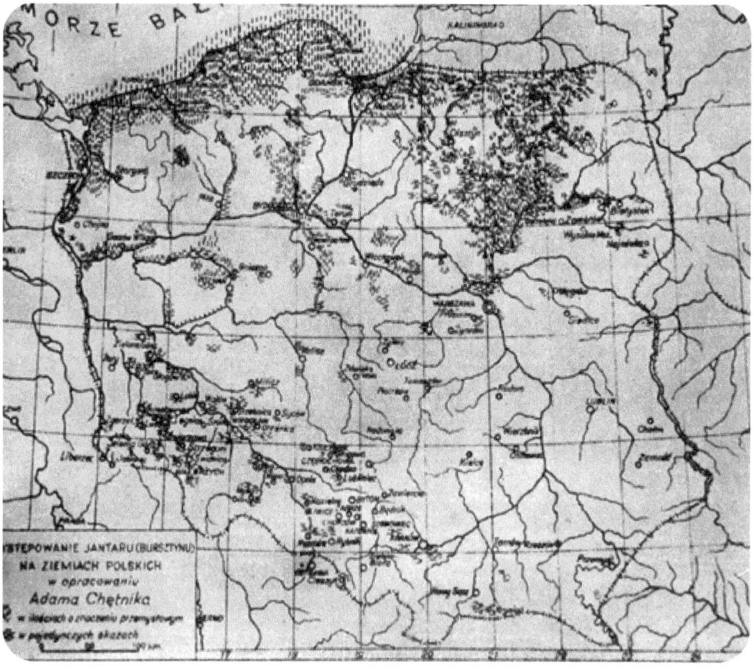

Fig. 55. An old map of the occurrence of amber in Poland. Source: Chętnik, A. (1961). „Polish Amber". Przegląd Geologiczny, 9 (5), 248. The shaded areas indicate where succinite is found.

Thanks to this natural hardener, the ambers from Jaroszów have survived for millions of years and have retained their jewelry qualities, even though they are a collector's rarity. They were not brought by the Eridan River, but by melting of unknown glaciers.

Photo 91. Amber from the Jaroszów region. Photo P. Barczak.

PYRITE IN AMBER

A naturally formed pyrite crystal

The mineral pyrite is often found in Baltic amber. It is an iron mineral from the sulfide group. It is the most common sulfide mineral found in marine sedimentary rocks. Often the process of fossilization of microorganisms in amber takes place with the participation of pyrite, although it is a poorly understood process. As a result of fossilization, insects or plants are sometimes filled with pyrite. Several insects were found in amber that were completely transformed by pyrite. The vast majority of pyrite replaces degraded tissues of plants and microorganisms found in amber by filling organic cavities. The formation and accumulation of pyrite in amber's surroundings is related to environmental conditions, including the availability of iron compounds, sulphates, organic carbon, and the existence of anaerobic conditions. These factors are of key importance for the development of sulfate-reducing bacteria, i.e. a group of microorganisms involved in the amber pyritization process. The fossilized resin, which is amber, traps the organic debris and stops the decomposition of organic elements. The next step is the mineralization that takes place inside the fossilized resin[Martín-González et al. 2009]. Pyrite-forming bacteria accumulate below the sea, and the formation of amber from the Cretaceous period in Alabama, USA, was due to carbonization, pyritization and compression [Knight et al. 2010]. In the Eocene Sea and the Sea of Paratethys, phenomena similar to those known today from the Black Sea may have occurred. There is a layer of hydrogen sulfide there, mainly because the sea is isolated from the open ocean. Hydrogen sulfide in the Black Sea is formed from the decomposition of organic matter in sulfate-reducing bacteria[Haklıdır et al. 2005]. Over the millions of years after the seas had dried up, deposits of sulfur and pyrite were formed from such layers, and in such an environment Baltic amber was preserved.

CHAPTER
III

3. HISTORICAL OUTLINE OF THE CREATION OF THE CONCEPT OF BALTIC AMBER

SUMMARY

Pliny, the Elder in "Natural History," quoted by Callistratus, recommended wearing amber beads for sore throats and tonsil diseases. In the Middle Ages, people with stomach or intestinal diseases were recommended to put pieces of amber in their beer, wine or water and use it for 14 days. On the other hand, putting amber into cow or sheep milk for one day was supposed to improve the functioning of the urinary system. In 1019, Avicenna (Ibn Sina) mentioned that amber has a positive effect on improving the work of the heart and reducing abdominal pain, fainting, fever, and nausea, as well as stopping vomiting. One of the recommendations was using amber to speed up the regeneration of broken bones.

In the Middle Ages, amber was considered one of the most effective drugs, along with camphor, gold, crystals, and beaver fat. In the work of Aricoli de Peste from 1554, using amber was recommended to fight pestilence by heating rooms with amber and camphor, which led to the rooms being disinfected. In the 16th century, Georgius Agricola, a German miner and doctor, carried out the dry distillation of amber and obtained succinic acid, amber, oil and rosin.

Baltic amber was used as a medicine by many pharmacists. In folk medicine, amber alcohol tinctures were used to treat lung disease. Another pharmacist, Paracelsus, described amber as a treatment for helping with head, stomach and intestinal problems. According to E. Fraquet, amber products could prevent colds and help treat rheumatism. According to Camillo Leonardo, amber administered orally strengthened the teeth.

Special properties were attributed to white amber, which was used against mass plagues. Matthäus Praetorius wrote in 1680: "During the plague, there were no deaths among the amber craftsmen in Gdańsk. Therefore, during the epidemic in 1709-1711, the Prussian authorities created an institution called the Health College, which recommended the fumigation of rooms with burnt amber."

In the publication "US Pharmacology" [1898], administering purified amber oil was recommended to treat spasms, and external rubbing was recommended for rheumatism. Amber was also valued as an ingredient in a drink called "Lammer-vin," a herbal concentrate in which amber was dissolved. **The drink was referred to as the elixir of immortality.**

In the 19th century, Baltic amber was subjected to a more detailed chemical analysis. It was found to contain succinic acid and terpenes. Consumption was recommended to help aging organisms. According to Jan Freyer, amber oil administered along with sugar had a good effect on the nervous system, accelerated blood circulation, alleviated skin ailments and alleviated lung diseases, which was a plague affecting all social classes in past centuries. It was also used as an anti-epileptic drug.

3.1. AMBER AND RESINS A UNIQUE COMPOSITION USED BY FOLK MEDICINE. CREATING MYTHS AND PHARMACEUTICAL KNOWLEDGE

The famous Greek philosopher Aristotle (384-322 BC), believed that the "electron" – what he called amber – was formed through hardening. He compared how it was made to myrrh, incense, and gum. He believed that amber was formed through solidification, and the trapped insects visible in it were proof of this process. Arab pharmacists also described the electrical properties of amber, and referred to it as Kahruba or Kahrabah, which literallyt means "a straw-absorbing substance" [https://referenceworks. brillonline.com/entries/encyclopaedia-of-islam-2/kahruba-SIM_3788]. Ibn al-Kabir noted that amber quickly and strongly attracted straw when it was rubbed lightly. The Arabs thought it was a Greek nut resin originating in Spain. It was mentioned that it comes from the Mountains Bulghār Dagh (now Taurus). Pliny – a Roman writer and philosopher of nature – mocked the Greeks who spread the idea that amber was the tears of Phaethon's slain sisters turned into poplars – tears frozen in the mythical Eridanos River. There were theories that it was fossilized horse manure or bird droppings (Sophocles) or that the raw material from which amber was made was lynx urine. Pliny, however, believed that amber was a solid pine resin collected by Teutons at the Baltic Sea and sold to ancient Pannonia. He probably also saw in person that villagers at the Po River in the ancient Roman Empire wore amber necklaces, which were prized for their healing properties. In his philosophical deliberations, however, he noticed that amber also occurs on the Adriatic coast as well as in the deltas of the Po and the Rhône rivers in their sediments.

Fig. 56. The Devonian period. From 419.2 to 358.9 million years ago. Painted by K.Doszla.

Pedanius Dioscorides (40-90 CE), a surgeon at the service of Nero's Roman army, in his dissertation entitled "De Materia Medica" wrote that amber is a resin from the aigeiros tree – that is, the poplar – a favorite delicacy of *Lepidoptera* caterpillars.

In later times, there were many bizarre assumptions. Amber was supposed to be the entrails of elephants. One of the authors, who lived in Prussia, doubted this version, however, as he noted that there were no elephants on the Baltic Sea coast – so how were these entrails to appear in that place? As such, amber was a secret of history. There were also some reasonable assumptions. Hartmann claimed that Baltic amber was made of pine and fir resins, which were formed due to fires, and that rivers from the ancient past carried this type of material into the sea. Particles of seaweed had been noticed in amber, so it was believed that these particles could indicate how the amber was formed. As a result of these old analyzes and deliberations, succinite (Baltic amber) was referred to as bitumen in Latin, i.e., the one flowing from the seabed [Hartmann 1677].

In the distant past, amber was seen in Egypt and Mauritania. It was found on the British Isles, China, and Japan. All hard resins were treated as amber. However, the advancement of societies and technology caused considerable differences to be noticed between particular fossil resins originating from different places.

Most of the succinite can be found on the Baltic Sea, in the Vistula River delta, and in the Kaliningrad region (Yantarny/Palmnicken), as well as in various places in Poland and Ukraine. It can also be found in the German town of Bitterfeld. However, modern chemical analyses using infrared can show the differences between Bitterfeld amber and Baltic amber. Bitterfeld amber was created by plants from the cypress family (*Sciadopityaceae*), such as the Japanese pine (*Sciadopitys verticillata*) – an evergreen coniferous tree. According to other concepts, Bitterfeld amber

was produced by pines and *Araucariaceae* trees [Mills, White, Gough 1984]. Baltic amber could also partly have been produced by this type of tree, but other species also participated in that process. We are still unable to establish the types of plants that formed Baltic amber, so there are still presumptions and hypotheses.

Fig. 57. An old engraving showing the amber collection in the Baltic Sea. Philipp Jacob Hartmann, "Succini Prussici physica et civilis historia demonstratione, etc", Francofurti 1677.

All in all, we know much more about the pharmacological properties of Baltic amber than about the plants that created it. Over time, a distinction was made between white and yellow amber. Nicolas Culpeper (1616-1654) referred to white amber as succinum [Culpeper 2006]. In ancient times, a Roman writer Callistratus claimed that amber is good at any age and that it is a prophylaxis in the treatment of disorders of consciousness accompanied by illusions (delirium), is a remedy for anguish, can be consumed in a liquid state, and is also an amulet that should be worn on the body. Claudius Galenus, the court physician of Emperor Aurelius, known in the Roman Empire as Galen of Pergamon (129-200 CE),, recommended amber-based lozenges to prevent digestive and respiratory problems [Kühn, Galeni 1965]. Galen's pill became quite popular and was used by Oribasius (c. 320-403 CE) – a doctor who was also from Pergamon – who was a personal physician of Emperor Julian the Apostate. They were used in the fight against dysentery, stomach disorders, coughing, and bladder disorders [Riddle 1973]. A Aecius Amidenus or Aetius of Amida, Mesopotamian doctor and court physician of the Byzantine emperor Justinian I, wrote that amber should be used in case of shortness of breath, pain in urination, and stomach problems [Drabkin 1973].

In China, amber has been known since ancient times. It was first mentioned in the work "The Famous Doctors" written by Tao Hongjing, who lived in the fifth century AD and who, while studying Taoism, sought the possibility of immortal life. The earliest known amber product in this vast country was found in the sacrificial pit no. 1 in Sanxingdui, Guanghan, Sichuan, and is 3000-5000 years old.

Chinese medicine lists three types of effects of amber. First of all, it works to calm the nerves and to support blood circulation and diuresis. Another Chinese book, "Compendium of Materia Medica," mentions amber in the context of strengthening the heart, improving eyesight, and relieving heart pain and epilepsy. According to Chinese medicine, amber eliminates blood stagnation and stops bleeding, eliminates insomnia, fights stomach aches, relieves headaches, strengthens the nerves, and relieves toothache. Nowadays, amber tablets known as "Domei" support memory processes and facilitate falling asleep and sleeping better.

Amber was a precious medicine in ancient China. It could be used only by emperors and wealthy people. In the tomb

of Princess Chen of the Liao Dynasty (907 - 1125), over two thousand amber accessories were found, some of which were set in gold. Most of the amber found in excavations and originating in the Ming (1368–1644) and the Qing (from 1644) dynasties are of the highest quality with clear and uniform colors, dense textures, and no contaminations. At that time, amber processing technology was already very sophisticated. The precious amber resin must have been extremely valuable and beneficial to humans because during the Tang Dynasty (771),the book "Cefu Yuangui" stated that the Chinese emperor sent his merchants to Persia to buy pearls and amber. The fossilized resin soothed anxiety, and helped with heart palpitations, insomnia, epilepsy, and amnesia. Amber was a symbol of status in ancient China. It was and still is appreciated in Buddhism, as amber Buddha statues can be found in many Chinese homes today.

As in China, Baltic amber became a recognized medicine in Europe. In the treatise "De Succini Natura" (1614) – published by Adrian Pauli (1583–1622), a professor of physiology from Gdańsk – one can read that "there is no part of the body that cannot benefit from amber" [Gelnitius 1614]. Baltic amber was considered one of the six best medicines in ancient times.

Alchemists also considered amber to be a concentrate of healthy energy. Felice Passera, a Capuchin monk from Brescia in Northern Italy (Cappuccino Infermiero della Provincia di Brescia), in the treatise "Il Nuovo Tesoro degl'Arcani Farmacologici Galenici, & Chimici, ò Spagirici" describes the effective healing properties of amber: "Scalda, dissecca, astringe leggermente, conforta, corrobora" ("It warms, dries, slightly tightens, soothes") [Ragazzi 2016].

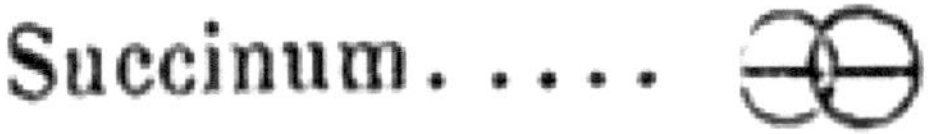

Fig. 58. The sign that alchemists used to mark amber.

Fig. 59. Source: Seeking amber in the Baltic Sea in the old days. Przyjaciel domowy: pismo zbiorowe dla gospodarzy, No. 47, Publisher Hipolit Stupnicki, 1856.

The healing properties of amber have been known in Europe for a long time. As early as the Middle Ages, small funnels were made of it, which were considered sacred, because they gave strength and contained a mysterious power. It was noticed that the oil obtained from amber positively affected "calming the nerves" and soothed epilepsy. In addition, it protected against poisoned air and other poisons [Andernacci 1571]. It was recommended to smell hot amber in the form of balls to strengthen the brain and chase away all ailments. Wearing amber beads protected against epilepsy [Boneti 1684].

Photo 92. White Baltic amber. Photo by. P. Barczak.

The following centuries of civilization development kept bringing more and more sophisticated ideas on using Baltic amber. Various herbs and valuable additives were mixed with amber. First of all, spices and herbs mixed with amber were beaten in a mortar. One such mixture came from England, and was the powder of the Countess of Kent. It consisted of ground elements of pearls, white coral, crab, white amber, and a rootstock of the tropical plant contrayerva. It was also mixed with deer horn jelly and musk. Saffron was added. Such a mixture was supposed to treat smallpox, measles, the plague, cancer fever, scarlet fever, and melancholy. Apparently, it was effective as the recipe was published many times. In turn, an English doctor, astrologer and herbalist, Nicolas Culpeper (1616-1654) [Duffin, Pymm 2015] invented the powder Pulvis Cardiacus Magistralis, which he recommended in all diseases associated with veins and also noted that the medicine helps with brain and heart diseases, as well as melancholy. The powder had a complex composition, for example, it contained deer horns, red and white coral, ground pearls, ground ivory, oregano, aloe, angelica and white amber [after: Ibid.].

In 1828 or 1829, the first attempts at fractionation via succinite extraction with ether and alcohol were made by Swedish chemist J.J. Freiherr von Berzelius (1779-1848). In addition to studying amber, he established the basic principles of inorganic chemistry in the 19th century. A series of solvent extractions were also undertaken by Swiss chemist Alexander Tschirch. He isolated four fractions referred to as resin acids from Baltic amber – "succinabietic acid" and "succinic silicic acid" [Tschirch, Aveng 1894].

In later centuries, an "Amber Tincture" was created based on amber and spirit, which was used against spasms and hysteria. Amber oil obtained through distillation (Oleum Succini) acted on the nervous system, accelerated blood circulation, and helped fight epilepsy. Succinic acid was used in asthmatic fevers and nervous apoplexy. Amber liqueur along with a small amount of alcohol (Amonii Succini) helped with the spasmodic suffering of lungs and nervous fever [Sylwan 1838].

Tinctura Succini (amber tincture) became a famous drug and was described in a book of pharmacology, "Pharmacopœia Officinalis & Extemporanea Or, A Complete English Dispensatory," in 1733. The book pointed to the important medical role of amber. The author, John Quincy, provided a recipe for Tinctura Puccini, which reads as follows: mix amber with wine in a vessel for five days, stirring two to three times a day.;when the tincture has an amber color, it can be used [Quincy 1733]. The tincture was used in hysterical ailments, i.e., neurotic disorders. It was also effective in weakness and rheumatic pains. Mass use was limited due to the high price of the tincture [Lewis 1761].

Another pharmacist, Nathanael Sendel, a doctor from Elbing (Prussia), indicated in his treatise on amber entitled "Electrologica" (1725-1726) that the amber infusion should be used against all types of brain diseases, such as strokes, epilepsy, paralysis, and also against illnesses of the respiratory tract. It was supposed to be an excellent antidote against swelling caused by insect bites.

More and more scientific studies on amber were completed. The University of Bologna, Italy, in a pharmacopeia developed in 1750, described the method of making amber tablets for ailments related to head diseases. Amber was also supposed to be an antidote for the plague epidemic. "Amber apples" had to be sniffed before falling asleep, which, according to the suggestions of people who lived during that time, helped fight the plague [Hinrichs 2007].

Amber has always been an expensive substance. Albert, the prince of Prussia, noticed this. Amber began to bring him a lot of income. His personal doctors, Andreas Aurifaber and Severin Göbel, were ordered to test the medical effects of amber, especially white amber. In the 17th and 18th centuries, pharmacies all over Europe were recipients of white amber. Amber was also donated to the court pharmacy in Berlin. In Prussia, amber was an indispensable medicine in every pharmacy [Aurifaber 1572]. It was also noticed that amber circumcision knives caused less infection in Judaistic rites than metal ones, so amber tips were often used instead of metal tips.

A study by the University of Dorpat (now Tartu in Estonia) found that amber stops hemorrhoidal and menstrual processes. It is useful against heart palpitations, and also strengthens the brain. Amber oil in the air stimulates the brain and heart [Kobert 1893]. Still, other healing properties were attributed to the amber tincture (Tinctura Succini), praised as a medication for curing chronic rhinitis, used as an antispasmodic, and in hypertension. Externally, it was used to heal wounds. In turn, amber oil (Oleum Succini), according to pharmacists of the time, influenced the functioning of the brain and certain parts of the nerves. It was especially praised for acting on weakened nerves which caused paralysis and spasms [Richter 1827].

In the 16th and 17th centuries, small amber tablets began to be produced. The ”amber pills” or “golden pills” contained aloe vera, rose petals, celery, fennel, anise, and saffron. Of course, the amber pills (Pillulaede Gentaro) also contained amber. Amber was mixed with sugar, and such mixtures were recommended for any type of disease. Balsamo europeo made in Italy, was a pharmaceutical known there [Donzelli 1737]. Amber was also used to stop bleeding [Passera 1688]. Another pharmacist – Averroes – recommended amber as an anti-fatigue agent in combination with cinnamon, Chinese cassia, and other aromatic spices. In turn, G. Agricola, a physician and a mineralogist (1494-1555), wrote that, because amber has a pleasant smell, it improves the heart's work and stops hand tremors, and the vapors of white amber prevent epilepsy [Polyakova 2013]. Pharmacists commonly combined amber and herbs in mixtures. Amber was mixed with ginger and rose water, thanks to which, a sore throat was soothed once applied. Succinum-based potions helped remove kidney stones. It was also noted that amber improved the functioning of blood and veins, which somehow confirms the effects of the compounds contained in amber, i.e., alleviating the symptoms of hemorrhoids, which is an illness connected to varicose veins [Aurifaber 1572]. Amber oil was given to treat all headaches and nerve disorders. Its effectiveness exceeded all other measures used for these types of diseases at that time [Untzer 1616]. Powdered amber was also one of the ingredients of the gem electuary – a mixture used for various ailments, such as heart and intestinal diseases [Manlio 1536].

Jan Baptist van Helmont (called Elmonzio) stated in his work "Opuscula Medica Inaudita," that there is no disease of the stomach, intestines, nerves, and mainly the brain, lest amber cure them when dissolved in wine [Donzelli 1737]. Other pharmacists and doctors of that time also noted the effectiveness of this remedy in fighting diseases of the brain and nervous system. It was the antidote to heal "darkness of mind." [Passera 1688]. An Arab doctor named Mesue believed that amber combined with musk and Spanish fly improves the work of the brain and heart. The preparation was also to act as a medicine against baldness [Mesue].

As already mentioned, alchemists were interested in amber, and also conducted dry distillation of amber [Passera 1688]. They were probably the first to notice that an acidic fluid containing succinic and acetic acids formed due to the anaerobic heating of amber in a laboratory retort. Over time, Barzelius described a phenomenon in which some succinic acid was deposited on the neck of the retort (crystallized pyratine). Amber oil was also produced in the chemical process, and a flammable gas was released.

In an English book from 1747, there is a recipe for an amber tincture, which, as the author writes, shows incredible effectiveness in all states, including ailments of the nervous system. It is recommended for use in hypochondria, hysteria, and similar disorders, including epilepsy [James 1747].

In his study, A. Aurifaber (a pharmacist and a scientist) gives a recipe for amber pills by Valery Cordus. In addition to crushed amber, they consisted of honey, pearls, and snail shells, as well as cinnamon, cardamom, saffron, and other substances. Several types of amber were used in the lozenges: white amber (Succinum candidum), falernian amber (Succinum falernum), and jet, i.e., very rare black amber (Succinum nigrum). V. Cordus recommended using amber lozenges (Trochisci de succino Cordi) in cases of epilepsy, hand tremors, hypothyroidism, and strokes.

Fig. 60. The Ordovician period – from 485.4 million years ago to 443.8 million years ago. Painted by K. Doszla.

Amber powder (Succinum pulveratum) was also used as a component of the drug called Rademacher Emplastrum Miraculosum, which was effective against ailments related to ulcers and skin lesions. An ingredient in this preparation was olive oil. Abraded amber (Succinum raspatum) was used in rheumatic diseases. Amber tinctures, in turn (Tintura succini) were applied in a dose of 20-30 drops of sugar. It was a European medicine. Succinic acid (Acidum succinicum) was administered mainly in combination with ammonia, and was considered a powerful remedy for fighting typhus, relieving apoplexy, and treating cramps and paralysis. An important condition that could be alleviated thanks to the amber tincture was pneumonia. Amber oil (Oleum succini) was used for whooping cough as an antispasmodic useful in asthma. It was also served in tea [The College Journal of Medical Science 1859].

Photo 93. Valchovite. Obora-Valchov, Moravskà Trevovà. Photo by P. Barczak.

In 1833 in Warsaw, amber tincture (Tinctura succini) was recommended for skin diseases or to alleviate hysteria. Amber oil (Oleum succini) also influenced the nervous system, accelerated blood circulation, and relieved lung spasms. It supported the treatment of hysterical suffering and epilepsy. In the old days, it was also used to alleviate tetanus ailments. It was known and used as a medicine for nervous apoplexy and paralysis. Liquor ammonii succinici was used in ailments related to pneumonia [Freyer 1833].

With the development of civilization, Baltic amber became more and more popular – since the Middle Ages, when it was considered hardened honey, lumpy kerosene, or a product of decomposition of wax [Stillé1868]. That is, until the time when more efforts were made to understand the structure and chemistry of this fossil resin. Amber was listed in the Dublin Pharmacopeia as a specific drug effective in rheumatism and paralysis. When administered internally, it acted on the nervous system. It was also effective in inhibiting the menstrual process. It was pointed out that succinic acid has urination-stimulating properties. Its use was noted to increase human sweating. It was also supposed to be effective in the case of gout disease [Pereira 1854].

In Russia and Scandinavia, amber was used as an antipyretic and an anti-inflammatory agent. In Arabian countries, it was known for its antibacterial and antiseptic properties, and for enhancing the body's natural immunity. It was also noticed that it reduces symptoms of colds and fever, and relieves rheumatic and muscular pains. In Europe and the Middle East, it was used to relieve stomach aches, rheumatic ailments, and viral infections.

Amber oil also made its way to pharmaceutical books in the USA. In a book published in 1849 in Philadelphia, entitled: "The Dispensatory of The United States of America," the properties of rectified oil made of amber were indicated. The oil had antispasmodic and stimulating properties, and it sometimes increased urine output. It was successfully used in the case of amenorrhea as well as of seizures and convulsive ailments. It was used in epilepsy, hysteria and childhood convulsions. It had a positive effect in the case of tetanus and intestinal irritation. When applied externally, it had an anticoagulant effect. Amber ointment was also noted to have positive effects on rheumatism. It was especially effective in rubbing the spine with a mixture of brandy, laudanum, and olive oil [Wood, Grigg 1849]. With time, amber oil has lost its importance and become a forgotten medicine. American publications refer to hard experiments and facts that indicate the effect of different compounds from Baltic amber. Legends are ignored. For some time, since around 1900, amber tinctures and amber oil have been sold by pharmacists as if for their own needs, often kept in the back of pharmacies. Amber has also disappeared from the British Pharmacopeia. At most, it alludes to amber beads that were used as teethers for children. Its homeopathic importance as a nerve calming agent is indicated in ailments related to head diseases – not as a medicine, but as an adjuvant. Amber oil and its effect on neurological diseases, hysteria, and hypochondria are described particularly positively [Clarke 1896].

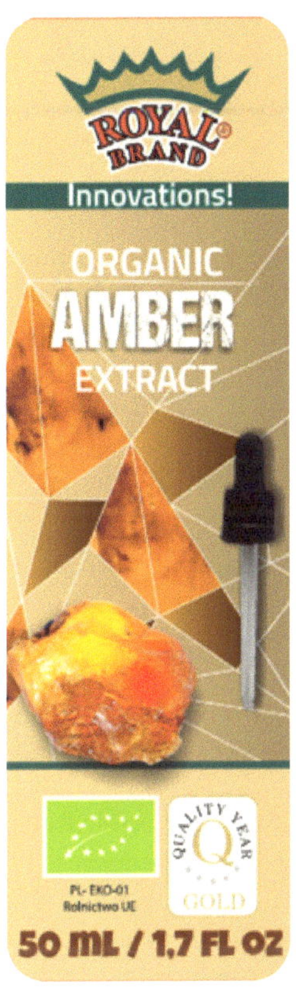

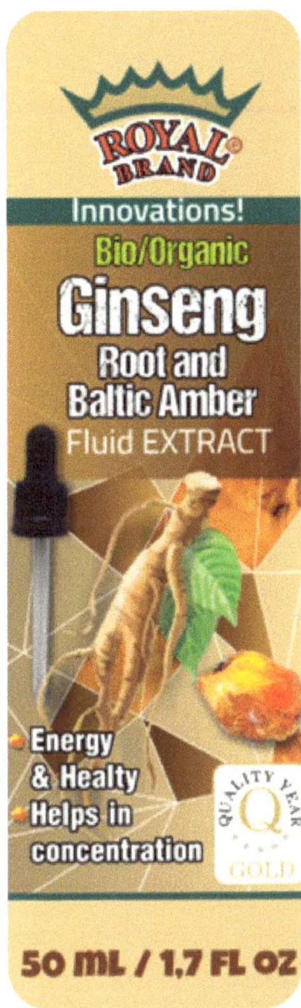

Photo 94. Amber extracts, in order from the left – amber extract, amber extract with ginseng, amber extract in aerosol. Photo: P.Barczak.

Various resins appear on the US market, and it is difficult to distinguish which come from the Baltic Sea and which are from other regions of the world. Amber has also been found in Maryland, on the Cape Sable, near the Magothy River. Associated with pyrite and lignite, it is similar to Baltic amber, but also contains different compounds [The Dispensatory of The United States of America 1849]. It seems that Baltic amber has somehow lost its brand because its composition and the way it was created are not exactly known, and, as a result, the recognition of its effect on individual diseases and ailments has been lost.

3.2. AMBER IN CHINESE MEDICINE

Amber has been known in Chinese medicine for a long time. Not only Baltic amber was used in China. Sometimes it came from other sources. Natural amber is known throughout China, and amber deposits were found in the provinces of Yunnan, Guangxi, Henan, Fujian, and Guizhou. Fushun amber from the Liaoning province was especially appreciated, and is currently a rarity in China. Some even claim that this amber is a myth. It is known to be from an opencast mine in Fushun, and was formed in the Cenozoic era (from 66 million years ago, to this day).

Fig. 61. The Cenozoic Era – from 66 million years ago till the present day. Painted by K. Doszla.

Chinese medicine has been using knowledge of amber for over 3000 years. Folklore describes the effects of taking amber as calming the mind and nerves, as well as improving circulation, and removing blood stagnation.

According to Chinese medicine, succinic acid has a positive effect on humans, strengthens the body and immune function, and balances the acidity of the human body. Amber prevents the aging of cells, and removes free radicals that have oxidant properties. Chinese physicians do not recommend taking amber before meals because it irritates the digestive tract. It is recommended that patients take it half an hour after a meal and discourages taking other medications at the same time.

An antibacterial effect of succinic acid has been pointed out. It has been noted that it has an inhibitory effect on the *staphylococcus aureus* and other bacteria. Injection of succinic acid into mice and rats caused an increase in the gamma-aminobutyric acid (GABA) content in the brain.

Chinese doctors also see the effects of Baltic amber in epilepsy, palpitations, and insomnia. Positive effects are achieved after consuming amber in states of depression and anxiety. In China, amber is prescribed to women after childbirth to stop pain and improve blood circulation. Amber-based ointments eliminate bruises and reduce swelling. Amber is also used for joint pain, and is known to activate the spleen, which is where white blood cells are produced, and old-damaged ones are removed. It is generally recognized that amber thins blood and improves circulation. In China, it has also been used to alleviate women's ailments.

In Chinese medicine, amber is a substance attributed with a diuretic effect, so it is used in diseases of the urinary system. Succinite helps to expel excess water. Amber and succinic acid

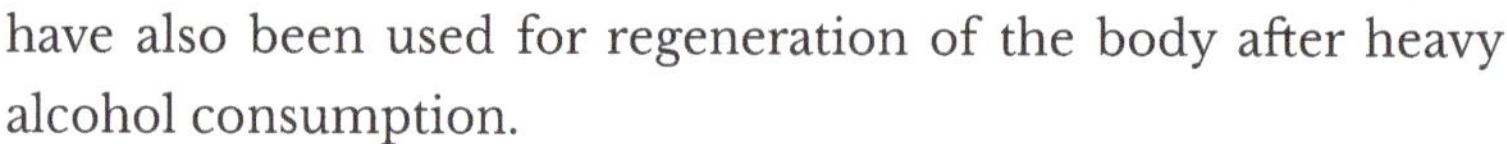

have also been used for regeneration of the body after heavy alcohol consumption.

Amber is also used in Chinese aromatherapy. It has a positive effect on breathing processes. It is known to affect viruses that are spread by droplets. According to Chinese pharmacists, amber can eliminate various pathogenic bacteria in the human body and prevent the virus from entering. According to Ming Li Shizhen's (1518-1593) "Compendium of Materia Medica", amber's scent is sweet, non-toxic, able to calm the five internal states of a person, stabilize the soul, and eliminate congestion. It improves five states in a person, strengthens the mind, improves eyesight, relieves pain, soothes nerves, and positively affects blood and vitality [Yuan et al. 1997].

To this day, Chinese doctors recommend the long-term use of amber because it benefits the skin. Amber can have a positive effect on skin diseases and support wound healing. It also protects against aging, and wearing an amber pendant can lower blood pressure, reduce stress, stabilize your mood, improve heart efficiency, and reduce hypertension.

Photo 95. *Taxus Chinensis*, *Taxaceae*, attenuates neurotoxicity and cognitive dysfunction [Zhang et al. 2020]

3.3. AMBER IN JAPANESE MEDICINE

In Japan, amber is called Kun no Ko. Kuji, in Iwate Prefecture, Japan is still known for producing "Kun no ko" amber, which is used as a herbal remedy. It is applied for abdominal pain and activates the liver and kidneys. A famous person who used amber was the widow of the Chinese Emperor Cixi of the Qing Dynasty. The Empress took a herbal medicine based on Kotaimaru amber in states of despondency.

Japanese scientists are working intensively on amber extracts. In 2019, patent number 654877 was obtained for an amber extract that lowers blood pressure. Another patent number 6601860 was obtained for an amber extract which reduces the absorption of sugar thanks to the use of a preparation. The Kohaku Bio Technology University is actively researching amber extracts. As part of the work, it was found that a properly prepared amber extract contributes to the smoothing of the skin (patent number 4953204). Amber extract also stimulates the secretion of hyaluronic acid, which increases skin moisture and leads to its renewal and fluffiness. As part of the scientific discoveries, it was confirmed that amber extract (with added mulberry leaf extract) inhibits the absorption of sugar into the blood. It can block the flow of glucose through the walls of the small intestine. Unabsorbed glucose is excreted to the outside. Amber extract was also found to affect hair growth by generating the VEGF gene [Shimizu et al. 2020].

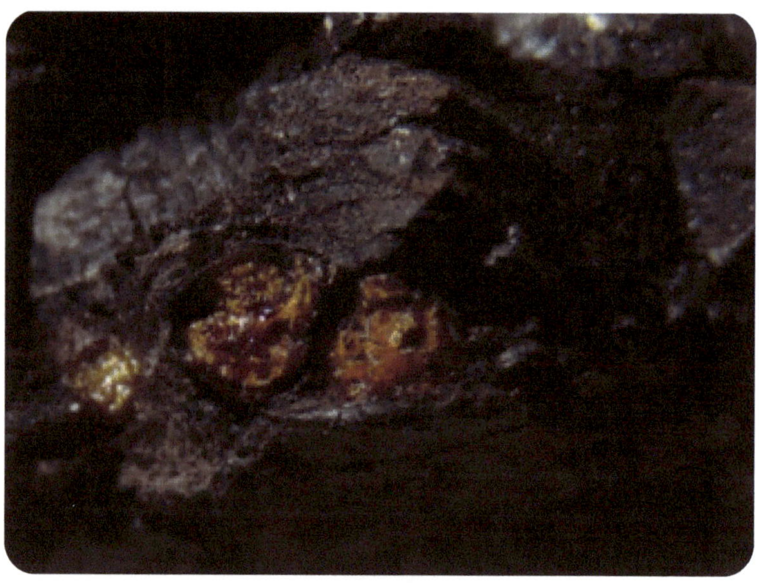

Photo 96. Amber Wyoming Hanna Formation, USA. Collected 7 Miles North of Hanna, Wyoming. 56 Ma. Collection Doug Lundberg, Colorado Springs.

CHAPTER
IV

4. CHEMICAL COMPOSITION

SUMMARY

In chemical terms, amber is a mixture of organic (terpenes, their derivatives, and resin acids) and inorganic (mineral) compounds. The group of terpenes can be classified as follows: the general formula for terpenes is $(C_{10}H_{16})x$, where x is 1; 1.5; 2; 3; 4, respectively. Digit 1 refers to monoterpenes (for example, pinene – $C_{10}H_{16}$), digit 1.5 to sesquiterpenes (cadinene – $C_{15}H_{24}$), digit 2 to diterpenes (cembrene A – $C_{20}H_{32}$), digit 3 characterizes triterpenes (squalene – $C_{30}H_{48}$), and digit 4 refers to tetraterpenes (carotene – $C_{40}H_{64}$) [Matuszewska 2012]. We can find various terpenes and terpenoids in Baltic amber, as each piece of fossilized resin could have been created by different plants and under different geological conditions. Carvacrol, isoborneol, borneol (6%), citral, carvone, para-cymene (49%), meta-cimene, fenchone, fenchenol (10%), limonene, α-pinene, β-pinene, 1,8-cineole (8%), camphene, and camphor (3%), pulegone all form part of the group of monoterpenes found in Baltic amber. The group of sesquiterpenes includes dihydro-α-curcumene, and calamene. Resin, communic, pimaric, and abietic acids can also be found in Baltic amber which, according to scientific hypotheses, constitute an insoluble polymer phase. In addition, carboxylic acids and simple phenols, such as benzoic acid, vanillic acid, and vanillin, have also been found in Baltic amber. The main compound is succinic acid, which often binds with terpene alcohols to form, among others, bornyl and dibornyl succinate, as well as fenchyl and difenchyl succinate. Another group of compounds, the least

known to be in amber, are aromatic compounds based on the structure of naphthalene, phenanthrene, and anthracene. The minerals iron, aluminum, calcium, magnesium, silicon, boron, silver, manganese, titanium, copper, chromium, and vanadium can all be found in succinite. The composition of amber varies and depends on its origin. It has been shown that Baltic amber has a composition similar to Siberian amber (which contains silver) but differs from Sicilian and Dominican amber (which does not contain silver). One of the chemical features of amber is the content of organic iodine compounds [Koziorowska, 1984; Matuszewska et al. 2002, Wawra, 1993, Poulin et al. 2014]. Among the volatile compounds, isoborneol, borneol, para- and meta-cymene, fenchone, limonene, α-pinene, and β-pinene have been detected.

In chemical terms, amber is a mixture of organic (terpenes, their derivatives, and resin acids) and inorganic (mineral) compounds. All terpenoids in plants likely arose from the transformation of the basic isoprene C_5H_8 units, which in turn, formed Baltic amber. Most mono-, sesqui-, di-, and polyterpenes are synthesized by the "head-to-tail" condensation reaction of the isoprenoid unit. The "tail-to-tail" dimerization of sesquiterpenes and diterpenes can produce triterpenes and tetraterpenes. Isomerization transformations harden the resin [Langenheim, Beck 1968].

$$CH_2 = C\,(CH_3)\,-CH = CH_2$$

Fig. 62. An isoprene molecule C_5H_8 and the chemical formula of isoprene

Baltic amber is a complex molecular structure that, as a result of millions of years of transformations, has acquired a unique structure enriched with plant compounds occurring on the Earth over millions of years. Succinite, or fossilized resin – as it is assumed – from the Eocene period, shows supramolecular structures, and each amber study reveals new compounds found in plants. Roughly 40,000 terpenes have been found to this day, and countless numbers of plant species have existed for the past 50 million years. Additionally, terpenes and terpenoids ("terpenes" with an additional oxygen atom) are transformed under the influence of high temperature, which could have occurred during geological processes when Baltic amber was deposited at a certain depth in the layers of peat and earth. Perhaps the fossilized resins interacted with plant oils, including terpenoids, which further complicates the analysis. The leading theory of the chemical structures found in amber indicates that the basic component of the succinite structure is abietic acid in the form of dimers, i.e., two identical molecules joined together.

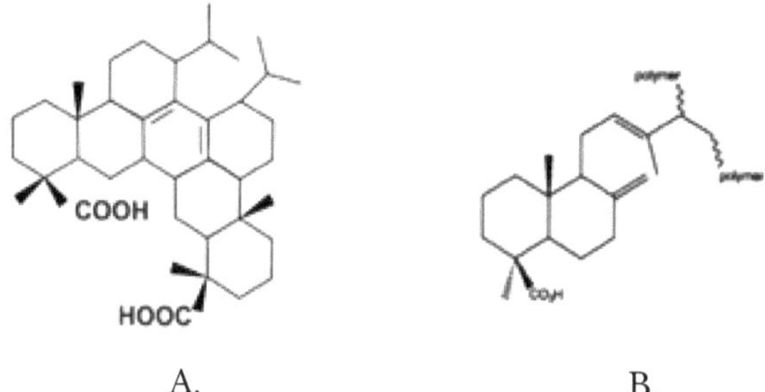

Fig. 63. A. Abietic acid dimer B. Communic acid-communol copolymer

Another theory, in turn, indicates that it is a polymer consisting of communic acid and communal [Tumiłowicz et al. 2016].

Succinite creates numerous chemical networks, but the classification is problematic due to the numerous types of amber, as well as the conditions for its formation and shaping.

The polymer is a "living fossil resin" that constantly exchanges its components with its surroundings.

Taking the chemical composition into account, amber is classified according to the following criteria:

1. Unsaturated monoterpenes, the main element of which is isoprene C_5H_8, which is described using the formula (C_5H_8)n, where n>2. Examples include camphene $C_{10}H_{16}$, limonene $C_{10}H_{16}$, α-pinene $C_{10}H_{16}$, and β-pinene $C_{10}H_{16}$.

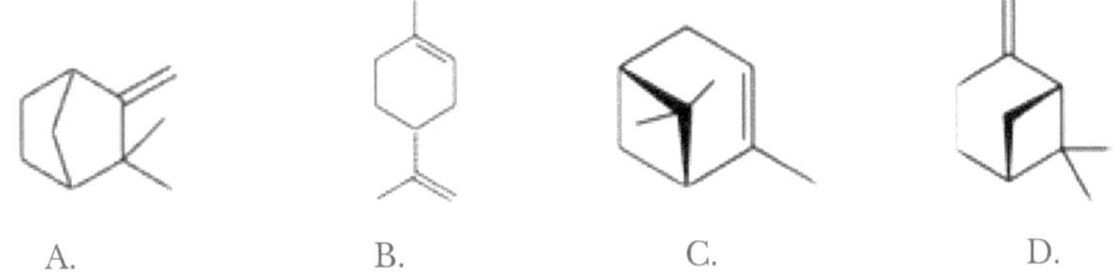

A.   B.   C.   D.

Fig. 64. Unsaturated monoterpenes found in Baltic amber. A. Camphene, B. Limonene, C. α-Pinene, D. β-Pinene

2. Aromatic monoterpenes. An exemplary aromatic monoterpene found in Baltic amber is p-cymene $C_{10}H_{14}$.

Fig. 65. P-cymene – a terpenoid found in succinite.

This compound is found in essential oils, such as camphor tree oil, *eucalyptus* oil, tea tree oil, and mint oil. Significant amounts of it are produced in the chemical processes of the effect of sulfur compounds on terpenes, which may have been important in the formation of Baltic amber.

3. Monoterpene alcohols and their esters. This group includes fenchol, isoborneol, borneol, bornyl acetate, and terpinen-4-ol.

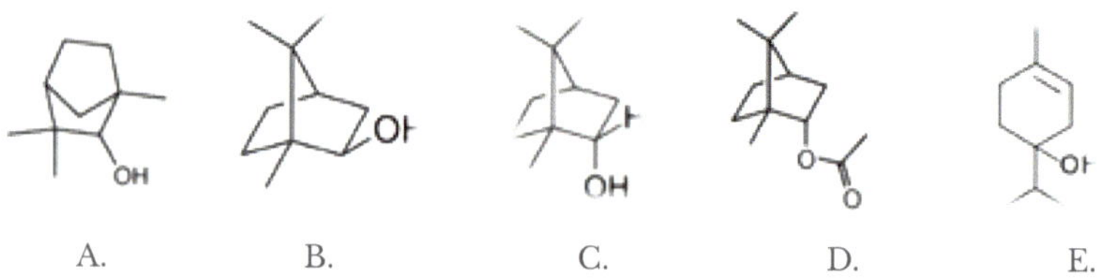

A.   B.   C.   D.   E.

Fig. 66. Monoterpene alcohols and esters contained in amber. A. Fenchol $(C_{10}H_{18}O)$, B. Isoborneol $(C_{10}H_{18}O)$, C. Borneol $(C_{10}H_{18}O)$, D. Bornyl acetate $(C_{10}H_{18}O)$, E. Terpinen-4-ol $(C_{10}H_{18}O)$.

4. Monoterpene ketones – linked by a double bond of carbon and oxygen.

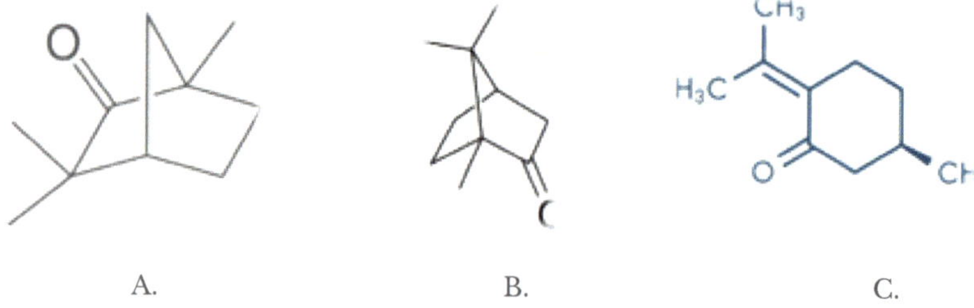

A.   B.   C.

Fig. 67. Monoterpene ketones contained in amber. A. Fenchone $(C_{10}H_{16}O)$, B. Camphor $(C_{10}H_{16}O)$, C. Pulegone $(C_{10}H_{16}O)$.

5. Monoterpene ethers with COC (eucalyptol) bonds.

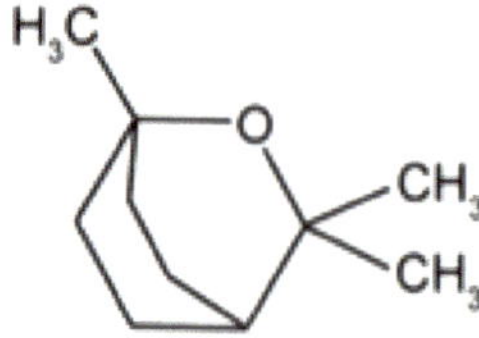

Fig. 68. Eucalyptol – $C_{10}H_{18}O$.

6. Polycyclic compounds and their derivatives, containing one or more cyclic systems. Polycyclic compounds and their

derivatives consist of abietic acid, communic acid, pimaric acid, isopimaric acid, abieta-7,13-diene, α-amyrin, and β-amyrin.

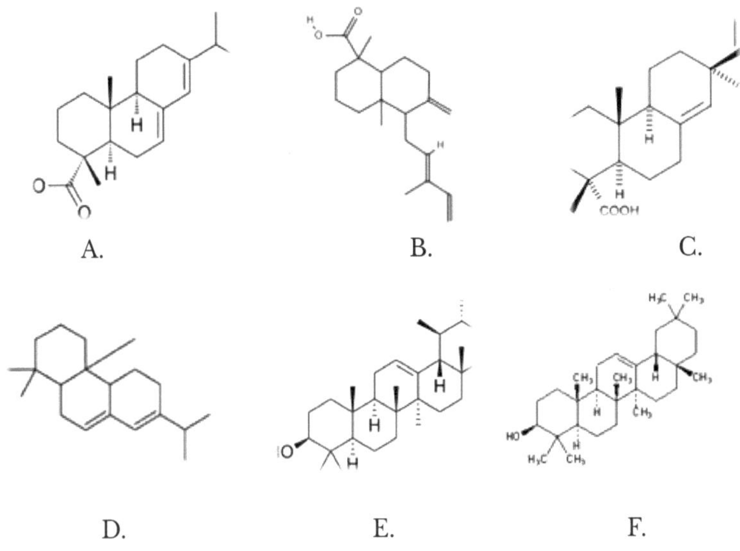

Fig. 69. Polycyclic compounds and their derivatives. A. Abietic acid ($C_{19}H_{29}COOH$), B. Communic acid ($C_{20}H_{30}O_2$), C. Pimaric acid ($C_{20}H_{30}O_2$), D. Abieta-7, E. 13-diene ($C_{20}H_{32}$), F. α-Amyrin ($C_{30}H_{50}O$), β-amyrin ($C_{30}H_{50}O$).

7. Succinic acid and fatty acid esters, including methyl palmitate.

Fig. 70. Succinic acid and esters of fatty acids. A. Succinic acid ($C_4H_6O_4$), B. Methyl palmitate ($C_{17}H_{34}O_2$). Based on: Paweł Tumiłowicz, Ludwik Synoradzki, Agnieszka Sobiecka, Jacek Arct, Katarzyna Pytkowska, Sławomir Safarzyński, "Bioactivity of Baltic amber - Fossil Resin", Polimery 2016.

As previously mentioned, numerous studies on Baltic amber constantly bring new results, and each one is slightly different, depending on the amber samples that are assessed and analyzed. One study shows a complicated number of compounds contained in amber. The five main compounds were terpenoids found in the amber extract, i.e., borneol, methyl ester, isopimaric acid, camphor, 2-fenchenol, and m-cymene, as well as camphane, camphene, and isoborneol. Residual amounts of agate acid, which can be found in kauri resin obtained from the Agathis tree from New Zealand, were also found, along with small amounts of abietic acid commonly found in pines. In addition, rosin and turpentine were detected. Baltic amber was also found to contain naphthalene – an organic chemical compound consisting of a mixture of gasses, liquid, and solid (bitumens) hydrocarbons. Trace amounts of benzene compounds were also found [Al-Tamini, Al-Saadi, Burghal 2020]. These substances arouse interest, as they are an integral part of crude oil deposits, which may indicate how amber is shaped and hardened by the sulfur compounds in this rock oil. This may be the reason why Baltic amber formed and hardened in the vicinity of bitumen and oil deposits. In coals from other geological eras, amber was not formed. It was not found in hard coal deposits in Upper Silesia in Poland or in lignite mines, such as Bełchatów, which were created during the Miocene period. (The amber deposits found in this area have been brought from other areas).

Amber, however, is mainly a mixture of terpenes and plant terpenoids. Among the chemical components found in succinite, monoterpenoids such as fenchol ($C_{10}H_{18}O$), isoborneol ($C_{10}H_{18}O$), and borneol ($C_{10}H_{18}O$) can be found. Sesquiterpenoids such as ®-caryophyllene-2,6-oxide $C_{15}H_{26}O$, dihydro-ar-curcumene $C_{15}H_{24}$, calamenene $C_{15}H_{22}$ have also been found, as well as diterpenoids such as 18-norpimar-8-ene ($C_{19}H_{32}$), 19-norabieta-8, 11, 13-triene ($C_{19}H_{28}$), 18-norabieta-8, 11, 13-triene ($C_{19}H_{2}$), dehydroabietane ($C_{20}H_{60}$), isopimara-8,15-diene acid

-18-oic ($C_{20}H_{30}O_2$), abiet-8(14)-en-18-ol ($C_{20}H_{34}O$), abiet-7-en-18-ol ($C_{20}H_{34}O$), pimar-8-ene-18-oic acid ($C_{20}H_{32}O_2$), dehydroabietol ($C_{20}H_{30}O$), pimar-9(11)-en-18-oic acid ($C_{20}H_{32}O_2$), pimar-7-en-18-oic acid ($C_{20}H_{32}O_2$), dehydroabietic acid ($C_{20}H_{28}O_2$), and abietic acid ($C_{20}H_{30}O_2$). As part of the chemical analysis of the micro components, the following succinates were discovered: difenchyl succinate ($C_{24}H_{38}O_4$), fenchylbornyl succinate ($C_{24}H_{38}O_4$), fenchylisobornyl succinate ($C_{24}H_{38}O_4$), dibornyl succinate $C_{24}H_{38}O_4$, bornyloisobornyl succinate ($C_{24}H_{38}O_4$), diisobornyl succinate ($C_{24}H_{38}O_4$) [Ibid.].

CHEMICAL STRUCTURE	FORMULA	COMPONENTS	PEAK NO.	CHEMICAL STRUCTURE	CONTENT (%)		RETENTION TIME
					BROWN	ORANGE	
5.686	1.09	1.25		$C_{10}H_{16}$	CAMPHENE		1
6.742	5.40	6.24		$C_{10}H_{14}$	M-CYMENE		2
7.543	1.45	1.71		$C_{10}H_{16}O$	L-FENCHONE		3
7.904	7.76	7.44		$C_{10}H_{18}O$	2-FENCHANOL		4
8.226	7.04	8.15		$C_{10}H_{16}O$	CAMPHOR		5
8.404	1.83	1.87		$C_{10}H_{18}O$	ISOBORNEOL (ISOCAMPHOL)		6
8.506	17.60	16.80		$C_{10}H_{18}O$	BORNEOL		7
9.169	-	0.72		$C_{15}H_{24}O$	SANTALOL, TRANS-BETA		8
10.902	-	0.93		$C_{15}H_{26}O$	VIRIDIFLOROL		9
11.460	0.92	0.97		$C_{13}H_{18}$	NAPHTHALENE, 1,2,3,4-TETRAHYDRO-1,6,8-TRIMETHYL-		10
11.718	-	0.72		$C_{14}H_{20}$	NAPHTHALENE, 1,2,3,4-TETRAHYDRO-5,6,7,8-TETRAMETHYL-		11
12.317	0.94	1.94		$C_{15}H_{26}O$	CARYOPHYLLENYL ALCOHOL		12
13.954	2.96	3.75		$C_{15}H_{26}O$	BICYCLO[2.2.1]HEPTAN-2-OL, 1,3,3-TRIMETHYL-, ACETATE, (1S- EXO)-		13
14.039	1.53	1.59		$C_{14}H_{20}$	NAPHTHALENE, 1,2,3,4-TETRAHYDRO-5,6,7,8-TETRAMETHYL-		14
14.149	2.66	0.86		$C_{13}H_{22}O_2$	1,3-DIOXOLANE, 2-(2,3-DIMETHYL-1-CYCLOPENTEN-3- YL)-2,4,5-TRIME-		15
12.317	0.94	1.94		$C_{15}H_{26}O$	ACETIC ACID, 1,7,7-TRIMETHYL- BICYCLO[2.2.1]HEPT-2-YL ESTER		16
14.810	1.95	1.67		$C_{13}H_{20}O_2$	ISOBORNYL ACRYLATE		17

15.911	-	2.60		$C_{38}H_{68}O_8$	L-(+)-ASCORBIC ACID 2,6-DIHEXADECANOATE	18
17.616	-	1.43		$C_{18}H_6O_2$	OCTADEC-9-ENOIC ACID	19
18.412	0.97	1.54		$C_{20}H_{30}O$	CRYPTOPINONE $$ DEXTROPIMARINAL $$	20
18.913	-	1.28		$C_{20}H_{30}O_4$	AGATHIC ACID	21
19.176	13.65	17.00		$C_{21}H_{32}O_2$	ISOPIMARIC ACID METHYL ESTER	22
19.427	-	1.67		$C_{21}H_{32}O_3$	PREGN-4-EN- 17(ALPHA),20(ALPHA)-DIOL-3-ONE	23
19.509	2.81	2.34		$C_{20}H_{32}O_2$	13ALPHA-DELTA(8)-DIHYDROABIETIC ACID (ABIET-8- EN-18-OIC ACID)	24
19.610	3.57	5.01		$C_{20}H_{32}O_3$	DIHYDROXYISOSTEVIOL	25
19.685	3.11	2.48		$C_{20}H_{30}O$	1-METHYLET ABIETA--8,11,13- TRIEN-18-OL	26
20.659	-	1.36		$C_{20}H_{30}O_2$	ABIETIC ACID	27
7.828	3.17	-		$C_5H_8O_4$	SUCCINIC ACID MONOME-THYL ESTER	28
10.903	1.11	-		$C_{12}H_{18}O_2$	EPIANASTREPHIN	29
12.316	2.24	-		$C_{13}H_{22}O$	5,5,8A-TRIMETHYLDECA-LIN-1- ONE	30
13.359	1.24	-		$C_{21}H_{42}O_2$	1,3-DIOXOLANE, 4,4,5-TRIMETHYL-2-PENTADE-CYL-	31
18.907	1.84	-		$C_{23}H_{32}O_5$	1.ALPHA-(ACETOXYME-THYL)- 7ALPHA.,8.ALPHA.--DIMETHYL-7- (2-(3-FURYL)ETHYL)BICYC-LO(4.4.0)DEC- 2-ENE-2--CARBOXYLIC ACID METHYL ESTERTHYL-	32
19.416	4.89	-		$C_{28}H_{48}O$	CHOLEST-14-EN-3-OL, 4-METHYL-, (3.BETA.,4.AL-PHA.,5.ALPHA.)-	33
20.881	1.62	-		$C_{16}H_{22}O_2S$	(2,6,6-TRIMETHYLCYC-LOHEX-1- ENYLMETHANE-SULFONYL) BENZENE	34
21.560	0.86	-		$C_{30}H_{51}O_4P$	PHOSPHORIC ACID, TRIBORNYL ESTER	35

Tab. 5. Source: Wijdan H. Al-Tamimi, Sahar AA Malik Al-Saadi, Ahmed A. Burghal, "Antibacterial Activity and GC-MS Analysis of Baltic Amber against Pathogenic Bacteria," International Journal of Advanced Science and Technology Vol. 29, No. 11, 2020.

Contemporary studies are more precise than those conducted in the 1970s, as they complement and extend them. New techniques and technologies bring us closer to the truth about Baltic amber. In the first laboratory studies on amber conducted by Mills' team in Great Britain, found mainly terpenes. The following compounds were found: camphene ($C_{10}H_{16}$), p- and n – cymene ($C_{10}H_{14}$), o-cymene ($C_{10}H_{14}$), 1,8- cineole, also known as eucalyptol ($C_{10}H_{18}O$) , dimethyl succinate ($C_6H_{10}O_4$), fenchone ($C_{10}H_{16}O$), fenchyl alcohol ($C_{10}H_{18}O$), camphor ($C_{10}H_{16}O$), isoborneol ($C_{10}H_{18}O$), borneol ($C_{10}H_{18}O$), carvomenthone ($C_{10}H_{18}O$), bornyl formate ($C_{11}H_{18}O_2$), bornyl acetate ($C_{12}H_{20}O_2$) , ionene ($C_{13}H_{18}$), dihydro-ar-curcumene ($C_{15}H_{22}$), calamene ($C_{15}H_{22}$) , and succinates ($C_{15}H_{24}O_4$). The compounds mentioned by Mills' team are terpenes and terpenoids found in plants. The skeletons of the discovered structures are similar to the compounds known from some of today's plants. The resin secreted by the trees had the same functions as today – protecting plants from intruders and attracting insects beneficial to the plant. The trees released the resin, creating their own defense mechanisms. Oils fixed in Baltic amber played a similar role. Mono- and sesquiterpenoids perform several important functions in injured trees, disinfecting, deterring, and transporting the resin to the wound site.

One of the phenomena visible in trees involving terpenes as electron transport units is the existence of an electrical potential difference in each tree (The experiment concerning this issue is presented later in the text). The voltage in microvolts is generated in the branches, trunk, and leaves in relation to the layer or earth. Terpenes are involved in the process of electron transmission in plants, keeping trees alive, and transferring electric charges from the leaf system to the root system.

Mills' team's research was groundbreaking and entered the history of amber research. For the first time, derivatives of terpenes were found in a Baltic amber extract – compounds of the podocarpatriene type (with a greater number of carbon atoms than terpenes and undivided by five). These types of chemical structures are found in numerous evergreen trees today [Devkota et al. 2011]. Other terpenes detected by Mill's team in Baltic amber are abietatriene chemical compounds ($C_{20}H_{30}$) and derivatives of pimaric acid. Pimaric acid is a diterpenoid with the formula ($C_{20}H_{30}O_2$). Pimarane diterpene derivatives were also found ($C_{20}H_{36}$). Another chemical compound – methyl palmitate ($C_{17}H_{34}O_2$) can be found in a number of plants, such as in oleander seeds [Deka 2011], in almond seeds [Ogunsuyi, Daraola 2013], walnut seeds, palm oil, as well as in the shrub Lantana camara often found in the tropics [Anwar et al. 2021]. It is therefore possible that these types of trees influenced the final composition of succinite.

British researchers have also discovered that terpenoid agathic acid ($C_{20}H_{30}O_4$)is a compound existing in Baltic amber. It also has the structure of labdane, and such structures undergo significant polymerization [Matuszewska 2017]. They can now be found in pine needles and pine oleoresins [Zinkel, Magee 1991].

One of the oldest studies on Baltic amber's chemical composition was the one carried out by a team consisting of A. Tschirch, E. Aweng, C. De Jong, and S. Herman in 1923, which is still quoted and described today. The study showed that Baltic amber consists of: 65% succinorezene ($C_{22}H_{36}O_2$), 12% succinabietic acid ($C_{41}H_{60}O_5$), 6% of succinabietol ($C_{40}H_{60}O_2$), 4% of succinosylvinic acid ($C_{24}H_{36}O_2$), 3% succinoresinol ($C_{12}H_{20}O_4$) 2% succinic acid ($C_4H_6O_4$), 0.5% succinoxabietic acid ($C_{20}H_{30}O_4$), 0.2% D - borneol ($C_{10}H_{18}O$), as well as 7.3% impurities and water [Tschirch et al. 1923]. Other, more recent, studies showed the following fractions in amber oil: ᴄ-pinene, endo-fenchol, fenchone, 5- methyl-3- [1-methylethenyl] - cyclohexene (m-menta-4,8-diene), 3,4-dimethyl-1- [1-methylethyl] -benzene (3,4-dimethyl-cumene), 1-methyl-4- [methylethenyl]

-benzene- (p-ʟ-dimethyl-styrene), 1,2,4a, 5,8,8a-hexahydro-4,7-dimethyl-1- [1-methylethyl] -naphthalene (®-kadinene), 1,2,3,4-tetrahydro-1,6-dimethyl-4-   [1-methylethyl] -naphthalene (1s, cis-calamene), 1-methyl-7- [1-methylethyl] -nafthalene , 1,2,3,4,4a, 9,10,10a-octahydro-1,1,4a-trimethyl-phenanthrene, 1-methyl-7- [1-methylethyl] -nafthalene, 1-methyl-7- [1-methylehyl] -phenanthrene (retene), 1a, 2,3,4,4a, 5,6,7b-octahydro-1H-cycloprop [e] azulene (ʟ-gurjunene) [Matuszewska 2004]. Compounds related to benzene and naphthalene reappear, again indicating the connection of the networks built in amber with hydrocarbon deposits.

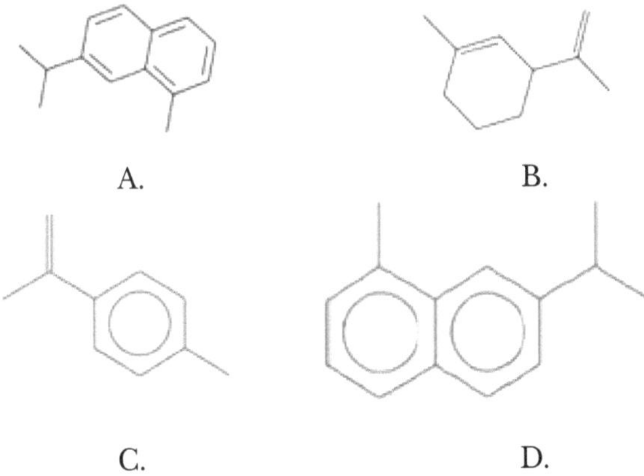

A.   B.

C.   D.

Fig. 71. Terpenes from Baltic amber connected to benzene and naphthalene. A. 1-methyl-7-[1-methylethyl]-naphthalene ($C_{14}H_{16}$), B. 5- methyl-3- [1-methylethenyl] - cyclohexene (m-menta-4,8-diene) ($C_{10}H_{16}$), C. 1-methyl-4-[methylethenyl]-benzene-(p-ʟ-dimethyl-styrene) ($C_{10}H_{12}$), D. 1-methyl-7-[1-methylethyl]-naphthalene ($C_{14}H_{16}$).

A fundamental characteristic of Baltic amber, perhaps the most important one, is the fact that it contains succinic acid (C4H6O4). Interestingly, in Bitterfeld amber (from Germany), more succinic acid was found in the tested samples in the form of free, unbound compounds than in succinite [Matuszewska

2012], even though it was originally believed that Bitterfeld amber did not contain this chemical compound.

Fig. 72. A molecule of succinic acid.

Despite the enormous amount of theoretical work, amber has not been intensively analyzed in the laboratory. On the one hand, it is understandable as research is costly. On the other hand, the number of fossilized resins found keeps growing, so the work becomes more scattered. There have been several basic chemical analyses over the past 200 years that the world of science currently refers to. Some of them are historical materials. The development of new techniques and technologies creates new opportunities to learn about the compounds contained in Baltic amber. Without laboratory tests, unverified hypotheses and theories remain. The experiment can make them more likely.

Photo 97. Araucaria resin leaking from a broken branch. Cyprus, Limassol.

CHAPTER V

5. BIOLOGICAL ACTIVITY OF AMBER

SUMMARY

Amber has been used in folk medicine for many years, serves as a medicine for the treatment of many ailments, and is also a component in many cosmetic preparations. Unfortunately, in most cases, there are no clinical trials that can provide a scientific basis for the use of succinite as an approved drug. Therefore, there is a need for research that would provide evidence-based knowledge. It has been suggested that amber's healing/health-promoting properties result from its chemical composition, including monoterpenes and acids. The succinic acid contained in Baltic amber is a widely used metabolite in pharmaceuticals (vitamin production, amino acids, anticancer drugs) and in the food industry. It is one of the main unions involved in the Krebs cycle. Its antioxidant, adaptogenic, and cardioprotective effects have been proven. From a pharmacological point of view, terpenes and their derivatives – such as borneol, cymene, abietic acid, and pimaric acid – are important compounds. They have antibacterial, antifungal, antiseptic, anti-inflammatory, and anti-rheumatic properties.

The list below describes a selection of succinite compounds and their properties.

Succinorezene Succinabietic acid Succinoabietol Succinosylvinic acid Succinoresinol	These are derivatives of succinic and abietic acid;compounds that are part of the so-called amber resin. They are also classified as polycyclic compounds. They create ester and, among others, cross-linked bonding with succinic acid or polyabietic acid. Abietic acid has antibacterial properties protecting from *Staphylococcus aureus*. This acid also has an anti-inflammatory effect in an in vitro and in vivo model. However, the activity of these compounds when isolated is not well known.
Succinic acid	In the food industry, it is used as an acidity regulator and flavor enhancer (E363). Three to eight percent of the outer mantle or "amber bark" of Baltic amber consists of succinic acid. Succinic acid is a natural catalyst for cellular processes, which can be beneficial to our health. Research has confirmed the acid to be a biostimulant. It stimulates the nervous system, regulates the functioning of the kidneys and intestines, and has anti-inflammatory properties. It may also be effective in treating diseases of the nervous system, bronchitis, rheumatic diseases, and asthma.
Succinyl abietic acid	It is an organic chemical compound belonging to the group of resins. It occurs naturally in pine resins (mainly in firs) or is obtained from rosin. It is a stabilizer and stiffener used in the production of cosmetics (including soaps and some face cleansers), as well as in the production of vinyl, varnish, and plastic. Abietic acid can cause allergic reactions and irritate the skin and mucous membranes. It is especially dangerous and harmful to aquatic organisms.

Borneol	It is a bicyclic organic chemical compound from the group of terpenes that is used as a camphor fragrance. Borneol is a component of many essential oils and is a natural insect repellent. It is also used in traditional Chinese medicine.
Para-Cymene	It is a component of many essential oils, most commonly cumin and thyme essential oils, and is a naturally occurring aromatic organic compound. It is classified as a monoterpene.
Fenchenol	It is an organic compound classified as a monoterpene. It is a colorless oily liquid with a camphor odor. Fenchenol is a component of absinthe and is extracted from fennel. It is also used as a flavor in foods and perfumes.
1,8-Cineole (eucalyptol)	It is an organic chemical compound from the terpenes group. It is a colorless liquid with a camphor scent, obtained from *eucalyptus* and contained in its essential oil, known from wormwood and rosemary. It is mainly used as an ingredient in fragrances and care products.
Camphor	It is an organic chemical compound from the group of terpenes. Camphor penetrates the skin thoroughly, causing a feeling of warmth. Ointments and alcoholic camphor solutions have a warming and anesthetic effect.
Isoborneol	Borneol isomer is a component of fragrances used in shampoos and detergents, as well as for the synthesis of camphor.
Communic acid	Communic acid has anticancer, antibacterial, relaxant, anti-inflammatory, and antioxidant properties. It has been found in plants of the genus Juniperus.

Pulegone	It is classified as a monoterpene. Pulegone is a clear, colorless oily liquid with a pleasant scent similar to peppermint and camphor. It is used in aromas, perfumes, and aromatherapy. The FDA withdrew approval for the use of pulegone as a synthetic food flavoring in October 2018.
α – Pinene β – Pinene	These are organic chemical compounds from the group of cyclic monoterpenes. They are the main ingredients of turpentine - obtained from pine resin. They are used as ingredients of perfumes and as raw materials for the preparation of fragrances.
Carene	It is a colorless liquid with a sweet smell similar to turpentine. It is an ingredient of cedar oil with antimicrobial properties. Carene is used in the perfume industry and as a chemical intermediate.
Limonene	It is an organic chemical compound from the group of monoterpenes, responsible for the smell of lemons (found mainly in their peel). It is used as a flavor and aromatic additive to food in the food industry. Additionally, it is also used as a fragrance component in cleaning products (detergents, air fresheners), perfumes and cosmetics (concentration 0.005 to two percent).

5.1. TERPENES IN AMBER AND THEIR EFFECTS

SUMMARY

From a pharmacological point of view, terpenes and their derivatives, such as borneol, bornyl acetate, cymene, eucalyptol, are important compounds gaining significance among agents supporting the functions of the human body. Also, polycyclic compounds and their derivatives, such as abietic and pimaric acids, have shown antibacterial, antifungal, antiseptic, anti-inflammatory and antirheumatic effects. [Zawietrzny et al. 2017, Tumiłowicz et al. 2016]. The antibacterial effect of amber is particularly noteworthy. There is not much research on the subject. The data available in the literature is based mainly on the antimicrobial effect of amber components [Mikucioniene et al. 2016]. Tumiłowicz et al. [2016] critically reviewed the knowledge of terpenes and phenols which are found in amber and act as antimicrobial agents. The described compounds are known and are commonly found in many essential oils of plants, including mint (Mentha piperita), caraway (Carum carvi), fennel (Foeniculum capillaceum), common thyme (Thymus vulgaris) [Pisulewska et al. 2008]. Many amber components, mainly terpene compounds, exhibit antibacterial properties, but their potency is lower than that of antibiotics, which precludes their use as a substitute for antibiotics [Załuski et al. 2015]. Synergism mainly takes into account the metabolites. Kaczmarczyk [2018] examined the effect of 20 ethanol extracts from various types of amber on gram-positive and gram-negative bacteria. The Baltic

amber extract showed antibacterial activity against *Staphylococcus aureus*. Perhaps the presence of communic acid, a compound commonly found in conifers with an antibacterial effect, influenced the bacteria *Staphylococcus aureus* and *Staphylococcus epidermidis* and induced antifungal activity against *Aspergillus fumigatus* and *Candida albicans*. Moreover, the amber extract was shown to have anti-inflammatory and antioxidant properties.

Juniperus rigida (Cupressaceae), Botanical Garden of the Polish Academy of Sciences in Powsin near Warsaw. The branches of the *Cupressaceae* family are found in amber from the Eocene epoch.

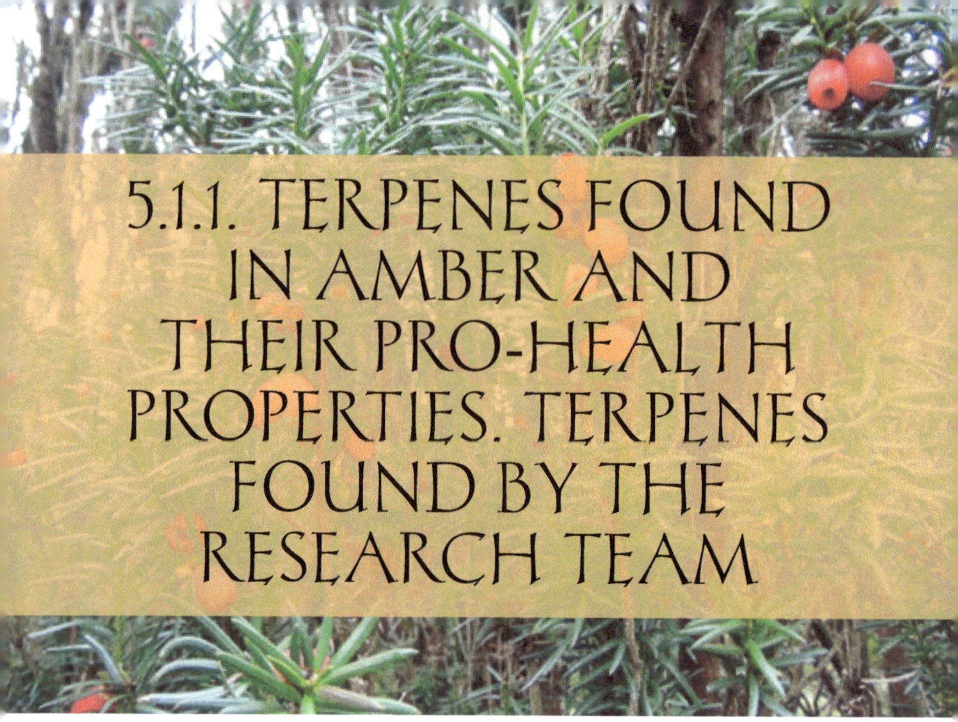

5.1.1. TERPENES FOUND IN AMBER AND THEIR PRO-HEALTH PROPERTIES. TERPENES FOUND BY THE RESEARCH TEAM

The key issue affecting the properties of the fossilized resin from 50 million years ago is the fact that it consists of compounds sometimes found in modern trees and shrubs, i.e. terpenes and terpenoids. Some of these compounds are contained in Baltic amber and are no longer present in plants, which makes amber extract an "elixir".

Essential oils are usually a mixture of terpenes and terpenoids. (Terpenoids are derivatives of terpenes – aldehydes, alcohols, ketones, esters, peroxides). According to our hypothesis, it was essential oils and resins that created the succinite structures. The trees that produced them protected themselves against microorganisms and viruses by secreting essential oils and influencing the environment in order to protect themselves against microorganisms and parasites. Sometimes trees and shrubs produced essential oils to deter insects. Amber has retained these substances, although terpenes are optically active – they are very reactive, they oxidize very easily in the air, when heated they easily change their chemical structure.

Terpenes are hydrocarbons, and terpenoids are terpenes that contain an extra oxygen molecule. For example, myrcene, with the formula $C_{10}H_{16}$, is a monoterpene, and when an oxygen atom is attached, a $C_{10}H_{18}O$ terpenoid called linalool is formed. Both chemical compounds are natural substances.

Terpenoids are a group of the most abundant components of plant oils. The basic building block of terpene is isoprene C_5H_8. Essential oils currently occurring in plants – including trees – contain two main groups of terpenoids: monoterpenes, with the chemical formula $C_{10}H_{16}$, and sesquiterpenes with the general formula $C_{15}H_{24}$. Some of the terpene components found in essential oils can be found in many oils. Components such as α-pinene, β-pinene, borneol, myrcene, limonene, linalool, 1,8-cineole, β-caryophyllene, and germacrene D [Malm 2021]. Most of these compounds can be found in Baltic amber.

Terpenes are compounds that form molecules consisting of carbon atoms divisible by five, namely:

monoterpenes (C_{10})
sesquiterpenes (C_{15})
diterpenes (C_{20})
sesterterpenes (C_{25})
triterpenes (C_3)
tetraterpenes (C_{40})
polyterpenes (C_{50})

In this group, there are also other molecules with the number of carbon atoms which is not divisible by five, for example steroids. They are formed at various times as a result of secondary reactions under the influence of temperature, pressure, external elements, and chemical compounds. Monoterpenes are also compounds contained in volatile oils, and their basic chemical formula is $C_{10}H_{16}$. The resins of the trees from which the Baltic amber originated were subjected to a specific maceration process in deep layers of peat and earth. Buried by volcanic ashes and sand brought by floods, proto-ambers shaped their chemical structure.

Fig. 73. Classification of terpenes. Based on: Najda, A. (2015). "Vegetal Volatile Substances–Essential Oils". Episteme, 2, 65-77.

From the very beginning, amber has had a polymeric structure that was open to other chemical compounds, and which – while it was laying in peat, brown coal, at the bottom of the ocean, or under glaciers – was constantly changing. This polymeric structure is also being formed today because succinite is constantly reacting with the outside environment. It has a unique structure, in a sense mysterious, with numerous chemical compounds derived from pre-trees.

In 1972, L. Goff and J. Mills showed that the bicyclic labdane diterpene is an important chemical compound in succinite, which constitutes as much as 77 percent of succinite's content. Thus, a diterpene was found, i.e. a substance from the group of terpenes, which contains 20 carbon atoms. However, that terpene can change.

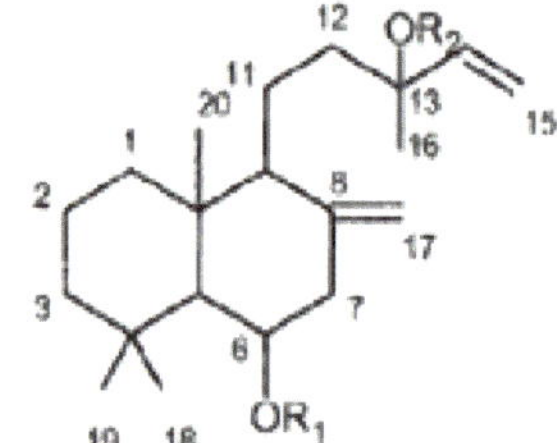

Fig. 74. Bicyclic labdane diterpene. In the places R1 and R2 it can be joined by numerous compounds.

It also occurred in Baltic amber. 20 percent of terpenes with the pimarane skeleton and 2.8 percent of terpenes with the abietane skeleton formed[Matuszewska 2012]. One of the trees rich in these compounds is *Agathis australis*, now growing in Australia and is abundantly resinous.

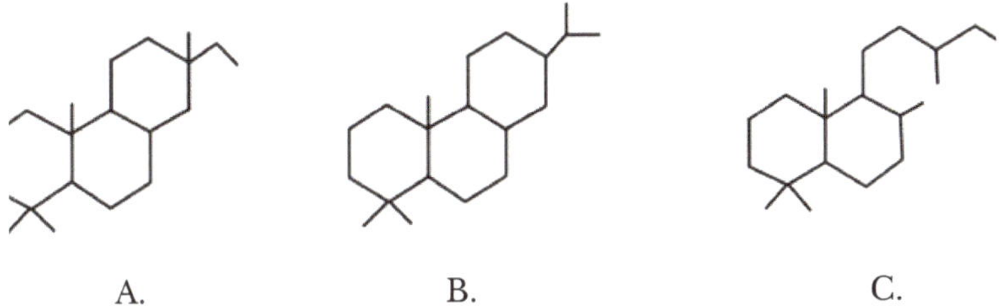

A.   B.   C.

Fig. 75. A. A compound with the abietane skeleton. B. A compound with the labdane skeleton. C. A compound with the pimarane skeleton

In the research by Goff and Mills, a number of compounds found in Baltic amber were established in the tested samples. The fossilized resin contains abietane diterpenoids (terpenes) made up of twenty carbon atoms. The analysis of trees containing these types of terpenes led to a hypothesis that Baltic amber was formed by gymnosperm trees – that is, trees producing seeds, which are literally called "naked seeds" or unclosed seeds. They can be pollinated directly by pollen from the circulating air (Quite a primitive form). Conifers, cycads, ginkgo, and gnetophyta are examples of this. Most of these plants are present in the world today, although some of them are endangered species. Cycads, for example, are similar to palm trees, although they grow very slowly. They can live up to 1000 years. Some species of cycads are pollinated only by certain species of beetles, which limits their possibilities of development.

Photo 98. *Cycads, Zamiaceae, Encephalartos ferox,* South Africa. Photo by P. Barczak.

Abietanes found in succinite are abietic acid derivatives, an organic compound commonly found in trees, especially conifers. A certain amount of the very rare abietic acid derivative – methyl abiet-8-en-18-mate – has been found in the Acridocarpus tree growing in New Caledonia, although some species can be found in Oman and the United Arab Emirates. This tree may have grown in the regions where Baltic amber was formed.

Acridocarpus orientalis is famous for its antioxidant and antiprotozoal properties. It is effective against lipid peroxidation and has anti-inflammatory and liver-protecting qualities. The oil soothes muscle aches and headaches. It also has an important function related to its influence on a carbonic anhydrase enzyme in cellular processes [Rehman et al. 2021.] (Amber has similar properties). The oil from this tree can play a significant role in human organic processes at the cellular level.

Photo 99. Duxite, Vršany u Mostu. Photo by P. Barczak.

Anhydrases are enzymes, which originated billions of years ago all over the planet. Man, like plants, uses these substances. Enzymes of this type (carbonic anhydrases) are found in mammals, as well as in plants and bacteria. One type of carbonic anhydrase can be found in hot springs, which may have been the place where life on Earth originated.

Photo 100. The eastern white pine containing labdane diterpenes [Zinkel,1987], Pinus strobus, Athens, Greece. Photo by P. Barczak.

These types of chemical structures are created by specific plants – including the oldest ones, which existed 50 million years ago. Sometimes essential oils of plants with labdane molecules have a protective function, other times they signal insects, and in still other times they stimulate plant genes. Compounds of this type can also be found in coal seams which, as it is known, are formed from plants.

Substances with the structure of labdane (terpenes) have been found in trees such as Araucaria, Agathis, and Wollemia. These are trees suspected of creating Baltic amber. Araucaria is currently represented by 19 species, Agathis by 21 species. Wollemia – discovered relatively recently – is known from one species. In the trees mentioned, growing mainly in Australia, diterpenoids are represented by bicyclic, tricyclic and tetracyclic compounds. It is worth mentioning, however, that bicyclic diterpenoids (they have two common carbon atoms in their structure) composed mainly of labdane compounds, are common in all conifers [Lu, Hautevelle, Michels 2013]. Therefore, it is not certain whether amber originated exclusively from trees such as Araucaria, Agathis, Wollemia or whether its chemical composition was influenced by other conifers and broad-leaved trees.

Photo 101. *Agathis* and *Araucaria*. Lima Botanical Garden. Peru. Photo by M. Barczak.

Labdane-type diterpenoids are found in all fossil resins present in Czechia. Czech ambers were made by Glyptostrobus and Quasi sequoia trees, and by the *Cupressaceae* family of conifers [Havelcová et al. 2018]. The content of labdane diterpenoids, however, does not determine the nature of amber, its appearance, durability or value as jewelry. Amber found in the Czech Republic has no use in jewelry, and its ethnopharmacological properties have not been found, although it contains labdane-type diterpenoids.

Photo 102. Petrified resin, Študlov, Třetihory, Eocen. Photo by P.Barczak

A study was conducted by a research team from Warsaw University of Life Sciences and the LLC Royal, financed – from EU funds – by the National Center for Research and Development which illustrated the content of terpenes in amber. The amber extract was analyzed with GC-MS gas chromatography. Only the main chemical compounds were shown in the laboratory-tested Baltic amber extract. By doing so, it is possible to identify the plants and trees that played a dominant role in building the fossilized resin. The research team discovered two main chemicals in the ethanol extract while extracting amber, namely succinic acid and 1,8-cineole (eucalyptol). There were also trace amounts of a little-known compound named plumbagin (C11H8O3), which is related to naphthalene. Let us discuss the discovered compounds.

Results of research

	Code of the sample	Described parameter	Method	Result	Unit
	SP 1925/20	Plumbagin	HPLC MS/MS	<10	mg/l
		Succinic acid	HPLC MS/MS	141	mg/l
		1,8-cineole	GC-MS	16	mg/l

Tab. 6. Table with the results of the analysis of the amber extract obtained in the extractor. The extraction with 96 percent ethanol alcohol had been performed.

The name of succinic acid ($C_4H_6O_4$) comes from the word succinum, which was used to describe Baltic amber in the past. In nature, this acid is in the form of a stable compound, the structure of the ester of succinic acid is also known. It is mainly produced by microorganisms. It has also been found in marine algae. It is common in nature, and fruits and vegetables often contain it. However, the amount of succinic acid that can be found in amber is hundreds of times more than what can be found in rhubarb for example, which also contains significant amounts of succinic acid. Succinic acid is soluble in water, alcohol-ethanol, and ether. It has been found to have no side effects as it is present in every human cell. All food substances, carbohydrates, fats, proteins are finally oxidized in the Krebs cycle with the use of succinic acid.

So it plays a key role in metabolism, and in cellular reactions. It takes the form of anions, is a metabolic intermediary, and, as a result, is an indicator of the metabolic state of a cell. Succinic acid plays an important role in this process, participating in the transformation of energy in the mitochondria. By accelerating the oxidation process in cells, succinic acid reduces the content of free oxygen, and thus reduces the likelihood of free radicals.

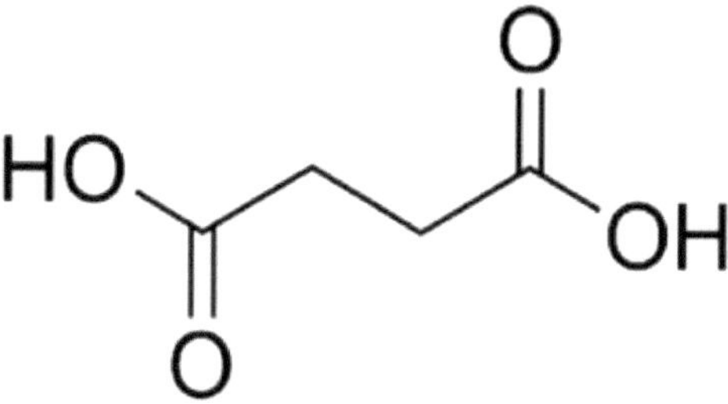

Fig 76. Succinic acid ($C_4H_6O_4$)

Succinic acid is a white solid with a very acidic taste. It ionizes very easily in an aqueous solution, forming succinate compounds. It is a well-known dietary supplement and food additive and has been recognized by the FDA in the USA as a safe chemical. Succinates play an important role in cellular metabolism. They are a component of many pharmaceutical products.

The first to carry out the dry distillation of amber in 1546 was Georgius Agricola. As a result, he released liquid chemical fractions, including succinic acid, amber oil and rosin. He described the taste of succinic acid as sour. The acid crystals that were obtained looked like salt, therefore succinic acid was later described as sal puccini (succinite salt), sal volatile (volatile salt) and also flores succini [Vávra 2009]. In 1728, Chambers described succinic acid as the spirit of amber [Matuszewska 2016].

In 1877, while conducting dry distillation, O. Helm found that amber consists of two to eight percent succinic acid. He attributed the ability to secrete succinic acid to amber from the Baltic Sea region, although it turned out that other resins also contain succinic acid, but contain much smaller amounts [Tschirch 1894].

Succinic acid is known to have a stabilizing effect on the human body. The stabilizing effect includes supporting the processes of cell reconstruction in the case of heart and kidney damage, improving age-related disorders of the nervous system, helping to increase the efficiency of muscle work, as well as reducing the effects of toxic substances in the human body. Our body produces about 200g of succinic acid daily, but the balance is disturbed in the case of stress or exertion. It has been found that succinic acid obtained, for example, from amber, penetrates into cells and modifies their activity by influencing the regulation of intracellular processes that restore the organism's homeostasis. Succinic acid supports mental and physical activity; restores vitality, and strengthens the body's protection. This compound is often used to relieve alcohol poisoning and reduces one's craving for alcohol. It slows down the aging processes of the body at the cellular level. Succinic acid is used to lower the content of cholesterol, fatty acids and ketone bodies in the blood, and also reduces the need for insulin as it stimulates the body to produce its own insulin. Succinic acid affects the body's efficiency during training, allows the body to quickly and easily adapt to physical exertion, and reduces pain in overtrained muscles. Succinic acid has been found to help prevent heart and blood vessels, bronchitis, and cerebrovascular strokes, and helps alleviate the effects of flu, colds, tuberculosis and allergies Succinic acid and its salts are used as diuretics, laxatives, anti-inflammatory and anti-anemic drugs.

The next compound detected was 1.8 cineole (eucalyptol).

1,8-cineole (eucalyptol) is an organic chemical compound from the group of terpenes with a camphor scent. 1,8-cineole is a toxin that protects trees against herbivores. Compounds of this type are clearly used to protect trees and other plants. Trees secrete terpenes, phenols, glycosides, alkaloids and tannins, which are used by humans as drugs or adjuvants. However, the overconsumption of terpenes - more than 1g per kg, causes serious poisoning of the organism [Tibballs 1995].

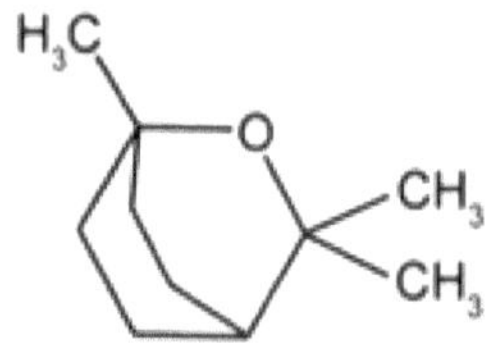

Fig. 77. 1,8- cineole (eucalyptol)

Eucalyptol may have a local pro-inflammatory effect. The pharmaceutical industry uses its properties to penetrate the skin [Santos et al. 2000]. This terpene is mainly obtained from rosemary, sage, wormwood and *eucalyptus* trees. It is the main ingredient of *eucalyptus* (75 percent) and rosemary (40 percent) oils. Eucalyptol is used to treat ailments such as coughing. It is antitussive and is useful in the treatment of bronchitis, bronchial asthma, sinuses, and rheumatism [McGilevery, Reed 1993]. It has also been shown to be useful because of its anti-inflammatory, antibacterial and analgesic effects and also influences bronchodilation. The anti-inflammatory action was proven to be effective in helping people suffering from

chronic lung disease. It was found that the substance may have a beneficial effect on people suffering from asthma. It has also been proven to improve lung function and the quality of life of people with this ailment. 1,8-cineole improved lung function, and reduced the incidence of dyspnea [Worth, Dethlefsen 2012].

Cineole influences the central nervous system and may be an antidepressant [Rao, Santos 2000].
The effect of eucalyptol in hypertension has been confirmed. The terpene has been shown to be useful in reducing aortic pressure, which may be due to its effect on vascular smooth muscle cells [Lahlou et al. 2002]. It relaxes blood vessels, and reduces heart rate and blood pressure. In the case of 1,8-cineole, in vitro and in vivo studies have confirmed its hypotensive effect [Santos et al. 2011].

Photo 103. Falling off *eucalyptus* bark. Limassol, Greece. Photo by P.Barczak.

Monoterpenes are known for their pharmacological properties in relaxing blood vessels. Sixteen monoterpenes have been tested for their effects on the cardiovascular system, including: carvacrol, citronellol, p-cymene, eucalyptol (1,8-cineole), linalool, menthol, myrtenal, myrtenol, α-pinene, rotundifolone (piperitone oxide), sobrerol, thymol, α-limonene, α-terpinen-4-ol, α-terpineol, and perillyl alcohol. They relax the blood vessels, and reduce heart rate and blood pressure. Some of these terpenes are known to be present in Baltic amber.

Inhaling essential oils containing 1,8-cineole and camphor has helped to lower blood pressure and provided relaxation and stress relief. Inhaling the oil increased the intensity of EEG waves in the human brain, namely alpha and theta waves, which are responsible for relaxation. On the other hand, beta and gamma waves related to brain activity, consciousness, and focused attention decreased while smelling the compounds. Only the inhalation of these compounds ensured the body's responses, including an impact on the heart and on blood pressure [Kim et al. 2018].

The search for new anti-cancer drugs is ongoing. Recent studies have shown that 1,8-cineole inhibits the proliferation of colon cancer in humans. The research was carried out on mice. The mechanism of action of 1,8-cineole is not fully understood. However, it has been noticed that the substance causes the death of diseased neoplastic cells in the case of colorectal cancer [Murata et al. 2013].

1,8-cineole protects the internal cell cycle against oxidative stress. It has been found to have antioxidant and anti-inflammatory effects in human cells. It is worth mentioning that Alzheimer's disease is characterized by an increase in pro-inflammatory cytokines at the cellular level. Degeneration of the nervous system in the brain causes inflammation of the cells. The administration of 1,8-cineole to the damaged cells resulted in reduction of the inflammation [Khan et al. 2014].

Eucalyptus oil has been shown to be effective against the bacteria responsible for causing periodontitis. Cineole is widely used in dental and oral care formulations.

One of the plants that contains eucalyptol is the holy basil (Ocimum tenuiflorum), which is very important in Hindu symbolism. The basil, also known as tulsi, contains 54 chemical compounds with different properties. It is also called the "mother of natural medicine. It contains three main terpenes: eucalyptol, camphor and eugenol. Tulsi extracts have proven effective in treating various types of poisoning, abdominal pain, colds, headaches, malaria, inflammation and heart disease [Pattanayak et al. 2013]. The eucalyptol content in the essential oil and leaves of this plant is estimated to be at 18.9 percent and 13.47 percent.

Cineole $C_{10}H_{18}O$ (eucalyptol) is the main ingredient of *eucalyptus* (75 percent) and rosemary (40 percent) oils. *Eucalyptus* globulus oil was found to consist of as much as 95 percent eucalyptol. It should be noted that the *Eucalyptus* genus, which is a significant supplier of eucalyptol, includes over 600 species of evergreen trees and shrubs. These trees thrive in both tropical and subtropical zones, therefore *Eucalyptus* is a significant contender to be indicated as an important creator of Baltic amber. The study confirms that the compound found in abundance in *Eucalyptus* is a significant component of the Baltic amber extract.

Due to the chemical composition, there are four types of *eucalyptus* oils: 1) cineole oil, containing over 60 percent 1,8 -cineole and is created by numerous species of *Eucalyptus*, 2) phellandrene-piperitone oil, which is contained in the twigs of *Eucalyptus* dives (only two to four percent), 3) citronellal oil with a lemon scent and slight rose note, containing 65 to 85 percent citronellal, geraniol, geranyl and eudesmol acetate, 4) geraniol oil, containing 65 to 75 percent geranyl acetate, and can be found in *Eucalyptus* macarthurii [Góra, Lis 2019]. 1,4-cineole is also found in juniper oil.

Scientific experiments confirm that 1,8-cineole, when heated with diluted sulfuric acid, turns into cis-1,8-terpin and then into α-terpineol [Wrzeciono, Zaprutko 2001]. The latter compound is known from succinite (Mosini, Forcellese, Nicoletti 1980). The existence of this compound may indicate that the processes taking place in the "sulfur sea" that is supposed to create Baltic amber were real. (This topic is covered in chapter 1.2.)

The final compound found by the research team from PH Royal LLC and Warsaw University of Life Sciences – although only a trace amount - was plumbagin ($C_{11}H_8O_3$), a compound with the chemical name 8-hydroxy-3-methylnaphthoquinone. It occurs in leadworts (*Plumbago*) – a genus of plants from the Plumbaginaceae family. Currently, they are most often found as ornamental plants due to their pretty blue flowers. Plants of this type occur in tropical and subtropical zones, which predisposes them to be included in the hypothetical succinite group. The climate in which *Plumbago* is present today is similar to that in which the amber forest grew. *Plumbago* plants probably grew there, but also possibly sundews (Droser and Nepenthes), containing plumbagin". It's worth noting, however, that plumbagin could also have recovered from naphthoquinone, a chemical related to naphthalene. According to the assumptions described in this work, naphthalene also influenced the Baltic amber produced over millions of years.

Another analytical study carried out in the GC-MS gas chromatograph by the PH Royal LLC and Warsaw University of Life Sciences research team revealed the main chemical compounds found in ground amber. The amber sample was not diluted with alcohol, and only ground amber was examined.

	Description of the samples by the Client	Description of the samples by the laboratory
.	Milled amber 50 g P.H. Royal	SP 1439/20

Results of research:

	Code of the sample	Description	Research method	Result	Unit
.		Camphene	Gas chromatography	5	mg/kg
.		Cymene	Gas chromatography	212	mg/kg
.	SP 1439/20	Borneol	Gas chromatography	42	mg/kg
.		L-fenchone	Gas chromatography	12	mg/kg
.		DL-menthol	Gas chromatography	3	mg/kg

Tab. 7. Compounds Shown in Ground Amber, GC-MS Gas Chromatograph Analysis.

The analysis showed a group of dominant compounds. First of all, a significant amount of terpene, called cymene, was found – 212 mg/kg. There was a high content of borneol and small amounts of L-fenchone (12 mg/kg) and camphene (5 mg/kg). The last of the compounds found was dl-menthol - 3 mg/kg.

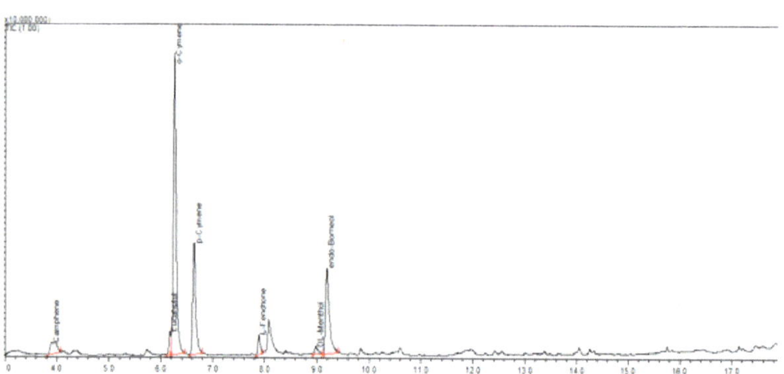

Fig. 78. Graphical diagram of the study carried out, as part of the project, with a GC-MS gas chromatograph.

Para-cymene (C$_{10}$H$_{14}$) is an aromatic organic compound, a monoterpene. It is miscible with organic solvents. Para-cymene is a component of many essential oils and a natural isomer. It is found, among others. in cumin and thyme.

Fig. 79. Para-cymene.

However, this compound is also found quite often in tree resins. Compounds from succinite have been found in American and Asian styrax, derived from the Liquidambar orientalis tree. Camphene, p-cymene, borneol, and terpenes have also been found [Fernandez et al. 2005]. Another essential oil which contains para-cymene can be found in trees of the species Protium heptaphyllum (*Burseraceae* family), which grow in the tropical zone of Madagascar, Papua New Guinea, and Brazil. Many plants contain it. It is found, among others, in the plant *Cymbopogon winterianus* (java citronella), which grows in warm tropical climates, but also in *Thymus pubescens* [Quintans-Júnior et al. 2013].

The oil protects the trees against insects, fungi and mold.

Para-cymene affects cardiological processes in mammals. The first signals concerning the influence of para-cymene on pressure reduction were noticed while testing black seed oil, which contains 5.8 to 11.6 percent of this compound. By conducting experiments on animals, its effect on lowering blood pressure has been noticed [El Tahir, Al-Ajmi, Al-Bekairi 2003]. These types of monoterpenes affect the cardiovascular system [Santos et al. 2011] and the central nervous system [Quintans-Júnior et al. 2010]. In mammals, para-cymene has antioxidant and anti-inflammatory effects. This terpene is the subject of many medical experiments. For example, it has been found to elicit specific antioxidant effects in the hippocampus of mice. Para-cymene may act as a neuroprotective agent in the brain. The administration of p-cymene has caused a significant (over 50 percent) decrease in lipids in the brain of mice, which is an important clue in the fight against degenerative diseases of the human brain. The administration of para-cymene also reduced the nitrite content and improved brainial energy processes essential for cellular processes. It should be noted that lipid deposition is one of the causes of Alzheimer's and Parkinson's diseases, atherosclerosis, and diabetic complications. Para-cymene may also contribute to the removal of free radicals derived from nitrogen groups. The decrease in the nitrite content in the brain observed in experiments on mice may be the result of para-cymene. This type of monoterpene contributed to the neuroprotection of the brain [Mendes, de Oliveira et al. 2015].

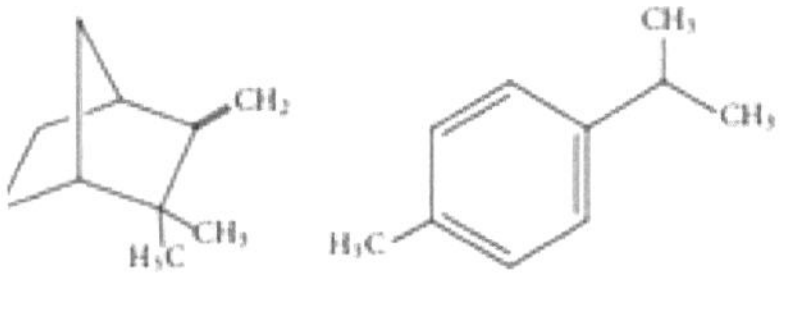

Fig. 80. Camphene and para-cymene.

A number of plants contain para-cymene, but this is a compound that is not well known and it is not widely used in medicine and pharmacology yet.

Borneol ($C_{10}H_{18}O$) is an organic chemical compound from the group of terpenes, a subgroup of camphanes. Borneol is a natural mosquito repellent and its smell is similar to camphor. It has been approved by the FDA (US Food and Drug Administration) as a flavoring food additive. It is a component of cosmetics, shampoos and toilet soaps. It is used as an anesthetic in traditional Chinese and Japanese medicine.

Borneol affects brain cells, including GABA receptors important in the nervous system. They influence the functions of neurotransmitters. Borneol improved the activity of neurotransmitters by 1000 percent [Renee 2005]. Folk medicine, especially Chinese medicine, widely uses borneol to relieve symptoms of nervous disorders such as insomnia, anxiety, and fatigue. Borneol manifested neuroprotective activity and reversed neuronal damage, including reducing the production of reactive oxygen species in cells that destroyed them. It inhibited anti-inflammatory effects in neuronal cells.

Its activity in the case of ischemic cerebral disease has also been analyzed. In China, this type of disease has been treated with borneol since ancient times. Borneol has been shown to help those with this type of ailment. This is most likely due to its anti-inflammatory characteristics [Dong et al. 2018].

It is also used as a food additive. According to the Chinese Pharmacopeia, borneol is present in 63 plant products. It has been proven to inhibit the growth of bacteria and fungi. The compound stimulates the body's immune functions, as well as antithrombotic and anti-platelet functions. Borneol has contributed to regulation of overpressure in rats [Kumar et al. 2010].

Borneol has special properties that protect DNA against damage and oxidative stress. It has anti-inflammatory, antibacterial and antioxidant properties. In rodent studies, borneol has had a protective effect on cells, although it is not fully understood. However, it can be assumed that this compound contributes to the protection of cells against aging [Horváthová et al. 2012].

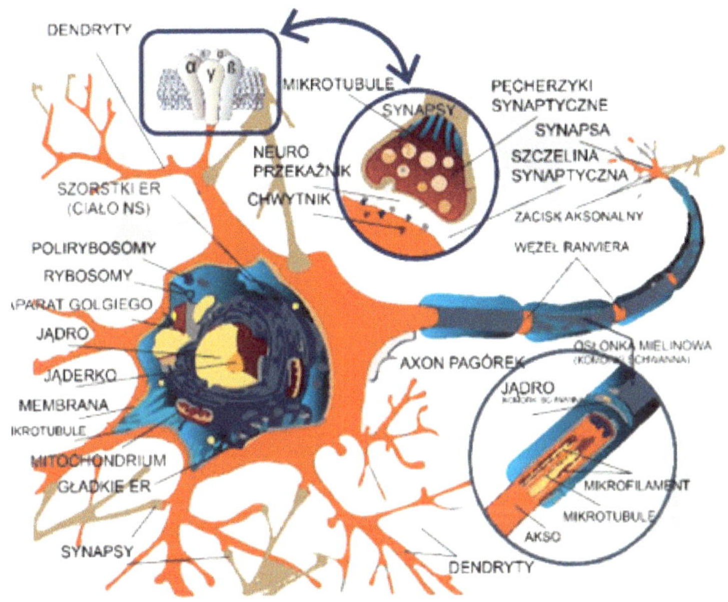

Fig. 81. GABA receptors in human tissue.

Borneol has been found in many plants and trees, including Dipterocarpus turbinatus. This tree grows in Northeast India, and its resin is internationally known as "Copaiba balm". Borneol occurs in both annual and perennial plants, such as in the following species from the Lamiaceae family: *Rosmarinus officinalis* (rosemary), *Salvia officinalis* (sage) and *Lavandula officinalis* (lavender), but also in trees growing in warm climates such as: *Lauraceae (Cinnamomum camphora), Myristicaceae (Myristica fragrans), Pinaceae (Pinus sylvestris), Poaceae (Cymbopogon winterianus), Apiaceae (Coriandrum sativum) and Verbenaceae (Lippia adoensis).*

One of the plants that contain large amounts of borneol is lemon balm. It is used to alleviate Alzheimer's disease, which is understood to be a defect of neurotransmitters. Short-term memory loss was found to be caused by deficits resulting from poorly functioning neurotransmitters [Francis et al. 1999]. Lemon balm extracts contain approximately 4 percent borneol.

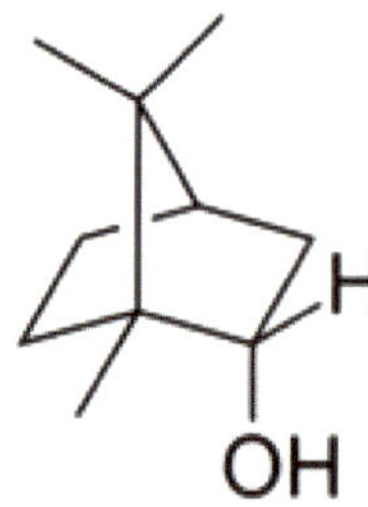

Fig. 82. A molecule of borneol

Lemon balm has been found to modulate mood and increase cognitive functions. The functioning of the nervous system in the human brain is mediated by cholinergic compounds, which can act as insect repellents in plants. These compounds affect the central and peripheral nervous systems of insects and plant-eating animals [Edelfrawi, Edelfrawi 1997], so it can be assumed that borneol also stimulates the human nervous system.

Lemon balm is especially valued in Iranian medicine. It is used for depression and anxiety. In small doses it strengthens the memory.

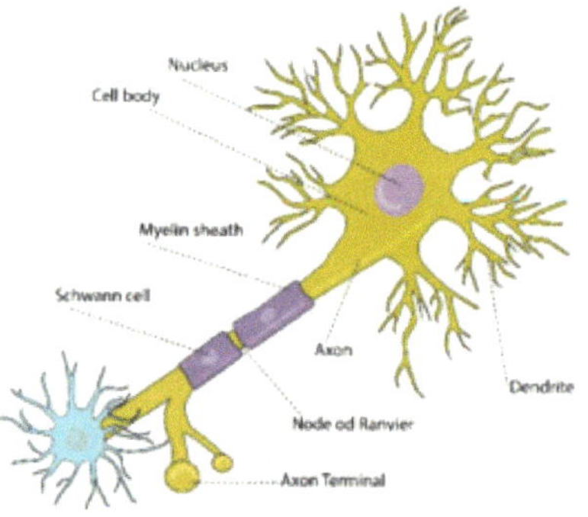

Fig. 83. Structure of a neuron, a cell stimulated by bioenergy.

Photo 104. Lemon balm extract. Photo by P. Barczak.

Melissa officinalis (Labiatae), at a certain dose, strengthens memory processes – especialy when it comes to neurons that are responsible for memory in mammals -- which are destroyed in the case of Alzheimer's disease. Lemon balm extract increases activity in neurons. It also improves one's learning abilities after administering a specific dose of lemon balm alcohol extract. Administration of the extract in the amount of 200 mg/kg resulted in the alleviation of memory loss in the examined rodents. However, too much extract did not lead to improvement, instead, memory processes did not change when the dose was increased [Soodi et al. 2014].

Another study has shown that lemon balm extract increased the flow of biocurrents in the subjects' brains. Brain waves imaged by the EEG signal increased, especially the alpha 1, alpha 2, and beta 1 waves. The central and temporal areas of the human brain were stimulated. The increase in activity in the parietal lobe, which is responsible for working memory is especially noteworthy (Dimpfel, Pischel, Lehnfeld 2004). Lemon balm oil alleviates cognitive deficits in Alzheimer's disease. By improving the accuracy of attention, it influences secondary and working memory. After four weeks of using lemon balm, Alzheimer's patients showed less agitation and less social withdrawal. It is predicted that it affects the human body by activating the acetylcholine receptor in the central nervous system [Kennedy et al. 2002].

There is much more borneol in the plant known as yarrow –

as much as 16.28 percent [García-Risco et al. 2017]. Therefore, it is worth getting acquainted with the effects of this plant. The yarrow is used in ethnomedicine for hemorrhoids, hypertension, blood circulation disorders and varicose veins, stomach ulcers and menstrual disorders. The yarrow is also used in treatment of colds and flu, as it effectively reduces fever and even inflammation of the throat and mouth. It is interesting that scientific analyses sometimes show a positive effect of amber extracts on hemorrhoids and hypertension.

Another plant containing significant amounts of borneol is sage. Research is ongoing to check to what extent the borneol contained in sage eliminates the β-amyloid protein in the brain, which is associated with Alzheimer's disease.

Camphene ($C_{10}H_{16}$) is a bicyclic compound – a terpene. It is characterized by a camphor-like smell. It sublimes quickly if it is not at room temperature. It is a component of many essential oils. It has been found in spruce essential oil, juniper twigs, valerian oil, and in one of the species of chrysanthemums (Chrysanthemum japonicum). D- camphene is found in cypress, nutmeg, ginger, and *eucalyptus* trees. It is also found in spices such as rosemary and fennel. When dissolved in alcohol, it replaced whale oil. It has a strong antioxidant effect, as confirmed in in vitro tests, and it significantly absorbs free radicals, including hydroxyl and suboxide radicals. This compound is also contained in the resin from the Bosswelia serrata tree and its essential oil. Camphene is used in hepatobiliary and kidney diseases.

Fenchone ($C_{10}H_{16}O$) is another terpenoid detected in the study that was conducted. Fenchone can be found in many essential oils. It is insoluble in water but soluble in ethanol. Fenchone tastes bitter and smells like camphor. It inhibits the growth of mold and fungi, and has a repellent effect on insects. In larger doses, it has a stimulating effect on the human body, and is used in the treatment of gastrointestinal diseases. It shows anti-inflammatory, antioxidant properties. The effect of fenchone on wound healing and the removal of inflammation has also been noted [Keskin et al. 2017].

Fenchone is a component of cedar (*Thuja occidentalis*) essential oils. In trees of this species α-pinene, camphene, sabinene, d-limonene, p-cymene, γ-terpinene and l-fenchone e have been found [Rudloff 1961]. Therefore, it can be assumed that trees of this type participated in the creation of Baltic amber. Most of the terpenes of this tree species have been found in Baltic amber. Fenchone has also been found in the needles of balsamic pine (*Abies balsamea*). Compounds of this type are secreted by trees in the form of volatile oils when the tree is attacked by insects [Caron 2013]. Plants of the Lamiaceae family also show a large amount of fenchone. 24.8 percent of fenchone has been found in the Mesosphaerum sidifolium extract. This species is especially abundant in the Mediterranean [Rolim et al. 2017].

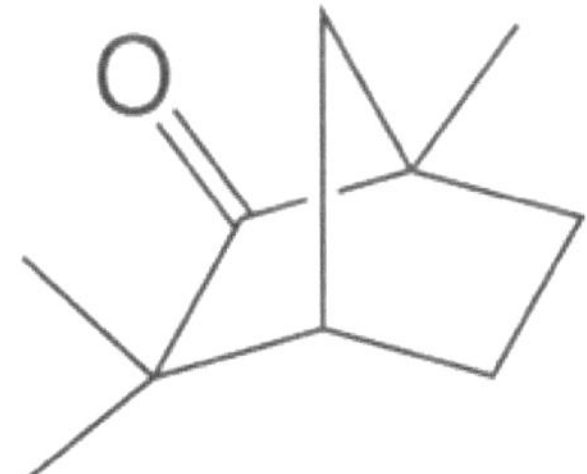

Fig. 84. Fenchone.

Fenchone is found in spices from seeds of the star anise (*Illicium verum*) [Kang et al. 2019]. In cellular studies, fenchone inhibited cancer cell growth, induced weight loss, and prevented the development of diseased cells [Rolim et al. 2017].

Fenchone occurs naturally in fennel. This spice has been used since antiquity in Asia Minor and Europe. It is used in the kitchen and in the treatment of ailments – especially indigestion, mild gastrointestinal ailments, and respiratory disorders. The compounds contained in fennel have an antispasmodic effect. Fenchone can constitute up to 20 percent of the fennel oil content.

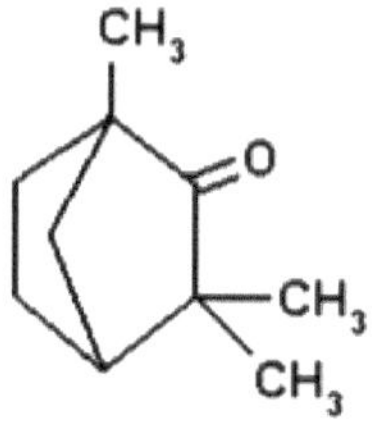

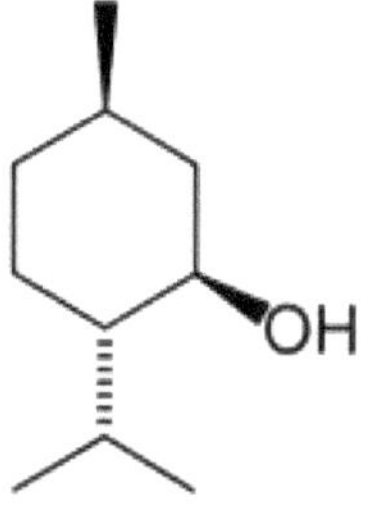

Fig. 85. A molecule of fenchone.

DL-menthol ($C_{10}H_{20}O$) is a cyclic, saturated alcohol from the group of terpenes with a mint scent. It causes a known feeling of coolness, which makes it a very popular addition to many cosmetic and utility products. It can come in eight varieties and four forms as menthol, neomenthol, isomenthol, neoisomenthol. It undergoes frequent changes in its structure. DL-menthol is a mixture of D and L-menthol. It has anesthetic properties, and reduces irritation of the mucous membranes. Menthol-containing essential oils aids in the treatment of colds and rhinorrhea. It is also added to ointments used to treat muscle pain. Rubbing substances containing menthol into your skin activates the receptors, increases the temperature in the muscles, and dilates the blood vessels, thus warming up the application site. One of its carriers are oils produced by *eucalyptus*.

Oil containing menthol facilitates digestion by stimulating secreting gastric juices and bile. It relieves flatulence, has a diastolic effect on the smooth muscles of the intestines, and helps in the treatment of inflammation of the gallbladder and bile ducts [Góra, Lis 2019].

Fig. 86. Menthol.

Peppermint oil also relieves headaches, migraines of gastric origin, and has a positive effect on nervous tension and fatigue. Menthol is allergenic, especially in children.

Oil containing menthol is effectively used for oral hygiene. It is antibacterial and helps fight against numerous bacterial strains accumulating in the mouth. Menthol is particularly useful in combating anaerobic bacteria that develop inflammation in the teeth, damage the enamel, and develop tartar. Oil containing menthol, even at a low concentration, contributes to killing more than half of anaerobic bacteria [Kusiak et al. 2010]. In addition to the aforementioned properties, it has a positive effect on inflammation of the mouth and gums, and on mouth ulcers and erosions. In combination with 1,8-cineole (eucalyptol), it can be an important element in the fight for oral hygiene and has anti-inflammatory and soothing properties.

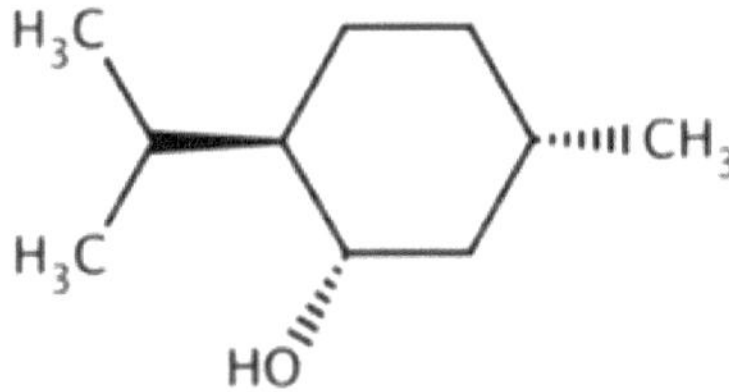

Fig. 87. DL-menthol.

In amber, this compound could undergo numerous transformations, transforming into isomers, so that trees do not produce it in excess. Menthol has been found in a species of the *Burseraceae* family – to which *Eucalyptus* belongs – specifically, in the tree Commiphora mukul. The Indian bdellium-tree, a resin-producing tree called also guggul, is used in Vedic medicine. The plant can be found in India, Pakistan, and Iran [Varamini et al. 2016]. It contains 1.77 percent menthol. Menthol is also found in the oil of another plant of the *Burseraceae* family – Boswellia sacra, a tree found in Somalia and the southern part of the Arabian Peninsula. 1.8-cineole (4.16 percent) and menthol (10.81

percent) have been found in the aqueous extract from its resin. Boswellia sacra resin is known all over the world, and is used in incense in aromatherapy and religious ceremonies. In Arab countries, the resin is known as Luban and is eagerly chewed, as it contains aromatic oils that are antiseptic and can soothe inflammation. Boswellia sacra resin helps heal wounds, ulcers, ulcers, hemorrhoids (amber alleviates these types of ailments – vide the Dorpat research) and inflammation. Boswellia essential oils have antiseptic, astringent, carminative, digestive, diuretic and calming properties. The water extract from this gum resin (incense) was traditionally used to treat coughs, gastric problems, and problems with the liver. The Boswellia water extract has been found to have the following composition: 1,8-cineole ($C_{10}H_{18}O$) – 4.16 percent, 1-acetate ($C_8H_{18}O$) – 6.91 percent, (-)-menthol ($C_{10}H_{20}O$) – 10.81 percent, 3-cyclohexen-1-ol ($C_{10}H_{20}O$) – 21.33 percent, octanoic acid ($C_8H_{16}O_2$) – 18.91 percent, thymol ($C_{10}H_{14}O$) – 6.14 percent, carvacrol ($C_{10}H_{14}O$) – 5.67 percent [Asad, Meshal 2015]. Thus, it contains a large amount of menthol. In turn, in Melaleuca cajuputi oil – made from a tree growing in Australia, Southeast Asia and Papua New Guinea belonging to the myrtle (Myrtaceae) family, has been found to contain a p-Menth-1-en-4-01, also found in Baltic amber. Menthol has also been detected in the oil [Pasaribu et al. 2021]. Myrtaceae could therefore have taken part in the formation of succinite.

Photo 105.
Melaleuca
Myrtaceae tree
flowers. Lima,
Peru. Photo by P.
Barczak.

5.1.2. POSSIBLE BIOLOGICAL ACTIVITY OF COMPOUNDS FOUND IN SUCCINITE IN TRACE AMOUNTS

There is a large number of terpenoids in Baltic amber, sometimes occurring in small percentages. Perhaps even minimal amounts of these chemical compounds can affect the human body, so it is worth mentioning them in the context of the use of amber in ethnopharmaceuticals. Their descriptions should be based on the systematics of terpenes and terpenoids.

A large group of terpenoid compounds formed from two C_5 isoprene particles (from 10 -C- carbon atoms) is contained in succinite. They are called monoterpenes or monoterpenoids. These substances have been found as a result of the work of several research teams. Mills' team (Mills, White, Gough 1984) found: tricyclene $C_{10}H_{16}$, camphene $C_{10}H_{16}$, p-cymene $C_{10}H_{14}$, 1,8-cineole $C_{10}H_{18}O$, fenchone $C_{10}H_{16}O$, camphor $C_{10}H_{16}O$, borneol $C_{10}H_{18}O$, isoborneol $C_{10}H_{18}O$. Yamamoto's group [Yamamoto, Krumbiegel, Simoneit 2006] found fenchol $C_{10}H_{18}O$, borneol $C_{10}H_{18}O$ and isoborneol $C_{10}H_{18}O$. Finally, a group from Basra in Iraq [Al-Tamimi 2020] confirmed the existence of borneol $C_{10}H_{18}O$, isoborneol $C_{10}H_{18}O$, camphene $C_{10}H_{16}$, m-cymene $C_{10}H_{14}$, 1-fenchone $C_{10}H_{16}O$, 2-fenchanol $C_{10}H_{18}O$.

Many of the monoterpenes have already been described. Their existence in Baltic amber is accompanied by a great variety of structures due to numerous possibilities of cyclization, the occurrence of various electrical bonds between molecules. Generally, however, they are liquid or solid compounds with a strong odor and are known to be volatile when in contact with water vapor. Oils containing monoterpenes are insoluble in water. They are widespread in the plant world, especially in the families Pinaceae, *Cupressaceae, Lauraceae,* Myrtaceae, and also *Apiaceae, Lamiaceae, Rutaceae, Asteraceae, Zingiberaceae,* and *Iridaceae.* Monoterpenes are compounds that irritate the mucous membranes and the skin, causing better blood circulation and redness of the skin. Internally, they are irritating to the mucous membranes in the digestive tract and the kidneys, therefore they have a diuretic effect, just like Baltic amber. They cause increased bile secretion. They also have an expectorant and disinfectant effect in aerosols by irritating the mucous membranes of the respiratory tract [Kohlmünzer 2007].

Photo 106. Resinating pine *Pinus (Pinaceae)*. The Botanical Garden of the Polish Academy of Sciences, Powsin. Photo by P. Barczak.

Tricyclene $C_{10}H_{16}$ – Camphene, which occurs in succinite has already been described. One of the teams that examined the chemical composition of amber [Yamamoto et al. 2006] did not detect this type of compound in succinite, which may indicate the complexity of research and the diversity of Baltic amber. Baltic amber is not a uniform fossil, therefore it is also important to know where and what geological layer it comes from. Each nugget can be of a different composition, although the basic compounds remain the same. Camphene ($C_{10}H_{16}$) is found in succinite, and was detected by other teams. Other monoterpenes having the chemical formula $C_{10}H_{16}$ are also known. The best known ones are: carene, limonene, myrcene, alpha-pinene, beta-pinene, terpinene, thujene, turpentine.

Alpha and beta-pinene are compounds which are quite common in conifers growing today. However, it has been noticed that Baltic amber very rarely contains them. There is a hypothesis that i α and β-pinene begin to evaporate when drying the resins. They dissolve about 20 percent of the time, and turn into a cross-linked polymer with a different chemical formula 80 percent of the time. The monoterpenes mentioned are common in the resins of presently existing trees, but it seems that these types of trees did not contribute to the formation of succinite [Armstrong et al. 1996]. The dominant compounds in amber are borneol and isoborneol, and in contemporary conifers α and β-pinene can be found. The color of samples that are yellowish – and seemingly have the same color as amber, but with a different chemical composition – may be misleading.

Fig 88. Alpha pinene and beta-pinene.

Alpha-pinene is the main monoterpene in Baltic pine (*Pinus sylvestris*) essential oils. The needles of this tree have a strong antioxidant effect. The extract from them has a high content of unsaturated terpenoids and flavonols. Flavonols, on the other hand, are a subgroup of organic chemical compounds found in plants – flavonoids – that act as pigments, antioxidants, and natural insecticides and fungicides. Flavonols were found in abundance in *Pinus sylvestris* needle extract. They have good antioxidant and anti-inflammatory properties. Alpha-pinene significantly influences the functioning of the nervous system of mammals by activating astrocytes and stopping epileptic seizures in mice. The compound protects astrocytes from oxidative stress [Ueno et al. 2020]. Administration of alpha-pinene prevents the depletion of neurons in the hippocampus while monoterpene reduces depression [Khan-Mohammadi-Khorrami et al. 2021]. Alpha-pinene significantly reduces the production of lipids by the brain and positively influences the motor activities of mammals with Parkinson's disease. It was found to have positive effects on memory processes [Goudarzi, Rafieirad 2017]. The administration of 100 mg/kg and 200 mg/kg of alpha-pinene resulted in an increase in motor coordination and an improvement in memory, and oxidative stress in the hippocampus diminished. However, lipid levels were significantly lowered by the 200 mg/kg dose. The preparation was administered to rats in which the chemical compound stopped Parkinson's disease. Similar phenomena were caused by the ginseng extract [Kampen et al. 2003].

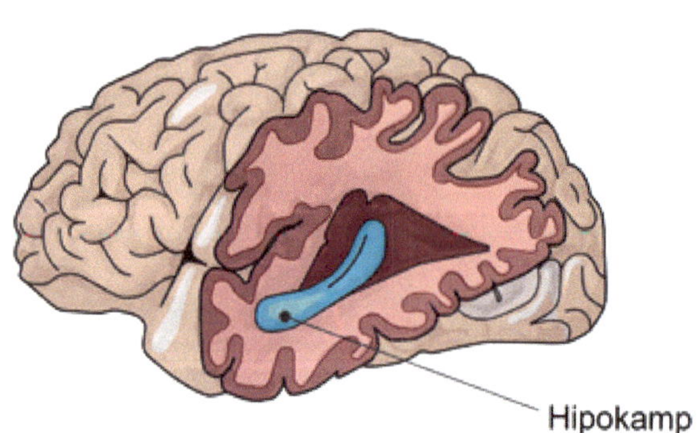

Fig. 89. Brain and hippocampus.

3-carene ($C_{10}H_{16}$) is another monoterpene found in amber. It is a component of turpentine (42 percent). It is the second, after alpha-pinene, component of pine essential oils. Under the influence of pyrolysis, it transforms into p-cymene. In a study carried out in 1976, succinite was found to contain: p-cymene (49 percent), fenchenole (ten percent), 1.8 cineole (eight percent), borneol (six percent), and camphor (three percent). 24 percent of the total mass of such terpenes consisted of: 3-carene, 1,4-cineole, isoborneol, o and m-cymene and other unidentified substances [Urbański, Glinka, Wesołowska 1976]. This monoterpene has been shown to positively affect human sleep. The compound also has anti-inflammatory, antioxidant and anti-stress properties. Its anxiolytic effect has been found [Woo et al. 2019].

Sesquiterpenes are an important group of compounds found in Baltic amber. It is a group of natural compounds with the general formula $C_{15}H_{24}$, forming a molecule half the size of monoterpenes. This group includes, among others, components of essential oils and resins with volatile properties. In total, more than 500 compounds of this type are known today. They can form chain structures, but also more compact structures. The combination of three C_5 isoprene units creates new complex structures with the base C_{15}. Most of the compounds are non-volatile or sparingly volatile, so substances may not show up in gas spectrometry analyzes. The pharmacological properties of the vast majority are unknown. Sesquiterpene lactones are potential contact allergens. One of the features is its influence on the development of the larvae of herbivorous pests. There is a suspicion that amber-bearing trees produced some sesquiterpenes as a kind of pheromone to perform definite functions related to insects. Nowadays, trees sometimes produce pheromones to protect pollinators, other times to attract their own pests' enemies. The best known sesquiterpenes include cadinene found in the *Pinus sylvestris* pine, the *Junipersus communis* juniper, the *Abies* alba *fir*, *Chamomilla recutita* and *the Mentha piperita peppermint.*

Yamamoto's team found dehydro-ar-curcumene ($C_{15}H_{24}$) and calamene ($C_{15}H_{22}$) in succinite [Yamamoto et al. 2006]. These compounds have been described earlier in the book.

Caryophyllenyl alcohol ($C_{15}H_{26}O$) is a sesquiterpenoid that is a component of many essential oils, including the most popular chamomile. It is best known for its anti-inflammatory effects. It also inhibits the action of gram-positive and gram-negative bacteria and fungi. Its variant is α-cadinol, which protects the liver [Tung et al. 2011] and has antifungal properties. δ-cadinol is a terpenoid found in many conifers, especially cedar trees. It works as a pheromone. Cedrol, another variation of this compound, is often found in Cupressus (cypresses) and Juniperus (junipers).

Fig. 90. Caryophyllenyl alcohol $C_{15}H_{26}O$.

The terpenoid has antiseptic, anti-inflammatory, antispasmodic, tonic, astringent, diuretic, sedative, insecticidal and antifungal properties [Jeong et al. 2014]. One of the compounds has been identified in cubeb pepper, a plant growing in Indonesia. Farnesol ($C_{15}H_{26}O$) is known to be present in lemongrass. Guaiol ($C_{15}H_{26}O$) in turn, has been found in *Callitris cypresses* currently growing in Australia, New Caledonia and New Zealand, predisposed by a system of seeds and roots that cause fires, that is, having a mechanism of fighting competition like the *eucalyptus*. The terpenoid-containing cypress Actinostrobus grows in Australia and Widdringtonia grows in the southern part of Africa.

Those trees also have a similar system of propagation as the *eucalyptus* – their cones open only after a fire, when the area has been "cleared" of competition. A sesquiterpene called ledol is commonly found in the *eucalyptus*.

Diterpenes constitute an important group that can be found in Baltic amber. Interestingly, they have not been found in the Sumatra glessite by Yamamoto's team. Diterpenes have the molecular formula $C_{20}H_{32}$ and consist of approximately 800 substances. Sometimes they are found in resins, growth substances and vitamins. They can have a linear, single, double or more cyclic structure. The joining of molecules most often leads to the formation of the labdane skeleton. Diterpenes are often found in the outer waxy leaves of higher plants. They are usually biologically active, often toxic. Sometimes they occur as glycosides. They are used to support the human body. Some of them are characterized by an outstanding influence on the activity of the heart and the circulatory system. Ginkgolides affect the cerebral circulation. Rosin, which forms yellow and yellow-brown nuggets, has a high content of diterpenes. When heated, it melts into a clear liquid. It consists of abietic acid and β-pimaric acid [Kohlmünzer 2007]. Most of the compounds in Baltic amber are not diterpenes but diterpenoids, which contain an oxygen molecule, confirming the oxidation processes taking place in the fossilized resin. In the Baltic amber the following diterpenes have been found, inter alia: abiet-8-en-18oic acid, pimar-8-en-18oic acid, pimar-9(11)-en-18oic acid, pimar-7-en-18oic acid and also dihydroxy-isostcviol in thc amount of 3.57 to 5.01 percent.

It has been known for a long time that isosteviols have significant pharmacological effects. There are a huge number of plants that produce steviol glycosides. They have an anti-inflammatory, antihypertensive effect that controls blood lipid levels. In the case of plants, perhaps they were produced by trees that inhibited the growth of competing plants [Ullah et al. 2019].

Pimarinal is a diterpene with the formula $C_{20}H_{32}O$ and has been detected in Baltic amber. It can be found in small amounts in the Japanese cedar Cryptomeria japonica. Oils of this type protect the tree against fungi, while abiet-8-en-18-oic acid, with the formula $C_{20}H_{32}O_2$, has been found in Baltic amber in the amount of 2.34 to 2.81 percent. This compound is not widely known.

Abietic acid $(C_{20}H_{30}O_2)$ is one of the terpenoids which are multifunctional natural compounds. It is also known as silicic acid. It is the main component of resin acid isolated from wood resins produced by the cells of coniferous trees in the temperate zone. Trees use this to defend themselves against external threats. It comes from various species of Pinus (pines, the Pinaceae family) as well as turpentine and rosin. It is also secreted by the grand fir (*Abies grandis*). It has antibacterial and antiviral properties, but is allergenic. Abietic acid has been shown to be fatal to fish by affecting their energy metabolism and respiration processes. Abietic acid derivatives are also fatal to rats, impairing their coordination and causing paralysis. The mechanisms of reactions of organisms to abietic acid are still unknown [Aranda, Villalaín 1997]. The terpenoid also causes positive reactions in the body when dosed correctly. It inhibits inflammation in organisms and regulates lipid metabolism, including stabilization of the development of atherosclerosis.

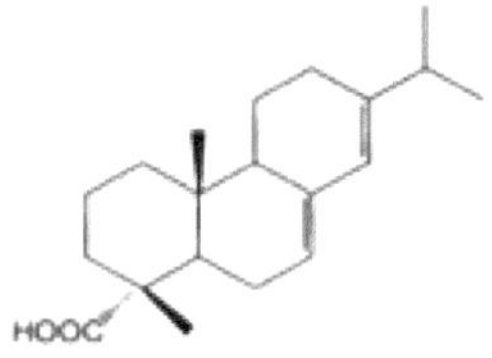

Fig 91. Abietic acid.

Some of the effects of abietic acid on the body are known. It activates macrophages, human cells tasked to destroy microorganisms, including abnormal, damaged cells. It also initiates the process of tissue regeneration. Abietic acid activates PPARQ (Peroxisome

Proliferator-Activated Receptor), which plays a major regulatory role in energy homeostasis and human metabolic functions. It affects lipid metabolism and affects processes related to atherosclerosis. Abietic acid inhibits the expression of genes involved in inflammation

Photo 107. *Pinus koraiensis* (*Pinaceae*). Photo by P. Barczak.

[Takahashia et al. 2003]. Dehydroabietic acid improved the metabolism of carbohydrates and lipids. Compounds based on abietic acid (catechols) showed anti-cancer activity. They inhibited the growth of cancer cells. On the other hand, abietol has proven to be effective in fighting herpes simplex. The acid-derived compounds have been found to be effective against some viruses and against certain fungi.

The compound with the formula $C_{20}H_{30}O_2$ can be converted into numerous forms in amber as a result of polymerization. Another form may be levopimaric acid. In some pine species it constitutes as much as 18 to 25 percent of the oleoresin content.

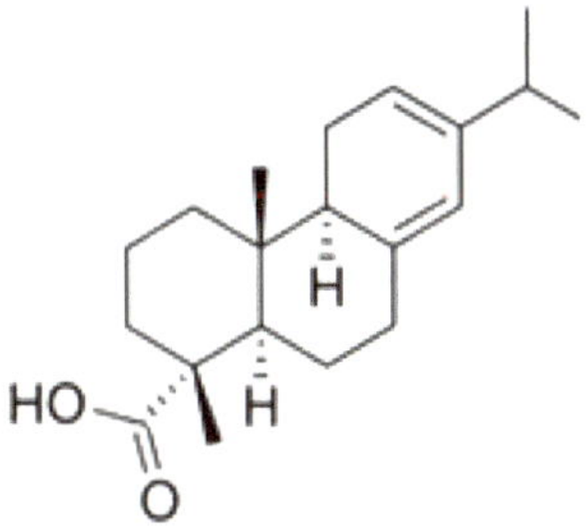

Fig. 92. Levopimaric acid.

A slight conversion causes abietic acid to change into isopimaric acid, which is a toxin, but acts as an opener of cell channels in the human body, which means it produces new properties. These transformations make Baltic amber a resin that can be researched endlessly, constantly discovering new formulas and compounds.

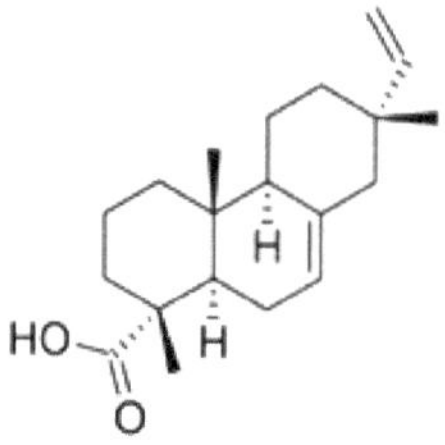

Fig. 93. Isopimaric acid.

Sometimes the compounds can act synergistically and the effects are not theoretically predictable.

Communic acid ($C_{20}H_{30}O_2$) is found in conifers, including cedars. It has been found in African cedars Widdringtonia [Sadgrove et al. 2020]. The acid has also been isolated from needles of the Korean red pine. It reduces blood pressure in rats, therefore it is predicted that the compound may be effective in medicinals for hypertension [Park et al. 2021]. This compound is contained in the plant Tetraclinis articulata – the sandarac tree, the resin of which consists of over 70 percent communic acid. This species of cypress grows in the Mediterranean and, like the *eucalyptus*, has adapted to outbreaks of fires by creating offshoots from stumps. The resin is commonly referred to as sandarac. The acid can also be found in juniper fruit.

The compound has properties of cross-linking with its aging. Complexes of polymers and saturated polymers with various structures are formed. Thanks to this, it can create biologically active chemical structures. Bruceantin, for example, has an anti-cancer effect.

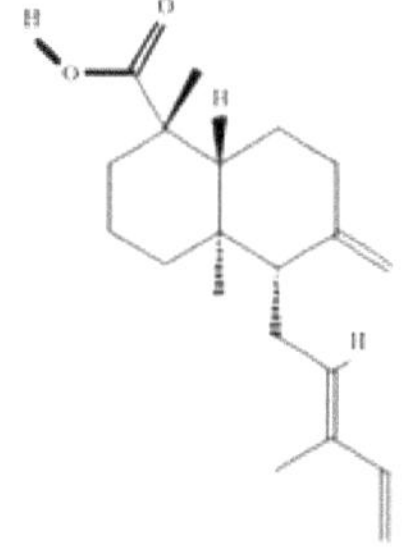

Fig. 94. Communic acid.

6β-hydroxy-trans-communic acid is a compound that is formed from communic acid, and can be found in sage (Salvia cinnabarina). This compound inhibits α-glucosidases. The number of chemical compounds based on communic acid can be very large, and each of them can have numerous properties of biological influence on the human body.

Dihydroxy-isosteviol ($C_{20}H_{32}O_3$) has been found in succinite in trace amounts (3.57 to 5.01 percent). The spatial arrangement of this chemical compound which is presented below may undergo changes and transform into separate molecular configurations. This diterpenoid also occurs in the form of HETE acid (hydroxyeicosatetraenoic acid), known for its properties supporting the vascular system, including blood flow [Bellien, Joannides 2013], as well as for the regulation of sodium and fluid flow in the kidneys. This type of reaction improving blood supply is observed when amber tinctures are applied.

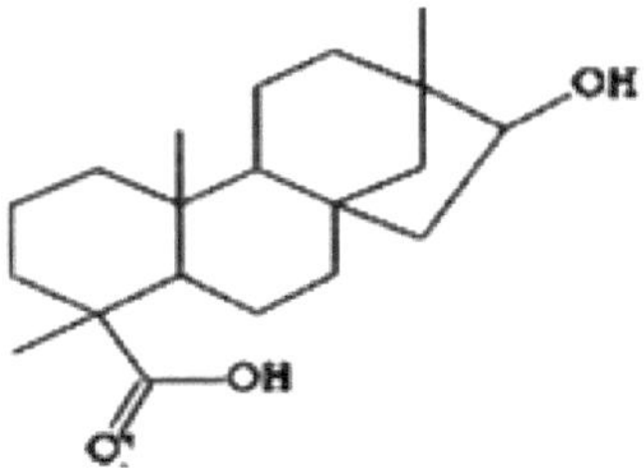

Fig. 95. Dihydroxy-isosteviol- $C_{20}H_{32}O_3$.

Another exemplary effect of $C_{20}H_{32}O_3$ in the form of EET acid (epoxyeicosatrienoic acid), with the same chemical formula as HETE acid, is the growth of axons in neurons and vasodilation. A properly converted diterpene (isocupressic acid) with the same chemical formula $C_{20}H_{32}O_3$ is known to come from needles of the Pinus ponderosa pine. It can also be found in Juniperus osteosperma juniper.

Dihydroxy-isosteviol is also obtained from the Stevia rebaudiana plant, which recently became a popular compound used in parapharmaceuticals and dietary supplements. An

important feature of the stevia is lowering blood pressure [Chan et al. 2000], which also somehow explains the properties of Baltic amber. Studies have shown that sodium isosteviol protects against persistent ischemia of the brain by inhibiting inflammation. The compound was effective within three to seven days after administration. It showed significant neuroprotective effects, including reduction of cortical damage, restoration of neuronal function, and an increased number of astrocytes. Sodium isosteviol reduced inflammation [Zan, Zhang, Lu 2018], protected the heart muscle, reduced the size of myocardial infarction, improved heart efficiency and left ventricular pressure, and stabilized the electrophysiological properties of the heart and ventricular fibrillation [Xu et. al. 2007].

Triterpenes have the molecular formula $C_{30}H_{48}$, i.e. they are made of 6 C5 units of active isoprene. In total, over 3,000 triterpene compounds have been found in the plant world. These are non-volatile compounds, quite easily soluble in water. They have a number of properties in common with steroids. Triterpenes are often found in tree resins as well as in the cork tissue of trees. They are found in lower plants, such as fungi and lichens, in which monoterpenes are usually absent. Alpha-amyrin and beta-amyrin are particularly widespread, although no such compounds have been found in previous research done on Baltic amber. Perhaps it is a matter of further research. On the other hand, traces of bornyl pimar-8-en-18-aete ($C_{30}H_{48}O_2$) and isobornyl pimar-9 (11) -en-18-aete, also with the chemical formula $C_{30}H_{48}O_2$, have been found in Baltic amber.

Alpha-amyrin is a triterpene found in plant resins, including the small tropical tree Bursera copallifera, growing, among others, in Mexico in mountainous areas. The resin of this tree has been found to consist of 21.1 percent α-amyrin. These types of plants are quite common in the tree family *Burseraceae*. This triterpene affects neurons, has anti-inflammatory, antipruritic, cell-

protective effects, and may protect against the effects of diabetes [Romerao-Estrada et al. 2016]. It has numerous pharmacological applications.

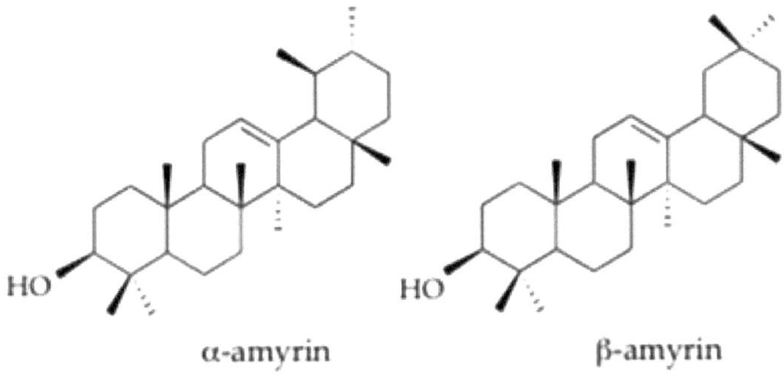

α-amyrin

β-amyrin

Fig. 96. Alpha amyrin and beta amyrin.

Amyrin has been found in numerous plants growing today, but also in Mexican ambers, in which 5 g/kg of this substance was found. The most important sources of amyrin are plants such as the lotus (Nelumbo nucifera) and bee pollen (3 g/kg). Amyrin has also been found in Boswellia carterii resin.

Beta-amyrin has been found in oleo-gum resin obtained from the Ferula gummosa plant growing, among others, in Iran, as well as in the Carpobrotus edulis succulent growing in South Africa. Beta amyrin has anti-inflammatory, antifungal and antiviral properties. This triterpene is used against some oral bacteria [Hernandez et al. 2012].

1- (+)Ascorbic acid- 2,6-dihexadecanoate ($C_{38}H_{68}O_8$) is a compound which has been found to have antioxidant, anti-inflammatory and antinociceptive properties. It was detected in 2.6 percent in Baltic amber, although only in one of the analyzed fractions. The terpenoid showed antibacterial activity against *Staphylococcus aureus*, *Escherichia coli*, etc. It also improves sperm quality and prevents spermatozoids' agglutination, making them more motile [Ogunlesi, Okiei, Osibote 2010].

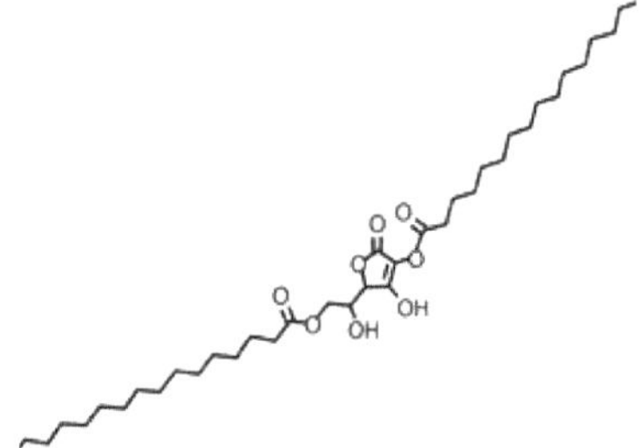

Fig. 97. 1- (+) Ascorbic acid- 2,6-dihexadecanoate $C_{38}H_{68}O_8$.

Acetic acid 1,7,7, trimethyl-bicyclo [2,2,1] hept-2-yl ester $(C_{12}H_{20}O_2)$ is a compound which is present in the oil obtained from cinnamon (2.28 percent). One of variations of this terpenoid is geranyl acetate, which is found in citronella oil in lemongrass. It is also an insect repellent. It is a liquid slightly yellow in color. Bornyl acetate, which has the same chemical formula, is the acetate ester of borneol and is a compound of essential oils found in pine needles (Pinaceae) [Garneau et al. 2012].

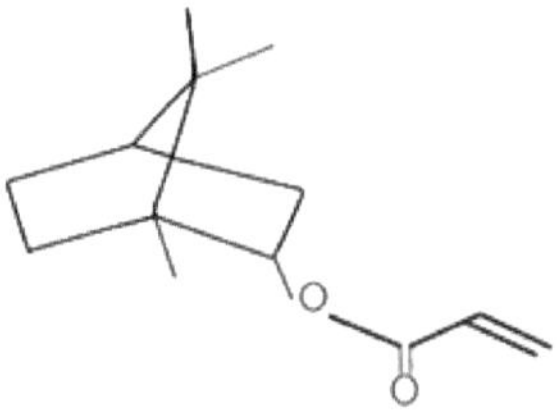

Fig. 98. Acetic acid 1,7,7, trimethyl-bicyclo [2,2,1] hept-2-yl ester- $C_{12}H_{20}O_2$.

Naphthalene, 1,2,3,4-tetrahydro-1,6,8-trimethyl $(C_{13}H_{18})$ is a volatile compound which has been found in black tea grown in the Wuyi Mountain region of China. Its content increased during roasting to one to three percent [Liu X et al. 2015], which suggests that if tea bushes also contributed to the formation of amber, the processes of heating up the matter took place in lignifying leaves that have made up the amber.

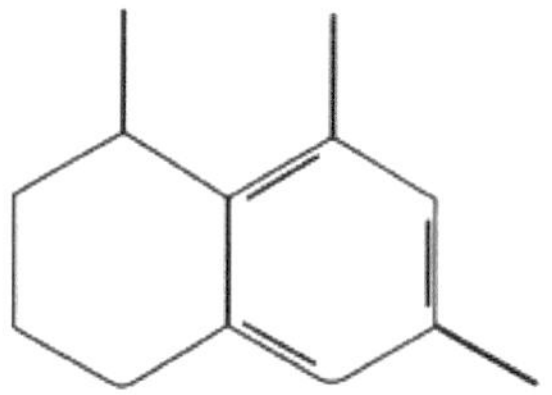

Fig. 99. Naphthalene, 1,2,3,4-tetrahydro-1,6,8-trimethyl – $C_{13}H_{18}$.

It is worth mentioning that naphthalenes are found in coal tar, crude oil and essential oils, which may indicate ways of creating Baltic amber. Plants use naphthalenes to repel insects, and one of the moth repellents contains naphthalene.

Naphthalene, 1,2,3,4-tetrahydro-5,6,7,8-tetramethyl ($C_{14}H_{20}$) is another naphthalene which may explain the way amber is formed. This compound has been found in bio-oil produced by anaerobic pyrolysis of wood mass. The pyrolysis takes place at a temperature of 400 to 900 degrees Celsius. During pyrolysis, flammable gasses are released, which is followed by carbonization, allowing the structures of carbon elements to obtain stable forms [Sichone 2013]. Diamantane is one of the forms of naphthalene. It is found in crude oil and is an important form of diamondoids due to its highly stable atomic structure.

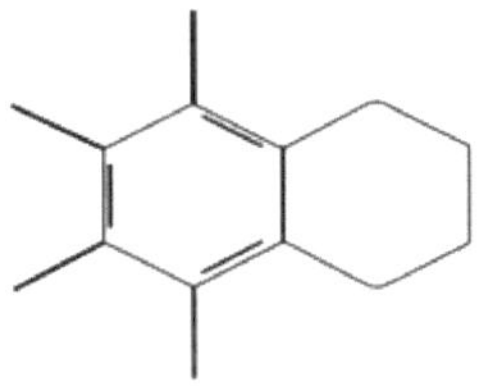

Fig. 100. Naphthalene, 1,2,3,4-tetrahydro-5,6,7,8-tetramethyl.

Bornyl formate ($C_{11}H_{18}O_2$) is a compound found by Mills' team (Mill, White, Gough 1984), one of the first to study Baltic amber with modern devices. Other research teams did not find this compound. It is an ester (esters are formed by condensation of alcohol and organic or inorganic acid), which is a combination of borneol and formic acid. It has a balsamic scent and is therefore used as flavoring and fragrance. It has poisonous properties in larger doses. Many esters are found in fruit, vegetables and spices. For example, ethyl acetate is found in apples and bananas, and menthol acetate is found in peppermint leaves. Significant amounts are found in Valeriana officinalis. Amounts of bornyl formate have also been found in the rhizome of a plant found in China called *Rhynchanthus beesianus*. The flower oil consisted of

21.7 percent bornyl formate, the leaf essential oil – 33.9 percent of bornyl formate. The stems consisted of11.6 percent of bornyl formate. Interestingly, this plant has several compounds found in amber: 1.8 -cineole (21.6 percent), camphene (7.8 percent), borneol (9.7 percent). The plant is known for its antibacterial properties and inhibits the production of the pro-inflammatory mediator NO (nitric oxide) in cells. In traditional Chinese medicine, flowers of Rhynchanthus are used to treat stomach ailments [Chen et al. 2021].

Photo 108. Tropical forests also made up Baltic amber. Peru. Photo M. Barczak.

6. IODINE CONTENT IN AMBER AND ITS ACTIVITY. FLAVONOIDS AND POLYPHENOLS

The iodine content in amber and its health-promoting effects depend on the presence of various mineral compounds. One of the minerals found in amber is organic iodine. Europe is a continent where chronic iodine deficiency exists and it has been suggested that 50 percent of the population is at its risk. According to WHO guidelines, daily consumption of 150 µg of iodine in food products is recommended, while pregnant and lactating women require a little more – 250 µg. The iodine content in the human body is about ten to 15 mg, while about 79 to 90 percent of this element is accumulated in the thyroid gland. Iodine deficiency causes diseases such as goiter and hypothyroidism. In 1997, a nationwide iodine prophylaxis program was introduced in Poland, consisting in obligatory iodization of table salt intended for domestic use in the amount of 30±10 mg KI/kg of salt. The greatest amount of iodine (45-60 µg/L), in the form of iodides, is found in seas and oceans. Iodides oxidize, then sublimate into the atmosphere and accumulate in water and soil in the form of rainfall. From here they are absorbed by plants and finally by humans and animals. Unfortunately, 40 percent of iodine is lost during technological and thermal processing. In order to replenish the deficiencies of iodine in products, new methods of providing people with iodine in food are being searched for.

One way is to biofortify vegetables and herbs with iodine by spraying leaves with it. Herbs from such crops contain much higher levels of iodine in leaves and stems, which increases the amount of this element in the diet [Domingues-Gonzalez et al. 2017; Limchowong et al. 2017]. Amber and its products may be an alternative to commonly used goods declared as a source of iodine. Baltic amber contains iodine of natural origin, but it is worth mentioning that no studies have been conducted so far on the iodine content found in amber extracts. In addition to iodine of natural origin, Baltic amber also contains other compounds that have a positive effect on the physiology of the body. For this reason, ROYAL products based on amber are the result of advanced research, the aim of which was, inter alia, determination of iodine content in the amber extract.

PROCESS USED EXTRACTION	DEADLINE HARVESTING	PROCESS EFFICIENCY (%)	TOTAL POLYPHENOL CONTENT (%)	THE CONTENT OF FLAVONOIDS (%)	IODINE CONTENT (UG/L)
MACERATION 70% ET.	I	1,22	4,82	0,37	38,90
	II	1,12	4,77	0,36	37,80
ULTRASOUNDS 70% ET.	I	1,14	2,86	0,35	35,25
	II	1,11	2,85	0,34	28,90

Tab. 8. The content of iodine, polyphenols and flavonoids in the analyzed amber extract. The presented data were obtained with two extraction methods – maceration and ultrasound in the Mars 6.

The study presented in the table above proves that the extract from Baltic amber contains 28.9µg/L to 38.9µg/L of iodine. Perhaps the iodine contained comes from ancient plants whose chemical compounds are found in amber, or perhaps it is a result of ion exchange between fossilized resin and sea water which contains iodine and some of it has been absorbed during long geological processes. The amount of iodine in the amber extract is not significantly high, similar to the content present in some vegetables, such as asparagus (42µg/kg), carrots (38µg/ kg), and broad beans (36µg/kg) [Jeznach 2012].

The amber extract was also tested for the content of polyphenols. It is a group of aromatic hydrocarbons found in plants, especially flowering plants. In higher plants, it is commonly found in lignin. It is also found in brown coal deposits, i.e. in the places where proto-amber was often produced. Due to the variety of structures, it is difficult to define specific physicochemical properties of polyphenols. These types of compounds are easily oxidized. They sometimes transform and turn into phenol and sugar. In the study presented (Table 8) the total polyphenol content can be considered as significant, oscillating about 0.3 percent.

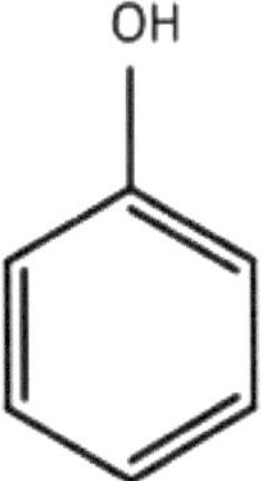

Fig. 101. Phenol.

For comparison, the content of polyphenols in some fruits and vegetables is as follows: blueberry – 56 percent [Gheribi 2011], cherries – 46 percent [Ibid. n.d.], strawberries – 26 percent [Fijoł-Adach et al. 2016] , pomegranate fruit – 12 percent [Kościuk et al. 2015], grapefruit – 8.7 percent [Ibid. n.d.], orange – 5.5 percent [Ibid. n.d.]. For example, the content of polyphenols in amber was similar to that found in oranges. On the other hand, vegetables rich in polyphenols include: tomatoes – 19 percent [Kowczyk-Sadowy et al. 2016], onions – 11 percent [Ibid. n.d.], carrots – 10 percent [Ibid., n.d.], celery root – 8.7 percent [Ibid. n.d.]. Low polyphenol content is noted in potatoes and spinach (5.4 percent) [Kowczyk-Sadowy et al. 2016]. A slightly higher content of polyphenols was found in the infusion of green tea (11 percent) [Kałwa i Wyrostek 2018] and olive oil (up to eight percent) [Szajdek, A., Borowska 2004].

As a result of biosynthesis of polyphenols, naphthalenes and anthracene compounds which are known to be in Baltic amber are formed. The antioxidant effect of polyphenols is used in pharmaceuticals, as these compounds are known for their anti-cancer properties and support the circulatory system.

The group of polyphenols includes flavonoids. They mainly occur as vegetal dyes, whose basic skeleton consists of 15 carbon atoms. They form a benzene ring. Flavonoids are most often yellow or colorless solids. As it is known, Baltic amber also has a yellow color, although the relationship between the color of amber and the presence of flavonoids has not been proven. Flavonoids are especially common in ferns (Filices). They have also been found in other plants, especially in the *Cupressaceae* family. Cypresses are one of the groups that created proto-amber. Interestingly, the flavonoids are rarely found in gymnosperm trees. However, they have been found in Viburnum, in honeysuckle (Caprifoliaceae), and in St. John's wort (Hypericum). They are also known to be in pines (Pinus). Flavonoids most often occur in the form of chemical compounds found in amber, i.e. quercetin and kaempferol.

Yellow and red vegetables, as well as green leafy vegetables have a high flavonoid content: capers – 32 percent [Gheribi 2011], parsley root – 31 percent (Ibid., n.d.), red onion – 3.3 percent (Ibid., n.d.), celery – 2.3 percent (Ibid., n.d.), broccoli – one percent.(Ibid., n.d.). A high flavonoid content has also been observed in fruit of various colors, such as: elderberry – 79 percent (Ibid., n.d.), apples – 20 percent [Sembratowicz and Rusinek-Prystupa 2015], cherries – 12 percent [Gheribi 2011], blueberry – 6.8 percent [Ibid. n.d.], grapefruit – 5.5 percent [Ibid. n.d.], and orange – 4.4percent [Ibid. n.d.].

Fig. 102. Neogene – a younger period of the Cenozoic era lasting from
23.03 to 2.58 million years ago.
Painted by K.Doszla.

Compared to plants currently existing, the flavonoid content in
Baltic amber is low, ranging from 0.34 to 0.37 percent. However,
the study proves that plants actively influenced the creation of
Baltic amber.

Flavonoids have a special feature in the cellular responses of
mammals. Most of the flavonoids are able to cleanse the cell
systems of excess amounts of oxygen that destroy cell structures
in the aging process of organisms. Flavonoids have a positive
effect on the cells of the arteries, which improves blood flow,
and this is one of the features of amber described in the texts
of ancient medicine. Former pharmacists appreciated this type
of impact to such an extent that during World War I amber
vessels were made to store blood, because the blood stored in
them did not deteriorate. In the case of the *Ginkgo biloba* extract

containing flavonoids, it was also noted that the extract caused increased activity of cells in the brain during its hypoxia. As a result, the cognitive properties of mammals improved, which inhibited the development of dementia-related diseases. The effect of flavonoids has been noted in the case of atherosclerosis and in the alleviation of diseases of the central nervous system. Flavonoids soothe neurological ailments such as tinnitus, dizziness, and headaches [Urata, Iijima 2000].

Adding the *Ginkgo biloba* extract to the amber extract can improve the functioning of the nervous system. Therapeutic agents used in China, as a rule, contain extracts from plants such as *Ginkgo biloba, Huperzia serrata, Rhizoma anemarrhenae*. Traditional remedies provide a multifunctional effect and few side effects. The *Ginkgo biloba* extract inhibits the development of amyloid-β. The two extracts combined into one liquid whole can support the nervous system using Baltic amber and *Ginkgo biloba*.

50 mL / 1,7 FL OZ

50 mL / 1,7 FL OZ

Photo 109. The Ginkgo extract and the Ginkgo extract with the amber extract. Photo by P.Barczak.

Baltic amber has the ability to reduce free radicals in the body, which can inhibit the development of many diseases. It is a specific antioxidant. Raw plant materials, which are ingredients of food, cosmetics and pharmaceutical products, are more and more often used as a substitute for synthetic antioxidants, which can often cause undesirable effects such as liver damage and initiate neoplastic processes. As part of the team's research, ethanol extracts were assessed for antioxidant activity using the DPPH and FRAP methods. The DPPH method consists in taking a standard chemical compound, which, being a chemical mixture, shows the rate of chemical reactions, including antioxidant activities. The FRAP method, in turn, shows the strength of the antioxidant effect by reducing iron ions in human cells.

For example, tests were also carried out on other plants to compare the potency of amber ethanol extract. The data is presented in Table 9 below.

Extract	% DPPH inhibition	FRAP [µmol TE/g DM]
Gypsywort herb	34.0	15.0
Basil herb fertilized with iodine	70.0	47.5
Summer savory herb fertilized with iodine	71.1	51.4
Amber	52.0	49.6

Tab. 9. Analysis of the influence of amber extract on antioxidant activity in comparison to other plant extracts.

The amber extract showed high antioxidant activity in comparison to the plant extracts, both in DPPH and FRAP tests.

6.1. ASSESSMENT OF THE ANTIMICROBIAL ACTIVITY OF AMBER EXTRACT COMPARED TO OTHER PLANTS

The evaluation of the antibacterial activity of the amber extract was carried out by standardized techniques, determining the minimum inhibitory concentration (MIC) and the minimum bactericidal concentration (MBC). The MIC is the lowest concentration of the extract that completely inhibited bacterial growth after 24 hours of cultivation. MBC, in turn, means the lowest concentration of the extract that completely inhibits bacterial growth. Ratios are expressed in mg/mL.

A strain of bacteria	Amber extract
Gram-positive bacteria	MIC (MBC)
Staphylococcus aureus	32 (32)
Listeria monocytogenes	32 (32)
Gram-negative batteries	MIC (MBC)
Escherichia coli	>32 (-)
Salmonella enterica	>32 (-)

Tab. 10. Measurements of the effect of the amber extract on gram-positive and gram-negative bacteria.

The research shows a rather weak action against bacteria. Interestingly, in the case of Gram-positive bacteria, the MBC and MIC was established at the level of 32 mg/mL. In the case of Gram-negative bacteria, the concentration of the extract which inhibits the growth of bacteria has not been demonstrated. It is therefore not clear why plants and trees, the precursors of amber formation, react differently to gram-negative and gram-positive bacteria. Could one type of bacteria be more tolerated by amber-producing plants than others?

Photo 110. Copaifera tree growing in tropical climates. It is over 1000 years old. The oil has unique antibacterial properties. Agathic acid has been isolated from Copaiba oil. Peru. Photo by M. Barczak.

6.2. LIVER CANCER CELL TEST

Amber extracts made on the basis of Baltic amber and 96 percent ethyl alcohol were tested at the National Medicines Institute in Warsaw. An MTT test was performed to assess the metabolic activity of human cells. The test was carried out on liver cancer cells, so it provides only some fragmentary insight into the diagnosis of the effects of terpenes found in Baltic amber. Full human studies have never been done. The only research on amber extracts (apart from the research in Poland) is currently being conducted in Japan in the context of research on fat cells. It shows the influence of amber extracts on energy phenomena occurring in human cells. The stimulation of the lipolysis process was observed, and as a result of the metabolic process in the body is accelerated This type of phenomenon can be used in the fight against obesity, atherosclerosis and diabetes [Sogo et al. 2021].

An in vitro laboratory test carried out in Warsaw proved that amber extract, with a small dosage, does not have a toxic effect on liver cells. However, above 0.4241 µl/ml half of the liver cancer cells were destroyed. Yet, it should be mentioned that testing on cells in vitro is not equivalent to the effects that it may have on the human body. Some terpenoids do not reach the body cells because they are transformed in the digestive tract. Therefore, an important element is the method of applying amber extract, and using time breaks during dosing.

Scientific papers do not mention any harmful effects of the use of terpenes in humans. Rather, their very positive role in eliminating a number of ailments is indicated. Herbs and medicinal plants are considered in the context of improving health, not its degradation.

The experiments carried out on rodents showed that the oils containing p-cymene, i.e. the compound found in amber, did not deteriorate the condition of mammalian cells and organs, despite the fact that they were subjected to huge doses of essential oils. The kidneys and livers of the rodents showed no changes [Tabarraei et al. 2019].

Dapatment of Medicines Biotechnology and
Bioinformatics
National medicines Institute
30/34 Chełmska Street
00-725 Warsaw

Name and address of the customer:
InnoNIL sp. z o. o
30/34 Chełmska Street
00-725 Warsaw

TEST RESULTS
The document describing requirements for the product:

No.	Tested parameter	Test method data identifying the method	Requirement	Result
1.	Evaluation of the cytotoxic effect of ethanol macerate of a plant-derived sample	MTT test	-	Tested product is not cytotoxic at concentrations ≤ 0.625 µl/ml*, at higher concentrations it shows increasing toxicity, IC50=4,241±0,076 µl/ml *

*concentration given as number of µl of product per ml of culture medium

Tab. 11. Test on liver cancer cells carried out at the National Medicines Institute in Warsaw.

Borneol has also been found in Baltic amber. The Chinese Pharmacopeia describes 63 plants that contain it. Natural borneol is different from the synthetic version. Natural borneol is chemically stable, while synthetic borneol decomposes and becomes harmful. The latter turns into camphor, which, in large amounts, can negatively affect living organisms. However, natural borneol is non-toxic. Research has been conducted on this substance and no side effects have been found in humans as a result of its use [Lei et al. 2011].

L-fenchone is not harmful to the body either. It is found in lavender shrubs – *Lavandula stoechas (Lamiaceae)*. 47.05 percent L-fenchone, 13.23 percent eucalyptol and 11 percent camphor were found in lavender oil [Sebti et al. 2020]. On the other hand, L-fenchone is effective in repelling plant-harming snails, and work is underway to use this substance as a natural pesticide [Abdelgalil et al. 2010].

In a study of Baltic amber, traces of 1,8-cineole (eucalyptol) were found. The study was conducted on mice subjected to very high concentrations of 1.8 cineol. As a result, doses of 21.38 mg/kg/ day and 64.15 mg/kg / day did not lead to any damage to the liver and kidneys, while pathological effects and blood vessel congestion were observed at a dose of 192.45mg/kg/day [Xu et al. 2014]. There have been cases of children up to 14 years of age drinking *eucalyptus* oil, which contains eucalyptol. Of 42 children, 80 percent were completely asymptomatic. None of the clinical cases required life support, and the complaints were minimal [Webb, Pitt 1993]. When children drank *eucalyptus* oil, 59 percent of them had symptoms of intoxication characterized by abdominal pain and vomiting, and coma occurred in four percent of the cases. Improvement in health was observed after 24 hours [Kumar et al. 2015].

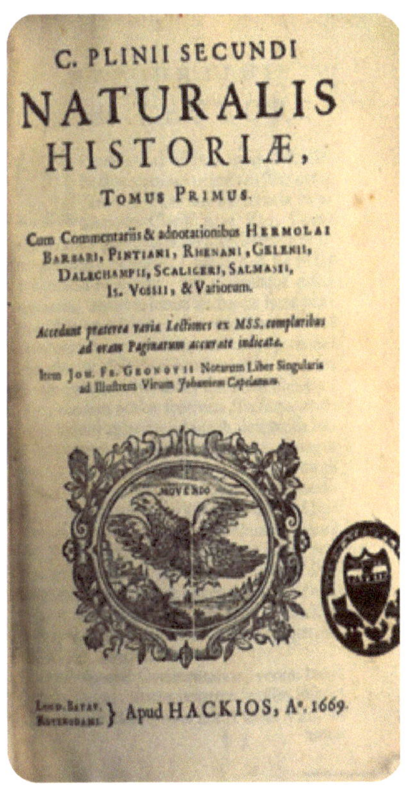

--Amber is called electrum because the sun is called Elector, which the poets used to say.

--Philemon claimed that it is extracted from the ground in Scythia in two places: in one it is white and waxy in color - this one is called electrum, while in the other it is reddish-golden - this one is sualiternicum.

--Sotacus believed that amber flows from rocks -in Britain, which he named Electridae.

--Asarubas relays that next to the Atlantic is Lake Cephisis, called Electrum by the Moors. When it is heated by the sun, amber flows out of the silt.

--According to Xenocrates, in Italy amber is also known as thium, and in the Scythians as sacrium, because it is also born there.

--Amber is formed in the islands of the Northern Ocean and is called glaesum from the Germanic language.

--Amber is the sap of a tree, and from this it was named sucinum.

--According to the evidence, the stone comes from trees of the pine genus, as it emits their scent when rubbed, and when set on fire it gives off the smell of a pine torch.

--Nero called his wife Poppea's hair amber. Since then, matrons began to desire the said color as the third in the hierarchy.

--Callistratus relays that at a certain age amber also helps with insanity and urinary problems.

--The stone is said to cure fevers and illnesses when hung as a pendant around the neck, while when used as a powder with honey and rose oil it helps with ear infections, and if rubbed with Attica honey - for poor eyesight.

--It is also used for stomach problems, either taken as a powder or taken with resin in water [Wasik, 2019], [Plinius C. Secundus].

CHAPTER
VI

7. POTENTIAL NUTRITIONAL AND PHARMACOLOGICAL EFFECTS OF AMBER AND SUCCINIC ACID

For several years, a new research concept has been observed, which combines the functions of groups of raw materials with high nutritional and therapeutic potential. These are the so-called nutri-pharmacological raw materials. They are the subject of advanced research due to their unique properties combining the characteristics of products used in pharmacology and nutritional raw material. According to Folkerts, this group includes functional food, which should contain ingredients with health-promoting effects [Folkerts et al. 2014]. Succinic acid is widespread in the animal and plant world, both in its free form and its derivative organic compounds. This compound participates in plants in the biosynthesis of amino acids. The succinic acid contained in amber is believed to be of secondary origin as a product of plant tissue fossilization. Succinic acid is a metabolite widely used in the pharmaceutical industry (in the production of vitamins, amino acids, and anti-cancer drugs) and food. In the human body, it is one of the main compounds involved in the Krebs cycle. Moreover, its antioxidant, adaptogenic, and cardioprotective effects have been described before. Due to its strong antioxidant effect, succinic acid is a component of dietary supplements, including those for athletes [Iqbal et al. 2016, Zarubina et al. 2012]. The results of tests carried out in the in vivo model showed the effect of amber on thyroid function. After adding amber to food, a decrease in blood TSH was observed. The activity of T4 (thyroxine) increased, thus proving amber's positive effect on hypothyroidism. Amber also had a positive effect on the functioning of ions. Aerosol and oral therapy containing amber allowed for the restoration of the normal levels of protein, lipid, and carbohydrate metabolism (the balance of thyroid hormones) when rats were experimentally induced with hypothyroidism. *We write more about succinic acid in Chapter 5.*

7.1. THE ANXIOLYTIC AND CALMING EFFECT OF AMBER

In a TCM (Traditional Chinese Medicine) pharmaceutical book, amber is prescribed to calm the mind and increase blood flow. The "Herbal Classic" by Shen Nong (25 BC) contains the first mention of its calming effect. The results of scientific studies carried out in in vitro and in vivo models confirmed its potential anxiolytic effect. In vivo studies in a mouse model have demonstrated the anxiolytic effect of succinic acid. There was a 50% decrease in animals' mortality compared with a control sample for the acid doses of 3.0, 6.0, and 12.0 mg/kg of body weight. Additionally, succinic acid at a dose of 1.5 mg/kg of body weight inhibited stress-induced hypothermia [Chen et al. 2003]. The research results, published in 2019, indicate the amber's calming effect as well as its potential use in the prevention of epilepsy. In the mouse model, amber was administered intragastrically for 14 days at a dose of 0.9 g/kg of body weight. According to the results obtained, there was a reduction in the frequency of seizures when compared to the control group. The authors suggest that the ambers' mechanism of antiepileptic action is similar to that of the drug phenobarbital [Zhu et al. 2019].

7.2. EFFECTS ON THE CARDIOVASCULAR AND RESPIRATORY SYSTEMS

Thanks to the use of amber as an atomizer, volatile substances are released, including terpene compounds and aromatic hydrocarbons. Amber resin has a very distinct smell. The aromatic hydrocarbons that are released are esters of succinic acid, which have a beneficial effect on diseases of the upper respiratory tract and lungs. In this case, the therapeutic effect is associated with inhaling anions that stimulate metabolic processes in the upper respiratory tract, trachea, bronchi, and alveoli, helping to relieve spasms, improve circulation, and eliminate edema and inflammation. In addition, they are said to have a bactericidal effect on microorganisms in the respiratory tract and lungs. The use of volatile compounds contained in amber for therapeutic purposes can be considered a kind of "amber aromatherapy." Amber therapy helps improve the valvular system of veins, relieves bronchospasm, and relieves inflammation of the respiratory tract and lungs. It was reported that people suffering from cardiovascular, endocrine, and neuropsychiatric diseases indicated improved health conditions after amber inhalation. However, there is no conclusive scientific evidence supporting this hypothesis.

SUCCINITE COMPOSITION

Carbon – C- 61-81%

Hydrogen –H- 8,5-11%

Oxygen- O- 8- 30%

Sulfur- 0,5% -7%

High sulfur content is found in the former region of the Paratethys Sea and the foothills of the Carpathians, in the regions of Poland and Ukraine [Kosmowska-Ceranowicz, 2012]. It contains trace amounts of elements: nickel, cobalt, copper, zinc, lead, manganese, iron, silver, sodium, calcium, titanium, chromium, tin, manganese, aluminum, silicon, iodine.

No.	Sample	N(%)	C (%)	H (%)	S (%)	O and other elements (%)
1	Madagascar copal	0.005	78.89	10.64	0.12	10.34
2	Colombia copal	0.005	79.05	10.45	0.15	10.34
3	Blue Dominican amber	0.071	81.99	10.93	0.20	6.81
4	Dominican amber	0.020	78.86	10.56	0.02	10.54
5	Mexican amber	0.053	83.65	11.24	0.37	4.74
6	Simetite	0.070	75.38	9.63	0.34	14.58
7	Lessini amber	0.010	79.60	10.91	0.06	9.42
8	Baltic amber	0.011	77.42	10.14	0.42	12.01
9	New Jersey amber	0.004	78.41	10.06	0.29	11.24
10	Cedar Lake amber	0.013	80.23	10.55	0.24	8.97
11	Triassic amber Dolomites	0.050	81.27	10.35	1.74	6.59

[Ragazzi et al., 2003]

Amber was found to increase the directed outflow of the cerebrospinal fluid from the ventricles in individuals whose cerebrospinal fluid dynamics in the brain's ventricles were somehow disturbed. The results of Khazanow's and Vengerovsky's research [Khazanov and Vengerovsky 2007] indicate the potential participation of succinic acid in the mitochondrial respiration process in brain cells, resulting in an increase of ATP (energy flow at the cellular level). Adenosine triphosphate (ATP) is an organic compound and hydrotrope that provides energy to drive many processes in living cells. It is an intermediary between energetic reactions in cells. ATP supplies cells with energy. One side of the process releases energy, and the other side absorbs it (see the picture). On one hand, energy is absorbed (from food), and on the other hand, energy is released (into the cell).

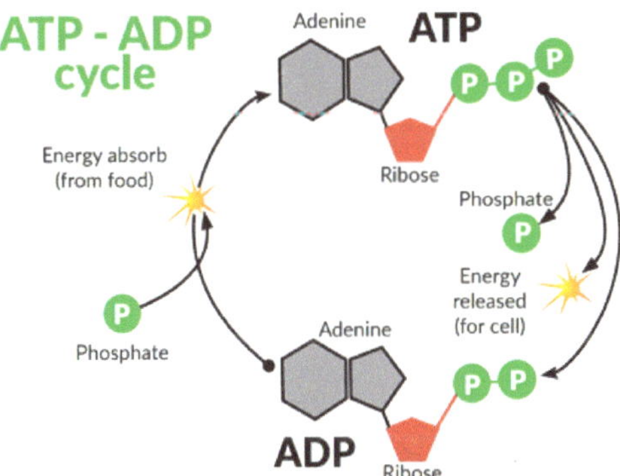

Fig. 103. A simplified cell cycle process in which succinic acid is involved.

7.3.1. DEMENTIA, ALZHEIMER'S DISEASE, AND CHEMICALS DERIVED FROM AMBER

Alzheimer's disease is a type of dementia. It involves neurodegenerative disorders, including progressive memory loss, difficulties in reasoning, and difficulties in using language. Personality changes can occur in some people. There are currently no effective pharmacological agents for treating this disorder.

The first descriptions of Alzheimer's disease appeared in 1906. The symptoms were described by German researcher Alois Alzheimer. This problem affects every fifth person over 80 in the world, and as the population ages, the problem of dementia increases as well. In 2025, the number of people aged 80-90 may double, which poses new challenges for medicine. Already today, about 7-8 million people in Europe and about 4-5 million in the USA suffer from this disease, which is almost 2 percent of the population of industrialized countries. To make matters worse, the number of cases is expected to triple in the next 50 years. The incurable disease is a big challenge for neurology, neuropsychology, and all medicine. Alzheimer's disease is the cause of most cases of dementia, but unfortunately, the mechanism of its formation is not fully understood. Abnormal protein aggregates appear in neurons (Lewy bodies), vascular diseases worsen, and a degenerative system that causes brain damage is formed. Inflammation of the nervous system in the brain is observed. Seniors lose their memory, and have problems remembering names and phone numbers. There is also a reduction in the brain's gray matter, which can be seen using the magnetic resonance scanning technique. During the aging process, synaptic connections in nerve cells in the brain are reduced. Abnormal proteins aggregate in brain cells, β-amyloid is formed in neurons, and α-synuclein in Lewy bodies. [Ahlskog JE]. Plaques and aggregates of the β-amyloid (Aβ) peptide consisting of 36-43 amino acids are formed in the nerve region of the cerebral cortex, in the extracellular regions. There are changes in neurons, in nerve cells, which are responsible for storing and processing information in the brain. Neurons disappear, inflammation of the peripheral and central nervous system arises, and the amount of free radicals destroying other nerve cells increases. Nervous system inflammation and abnormally formed peptide aggregates (Aβ) damage synapses and neurites in the brain. The "lines of communication" in the brain are disturbed. The human information system is disrupted. Neurons that function as neurotransmitters seem to be particularly damaged. As a result of the destruction of neurons that affect the production of serotonin and norepinephrine, the body's homeostasis in the endocrine system is disturbed. Disturbances in the flow of serotonin cause insomnia, depression, aggressiveness, and fatigue. This is due to the fact that damaged neurons show increased oxidative damage, disturbed energy metabolism in the brain, and disturbed calcium homeostasis of cells. Altered metabolism in cells and in the brain increases oxidative stress in cells, impairs energy metabolism, and disrupts ion homeostasis in cells. [Bartus 1982], [Ashwani Arya 2021].

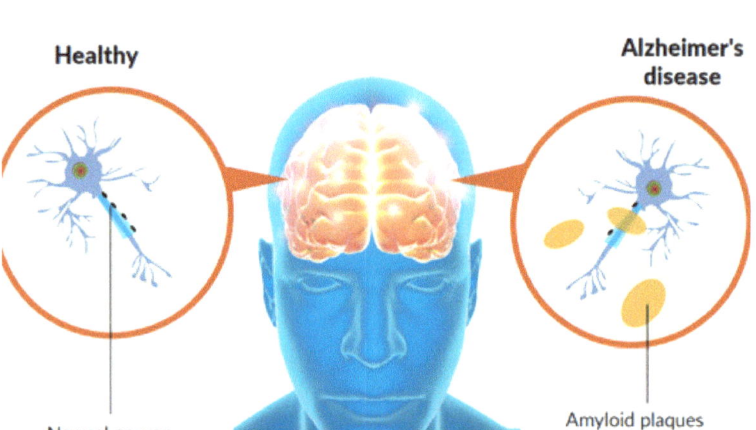

Fig. 104. Alzheimer's disease and neurons. Those on the right are contaminated with tau proteins and amyloid.

Elements of the human nervous system include mitochondria – nerve cell components. There is a concept that mitochondria descended from a common ancestor, α-proteobacteria, which came into being about 2 billion years ago, resulting in the anaerobic cell symbiotically absorbing the proteobacteria that needed oxygen. This allows the cells to produce more energy together in the human body. However, this cellular system is quite fragile. In the case of disturbances in the aging process, free radicals damage the mitochondria, which negatively affects the energy production process. There are changes related to electron transport and disturbances of the Krebs cycle, which is the basis of cellular respiration. The aging processes and mitochondrial damage occur similarly in all living organisms. [Kijowska-Oberc, et al. 2021]. Succinate, found in amber extract, can be considered a signal transducer with a hormone-like effect [Castro Fonseca, et al. 2016]. It activates signal pathways in the human nervous system, thanks to which the body tries to eliminate damage. It has been found that after the application of succinate, the pathways of nerve cells were stimulated. Succinate is a compound that activates signaling pathways through intracellular regulations in the event of disorders. Observations regarding succinate and its association with vascular diseases have been known since the 1950s when succinate infusions were used to treat barbiturate poisoning. The role of succinate in blood pressure regulation is known and has been documented in vitro and in vivo [Pluznick 2013]. Succinate has also been detected in the most important places regulating the functioning of the circulatory system [He, Miao 2004]. It has been suggested that succinate is a compound that informs the body of cellular stress. Succinate positively affects the work of neurons and astrocytes (a component of the nervous tissue), and accelerates their regeneration. The brain as a whole also can repair itself.

Succinate has vasodilator properties, and it also improves the functioning of the cerebral vessels. Thanks to this, they restore the functions of neurons and regulate energy production. The blood supply to the brain is restored, reducing potential inflammation. Positive changes occur mainly in astrocytes and nerve cells. The results of research on the effect of succinate and the mechanisms of treatment of ischemic brain diseases have been patented in the USA. (Patent No. US 2010/0267826 A1) [Hamel 2013].

7.3.2. ACTIVE PHARMACOLOGICAL COMPOUNDS CONTAINED IN SUCCINITE ACTING ON THE NERVOUS SYSTEM. THEIR HYPOTHETICAL IMPACT ON DEMENTIA AND ALZHEIMER'S DISEASE

Para-cymene, the substance described earlier, is an aromatic organic compound, a monoterpene. Para-cymene is known to affect the cardiovascular system [Santos, others 2011], as well as the central nervous system [Quintans-Júnior 2010]. Several experiments have been carried out with para-cymene. Administration of para-cymene resulted in the regulation of lipid metabolism processes in mammalian brains. It acted as a neuroprotective factor in the brain, and its administration caused a significant decrease (over 50%) in the lipid content in the brains of mice [Mendes de Oliveira et al. 2015].

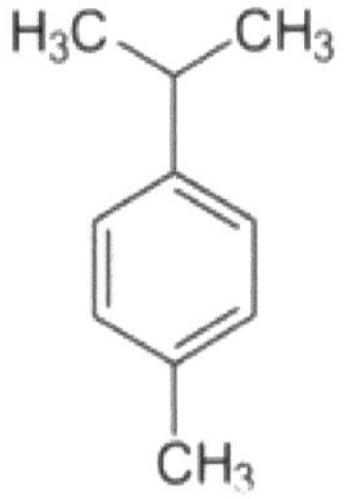

Fig. 105. Para-cymene.

Borneol ($C_{10}H_{18}O$) and isoborneol ($C_{10}H_{18}O$) are organic chemicals from the group of terpenes which are also described above. Borneol affects brain cells, including GABA receptors that are important in the nervous system. Borneol showed neuroprotective effects and regenerated neurons, including limiting the production of reactive oxygen species in cells [Dong 2018].

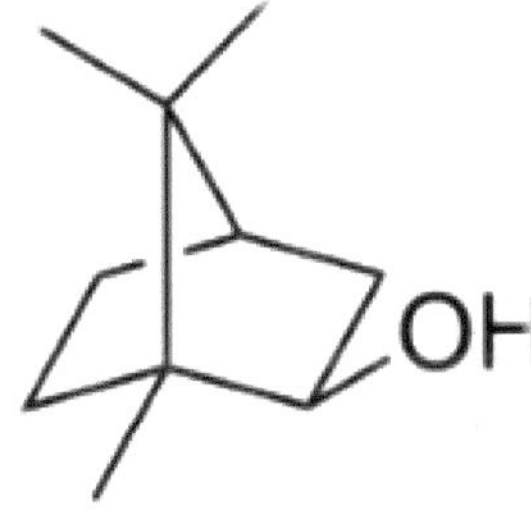

Fig. 106. Isoborneol.

Folk medicine, especially Chinese, extensively uses borneol to relieve anxiety and symptoms of nervous disorders, such as insomnia, anxiety, and fatigue. This monoterpene is used as a food additive.

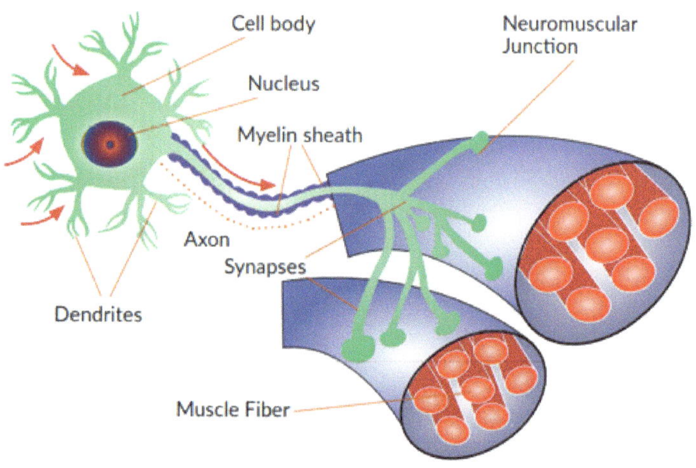

Fig. 107. A nerve cell and connection to the veins in the muscles.

Dihydrocurcumin ($C_{15}H_{24}$) is an aromatic compound found in succinite or Baltic amber. Dihydrocurcumin is not abundant in modern plant material but has been well studied in the fossil material – amber [Ellis 1995]. Dihydrocurcumin occurs mainly in a modern tree known as *Agathis australis*, as well as in a spice called Curcuma obtained from the Curcuma longa plant, though its content is lower in other types of resins and plants [Lu 2013]. Curcumin plays a vital role in pharmaceuticals due to its anti-cancer effects [Ellis 1995]. The properties of turmeric rhizomes have been studied through many experiments, and it has been established that they reduce free radicals, including hydrogen peroxide. Dihydrocurcumin inhibits the enzyme acetylcholinesterase, which is involved, inter alia, in the development of neurodegenerative diseases [Fujiwara 2010]. The compounds found in turmeric have reduced the amount of tau protein deposited in the human brain, specifically on

neurons, which is likely one of the causes of loss of consciousness in Alzheimer's disease. Consuming curcumin for 18 months improved memory and made it easier to concentrate. Curcumin improved the functioning of synapses in neurons, resulting in improved memory and the regulation of metabolic functions of the mitochondria [Wu 2015]. Turmeric has been shown to have antioxidant, anti-inflammatory, anti-amyloidogenic, and cognitive enhancing properties, which may suggest a potential neuroprotective nature [Apetz 2014]. On the other hand, curcumin reduced the number of Aβ amyloid plaques. It also influences the improvement of cell polarity by regulating intracellular processes [He 2009], [Sánchez-Ferro 2013], [Massoud 2010]. However, the role played by dihydrocurcumin has not been proven, although we know that it occurs in the plants mentioned previously.

Dihydrocadinene - cadinenes are a group of hydrocarbons found in various plant essential oils. This compound has also been found in succinite. Gamma-cadinene –

$C_{15}H_{22}O$ – belongs to the group of sesquiterpenoids. It is a component of Vetiveria zizaniotles grass oil. Cadinene is present in the fruit of the juniper (Juniperus *communis*), which grows in northern Europe.
Juniper oil has been used in folk medicine for centuries. It has diuretic, anti-inflammatory, antifungal, analgesic, antidiabetic, and antioxidant properties and is also used as a neuroprotective in Parkinson's disease [Rana N. 2014].

Cadinenes have also been found in some types of marine coral. A fairly large amount of them is contained in black pine needle oil. Black pine essential oil contains 53 components (93.4% total oil), among which the dominant ones were α-pinene (24.6%), β-pinene (10.9%), γ-cadinene (9.9%), β-fellandren (6.3%), manoyl oxide (6.2%), trans-caryophyllene (4.2%). Such a large amount of cadinene may indicate a specific action of this oil. Significant effects have been reported in inhibiting dementia and Alzheimer's disease [Bonesi 2010].

Calamene ($C_{15}H_{22}$) is a pharmacologically active sesquiterpene found in succinite. Calamene-stimulated dendritic cells increase the ability to stimulate lymphocytes while increasing the concentration of intracellular Ca (2+). Dendritic cells play a fundamental role in stimulating lymphocytes [Takei 2006]. Calamene has been isolated from the Cryptomeria japonica tree, a species of coniferous tree from the cypress family. It occurs naturally in humid mountain forests in Japan [Takei 2008].

Dimethyl succinate ($C_6H_{10}O_4$) – succinates are involved in many cellular processes. It has been noticed that in ischemic processes of the heart and in strokes, when cells die, reactive oxygen species are formed, and this succinate accumulates in human cells. After blood flow is restored in cells in the damaged organ, succinate is released again, regulating the flow of electrons in the cells [Chouchani 2014]. It is assumed that reactive oxygen species (ROS) are important in cell signaling, and their reduction has a positive effect on the functioning of the body. If levels are too high, it may cause oxidative stress and damage to cell membranes. ROS can damage macromolecules such as lipids, nucleic acids, and proteins [Hekimi 2011]. The administration of succinates has proven to be an effective remedy for brain damage. Succinate may support endangered mitochondria.

Succinates are derivatives of succinic acid that induce changes in the mitochondria of cells in rats. Succinates increase the potential of the mitochondrial membrane. They restore the functions of mitochondria in states of cerebral ischemia. Mitochondrial dysfunction is one of the components of the brain's ischemic cascade.

Succinates can support damaged mitochondria. The use of doses of 100 mg/kg and 200 mg/kg have contributed to more than a twofold (2.25) increase in ATP production. Under conditions of cerebral ischemia, the number of anaerobic processes increases, and the potential of the mitochondrial membrane changes. It should be emphasized that brain damage is largely related to a lack of ATP, and these deficiencies result from disturbances in the electron transfer process in the respiratory chain of cells [Kumar 2016].

In cellular disorders, mitochondria play the role of ATP consumers [Pozdnyakov, Dmitry 2021]. Administration of succinates helps to maintain the proper level of ATP in damaged cells, as well as helping to reduce the amount of lipids in cells. Succinate has been found to alleviate mitochondrial dysfunction and energy deficiency in the brain. Supplementation with succinate may turn out to be therapeutically valuable. Succinate uses an alternative pathway of mitochondrial energy metabolism and reduces oxidative stress. It alleviates energy shortages. Disruption of the metabolic state of neurons leads to a state where mitochondria, instead of being producers of ATP, become its consumers [Nowak 2008].

The effect of amber extracts on peptides in the brains of patients suffering from Alzheimer's disease has also been investigated. It was found that amber extract positively influences the process of cell apoptosis in the body and protects neurons against the harmful effects of excessively secreted β-amyloid – excess β-amyloid can lead to the destruction of neurons.

Medicines used in China to support Alzheimer's disease therapy typically include *Ginkgo biloba*, and *Huperzia Serrata*. For example, *Polygala Tenuifolia* extracts and *Ginkgo Biloba* extracts inhibit the growth of β-amyloid. Succinates also activate the production of collagen. It has been found that amber extract protects neurons against the harmful effects of β-amyloid and accelerates the removal of unnecessary degraded cells. It has been shown that amber compounds regenerate neurons, causing reconstruction of damaged cerebral cortex and hippocampal neurons located in the brain, for example. Amber extract also affected gamma-aminobutyric acid (GABA) in the cerebral cortex. It regulated the level of this acid [Zhu 2019]. The combination of Ginkgo and amber extract can positively affect the functioning of the human brain.

Succinic acid occurs in succinite (Baltic amber), which has already been mentioned in this paper. This is a special feature of amber, so the topic is worth expanding it a bit in the context of neurodegenerative diseases. The functions of mitochondria in human cells degrade with age, iron builds up in the mitochondria, and the functions of the human brain and the cerebellum are disrupted. Succinic acid in the brain prevents damage to mitochondrial membranes, and regulates metabolic processes, including energy synthesis. In addition, there is a significant decrease in the degeneration of Purkinje cell nuclei (cells of the nervous system) and nerve damage in the cerebellum. Succinic acid has been suggested to improve the functioning of the cerebellum in mice. After administration of succinate, the nuclei of Purkinje neurons in mice were not damaged, unlike those in the control group, which did not consume succinic acid.

CHAPTER VII

8. THE INFLUENCE OF AMBER IONS ON THE HUMAN BODY

Air therapy enriched with negative ions obtained from amber is used in some spa centers. It has been calculated that in rooms covered with amber there are produced about 6,000 ions per cubic meters per second, while in unprotected rooms the number of ions was estimated at about 100-200. Negative ions have a very positive effect on the human body, unlike positive ions that negatively affect the functioning of the body. Therefore, inhaling the amber aroma containing negative ions can contribute to achieving positive states of the human body. Inhalation of negative ions from amber may have a positive effect on cardiovascular diseases, hypertension, diabetes, varicose veins of the lower extremities, joint and spine diseases, bronchial asthma, gastrointestinal diseases, liver cirrhosis, neuroses, consequences of injuries and diseases of the central nervous system. Each patient feels the influence of amber ions individually, and this type of practice can support the achievements of modern medicine.

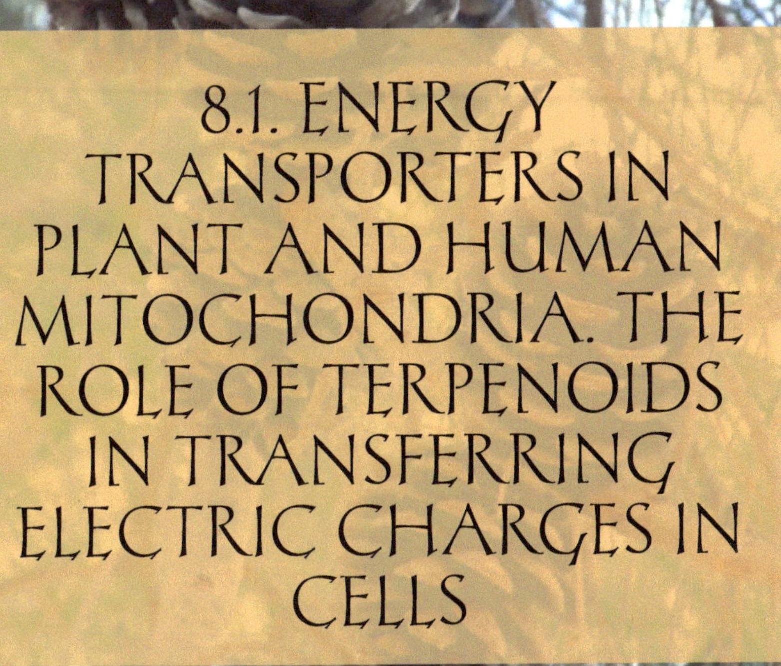

Cellular mitochondria in humans and plants have some similar functions. Mitochondria in humans and chloroplasts in plants are "cell powerhouses" that take up nutrients, break them down, and supply cells with energy. Constructed similarly, they probably stem from one common form from a billion years ago. They take energy from sugar or glucose and convert it into simpler chemical forms that can be used more easily by cells. The similarities between human mitochondria and chloroplasts include the production of energy by both types of mitochondria and having their own DNA. Oxygen O_2 and carbon dioxide CO_2 take part in the processes. The fundamental difference between the human and plant mitochondria is that chloroplast converts solar energy into chemical energy (sugar), and mitochondria in human cells convert chemical energy (sugar) into another form of ATP chemical energy that can be used by the cell. In plant mitochondria (chloroplasts) there is oxidative phosphorylation, the citric acid cycle, and oxidation of fatty acids. Mitochondria are involved in the multiplication of plant cells, they support the growth, development and death of plants [Millar et al., 2003]. Chloroplasts are the main producers of ROS – reactive oxygen species. They are the main complexes of the ETC – electron transport chain. Higher amounts of ROS in plants can cause oxidative stress and damage plant cell membranes. Thus, the process is similar to the processes taking place in human cells. Excessive ROS also damages human cells. In plant cells, as in human cells, nitric oxide is also produced, which, if excessive, can damage human and plant cells [Lázaro et al., 2013].

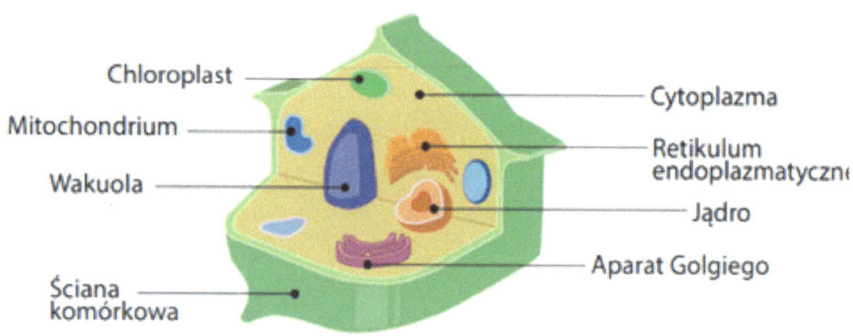

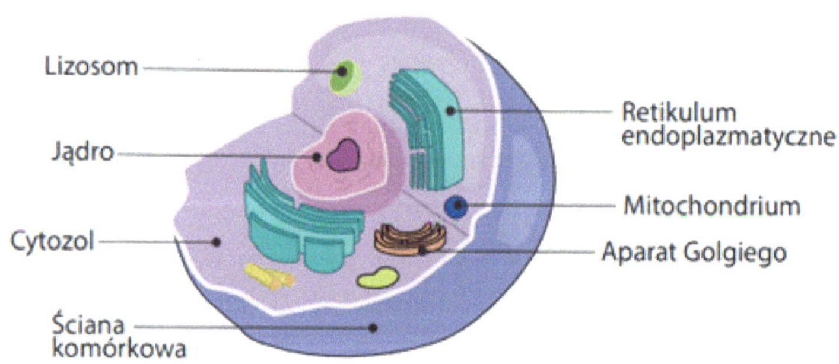

Fig. 108. Plant and human mitochondria.

The compounds compatible with mitochondria in energy transfer are terpenes and terpenoids. The compounds consisting of carbon atoms C_{10}, C_{15}, C_{20} are used in internal and external communication of plants. It has been found that all terpenoids are derived from the five-carbon building block isopentenyl diphosphate (IPP) and the dimethylallyl diphosphate (DMAPP) isomer. From this basic unit, all the other types of terpenoids, i.e. about 40,000, were formed. It proves the dynamics and possibilities of plants. Terpenes are also involved in the transport of electrons in plants, including the known role of ubiquinone and plastoquinone, which are organic chemicals from the group of quinones found in plant and animal mitochondria, responsible for the transfer of electrons in the respiratory chain [Tholl, 2015]. Nerve cells of animals and plants have excitable membranes through which electrical excitations can propagate in the form of active electrical potentials. Plants produce biochemical signals that resemble nerve impulses and are present at all evolutionary levels. Plant cells, tissues and organs transmit electrochemical impulses over short and long distances. A change of the potential on the surface of the cell occurs when an impulse is received. Such an impulse for plants can be a wound or a thermal shock. Electric impulses travel from stems to roots and vice versa. Depolarization waves are formed. Phloem is a potential way of transmitting electrical signals [Volkov, Ranatunga 2006].

Other terpenes are also known to participate in the transfer of electric charges in plants and human cells. α -pinene influenced mitochondrial respiration by inhibiting the electron transport chain, artemisinin (a sesquiterpene) metabolic compounds inhibited photosynthetic electron transport [Verdeguer,, Sánchez-Moreiras, Araniti, 2020]. Therefore, terpenoids serve a chemical transformation, as a result of which energy flows in plants [Pichersky, Raguso, 2018]. Terpenoids also form pathways in biosynthesis in plants [Siedow, Umbach, 1995], which results in current flow through plants being a kind of indicator of their internal and external situation.

Photo 111. Measurement of bio-voltage in a 20-year-old mahaleb cherry tree. Photo P. Barczak.

Terpenes also influence the functioning of the nervous cell systems in bodies of mammals. Treatment with Ginkgo extract improved electron neurotransmission, especially in mitochondria. The experiments eliminated defects in electrical metabolism in mice caused by tau proteins, which are also secreted in the brains of patients with Alzheimer's disease. A significant improvement in mitochondrial function was observed [Eckert 2012]. Brain work shows that terpenes are active in electric charge transfer, which indicates some common features of terpenes in both plants and humans [Ivic, 2003].

Photo 112. Bio-voltage observed as a result of the measurement in the mahaleb cherry. Photo P. Barczak

A simple experiment with the measurement of bio-voltage in a mahaleb cherry confirms the described mechanism of biocurrent flow in plants. One of the meter electrodes was poked into the phloem of the tree, while the other was driven into the ground. The meter showed a potential difference of -181.8 mV (measurement – 06.2022). The similarity of mitochondrial mechanisms in plants explains the effects of the actions of amber extracts. The structures of plant terpenoids, which are electron carriers in plants, have been preserved in them. They play a similar bioelectrical role in human cells, including neurons and the human brain. Hence probably so many descriptions of the effects of fossil resins on human organisms since ancient times.

Photo 113. *Pinus halepensis* essential oil attenuates the toxic Alzheimer's amyloid beta and induced memory [Postu et al., 2019]. In the photo *Pinus koraiensis*. Photo by P.Barczak.

8.2. TERPENOIDS AND SUCCINITE IMPACT ON THE BRAIN ACTIVITY

Proper human nutrition is the basis of the proper functioning of the brain. Its work is based on electrical signals. Neurons – the cells of the nervous system – make up the control system of the entire brain and body. A biocurrent of 1.23 volts constantly flows in the brain. The entire central nervous system is essential in regulating the body's homeostatic function (equilibrium) and the brain is also stimulated by electrical impulses. The body needs energy from food all the time, and the main energy consumers are neurons. Nerve cells respond rapidly to fluctuations in nutrients and metabolic hormones. They are innervated by axons that contain major neurotransmitters. Both carbohydrate and adipose tissue stores are monitored by the brain with metabolic and neural signals mediated by neurons. Glucose, which is the body's basic fuel and energy source, is of particular importance [Levin, Dunn-Meynell, Routh 1999].

Neurons need continuous energy supply by the body. Information flows between neural cells when electrical ions pass between cells in the neural system. However, if the neural cells are damaged or less efficient as a result of aging processes, stress, bad eating habits, the brain starts to have a problem. The work of mitochondria, which are an important part of cells, is disrupted.

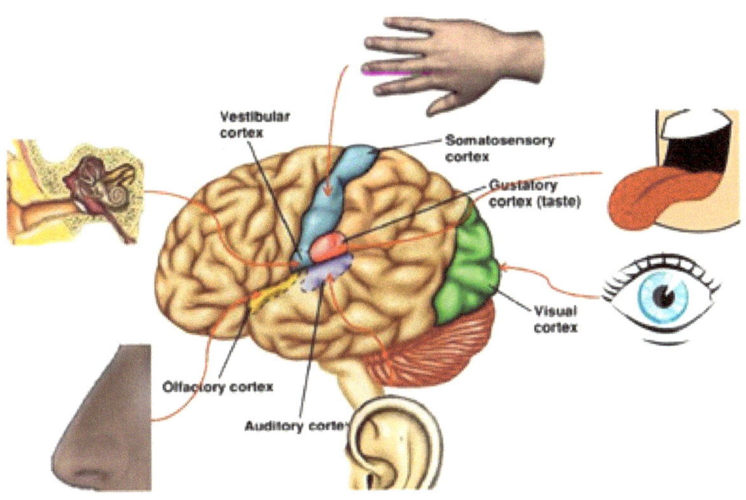

Fig. 109. Organs controlled by the "electric" brain.

According to one of the hypotheses, 2-4 billion years ago, archaic anaerobic cells incorporated mitochondria into their structures (they are ubiquitous in the world of plants and animals), thanks to which they obtained higher energy efficiency. Mitochondria, which are part of nerve cells, are oxygen dependent. Previously, all living organisms on Earth had been anaerobic, because the content of oxygen in the planet's atmosphere was very low. Mitochondria are therefore a common element between plants and humans. Thanks to them, life processes take place in all organisms that use oxygen on Earth.

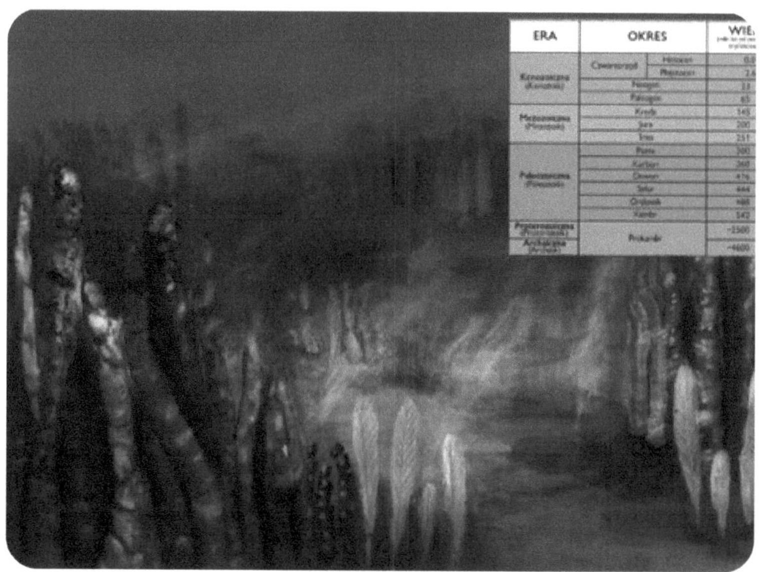

Fig. 110. The Precambrian – from 4.6 billion years ago to 538.8 million years ago.

The primary reaction that occurs in the human mitochondria is the oxidation of hydrogen into water. As a result of this, energy is absorbed by the human cells. In addition, oxygen is needed for every cellular reaction. It is taken from the air. The main energy carrier in cells is ATP (adenosine triphosphate). This compound is used particularly intensively in the brain. The brain uses a large amount of resources, including 20 percent of oxygen, despite the fact that its mass is only 2% of the human body weight. The brain is especially sensitive to hypoxia.

The entire neuronal cycle is governed by the proper functioning of the mitochondria. Microscopic mitochondria are like tiny power plants. They take nutrients from the provided food, break them down into prime factors and, after transformation, supply all organs with energy in the process of cellular respiration. Mitochondria are cellular signalers that indicate errors in the neurological system. So if the mitochondria are damaged, the work of the entire brain becomes disorganized, which can result in dementia and other neurological conditions.

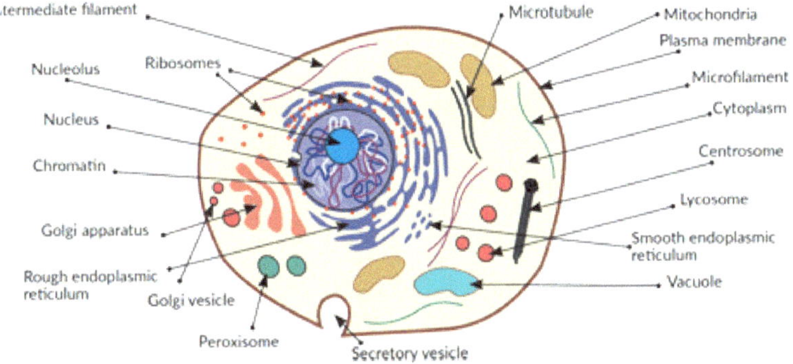

Fig. 111. The cell and its organs.

Oxygen deficiency in the cellular environment reduces oxygen synthesis and the content of high-energy compounds (ATP, CP), and it also weakens the electrical potential in cell membranes. The balance in the penetration of ions into nerve cells is disturbed, as a result of which calcium ions penetrate them in excessive amounts and the cells are subject to degradation and necrosis. Destruction of nerve cells leads to the development of diseases of the nervous system and, as a result, to human death.

Terpenoids are an element of the life system, mainly of plants. They take part in changes in the mitochondria, which is why scientists increasingly see their role in improving the functioning of human mitochondria. The influence of terpenoids is still in the realm of research and cognitive activities. Suffice it to mention that there are over 40,000 of them, which indicates a tremendous development of this natural group of chemicals. Some processes take place in the mitochondria of plants. It is known that various terpenes are produced by plants and microorganisms. They participate in biological processes in plants, mainly in biosynthesis of cell walls, transport of electrons and conversion from solar energy to chemical energy. They are also used by plants as a beacon informing about dangers. **Terpenoids are also electron carriers in plants** [Tholl 2015]. Higher levels of cellular respiration increase the conversion of terpenoids [Zhang et al. 2021].

Photo 114. Oleander, Athens, Greece. The chemical compound methyl palmitate ($C_{17}H_{34}O_2$), found in Baltic amber. is also known from a number of plants. It is found in a number of seeds [Deka, Basumatary 2011].

In plants, substances are transported according to specific patterns. It takes place through ion channels in plant cells and through the use of proteins by ABC transporters (an abbreviation for the ATP Binding Cassette). Due to energy contained in cell bonds, electrons are transferred from one side of the biological membrane to the other side. In this way, the molecules travel in the microworld in the living organisms of plant and human cells. In vertebrates, energy is also produced in the mitochondria. The synthesis of ATP – a compound involved in cellular transformations – is closely related to the flow of electrons in cells. The process of consuming glucose in cells leads, among others, to formation of succinate (a compound contained in Baltic amber), which supplies electron beams to the respiratory system of cells. Thus, damage to mitochondria can cause many

human diseases, heart failure, diabetes, neurodegeneration, and it is responsible for the aging process [Trifunovic, Larsson 2002]. Mitochondria are factories of human energy and producers of reactive forms of cellular oxygen, which sometimes destroy the cell system [Marreiros et al. 2016]. Providing the human body with succinates can therefore positively affect the regeneration of the damaged cellular system.

The compounds contained in Baltic amber can be used by the human body for cellular energy processes. One of them is succinic acid, which removes energy deficits in cells. It reduces damage to mitochondria [Nowak, Clifton, Bakajsova 2008]. (We write about succinic acid in the chapter 5.1.) Therefore, succinates play an important role in living organisms, including life processes in humans. Numerous experiments are currently underway.

Succinic acid derivatives, which are terpenoids, caused changes in the mitochondria of rats. After administration of the substance, the potential of the mitochondrial membrane of cells increased. The succinates restored the functions of the mitochondria in states of cerebral ischemia. Mitochondrial dysfunction is one of the components in the brain's ischemic cascade. Succinates (terpenoids) may support damaged mitochondria. The use of doses of 100 mg/kg and 200 mg/kg contributed to an over twofold (2.25-fold) increase in ATP production. Under conditions of cerebral ischemia, the number of anaerobic processes increases, and the potential of the mitochondrial membrane changes. It must be pointed out that brain damage is largely related to the lack of ATP, which develops during the disruption of electron transfer along the mitochondrial respiratory chain and causes a decrease in the driving force of protons [Kumar et al. 2016]. Administration of succinate maintains ATP levels in damaged cells, and the reduction of lipids is maintained at a better level in cells treated with succinate than in damaged cells. Succinate has

been found to alleviate mitochondrial dysfunction and energy deficiencies in the brain. Supplementation with succinate may turn out to be therapeutically valuable. Succinate uses an alternative pathway of mitochondrial energy metabolism and reduces oxidative stress. It alleviates energy shortages. The breakdown of the metabolic state of neurons leads to such a state in which mitochondria, instead of being producers of ATP, become its consumers [Nowak, Clifton, Bakajsova 2008].

These statements can be confirmed by an experiment with the use of brain EEG. A NeuroSky MindWave Mobile 2 device was used. The brain was imaged before administering the amber extract and after administering the extract. Gamma 2 EEG rhythms were analyzed. The left image shows the emitted EEG brain currents before administering the extract, the right image – after administering the extract, about 10 minutes later. The two EEG wave images are different. Additionally, the subject experienced changes in the perception of the state of the organism. After administration of the extract, he experienced a feeling of warmth and an increase in tinnitus, which indicates changes in the current flow through the nervous system.

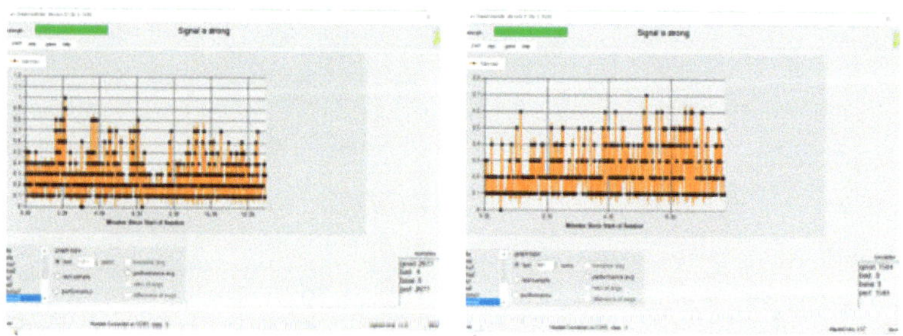

Fig 112. EEG rhythms. -- after administration
of the amber extract (Left). before administering the extract. (Right)

Succinate activates signaling pathways in the human nervous system, pointing the body damaged sites, thanks to which the body tries to eliminate the damages homeostatically. Succinates enhance electron transfer and energy flow. Stimulation of, among others, neuronal cell pathways, has been observed. It is a compound that, **through intracellular regulations, activates the body's signaling pathways** in the case of disorders. Succinates were detected in the most important places regulating the work of the circulatory system [Miao et al. 2004]

For example, succinate accumulates in the kidneys of diabetic mice. It is an indicator of disease states that the body fights by triggering homeostasis. Information flows by electrical impulses. In human cells, succinate inhibits the secretion of free fatty acids. It can be presumed that succinic acid (a terpenoid) is an important element in diseases such as hypertension, overweight and diabetes. It can therefore be an indicator of metabolic damage to the body. Weight loss was also noted early in the initial stage of the study [Gilissen et al. 2016]

Photo 115. Baltic amber extract with ginseng and pure amber extract.

Succinates are also active in the case of serious human diseases. In ischemic diseases – strokes, myocardial infarctions and acute kidney damage – there is an increase in the succinate content in the cells. Succinates inform the body about the pathologies taking place.

Contrary to cellular signals from the body, in which succinate is an indicator of cellular stress, in other situations succinate positively affects the work of neurons and astrocytes. **It oxygenates the brain and regenerates it.** The brain has repair abilities, and succinates are "grease" which regenerates the entire human system through blood vessels. Succinates have been found in brain nerve cells – neurons and astrocytes.

Hypoxia leads to the deposition of metabolites in cellular systems. On the other hand, succinates increase the density of blood vessels in cells and minimize the phenomena of brain damage. As a result of its action, the density of cells increases.

By means of electrical functions and the transfer of electric charges, succinate blocks inflammation in the brain. It improves the functioning of cerebral vessels and restores the functions of neurons in supplying energy to the neurological system. Succinate triggers specific cell functions. The blood supply to the brain tissues is restored, thanks to which the body is able to determine in which cells the inflammation arises. There is an increase in the density of vessels in the brain. This takes place mainly in astrocytes and neuronal cells. The mechanisms used in the treatment of ischemic brain diseases have been patented in the USA. (Patent No. US 2010/267826 A1) [Hamel et al. 2013].

Astrocyt

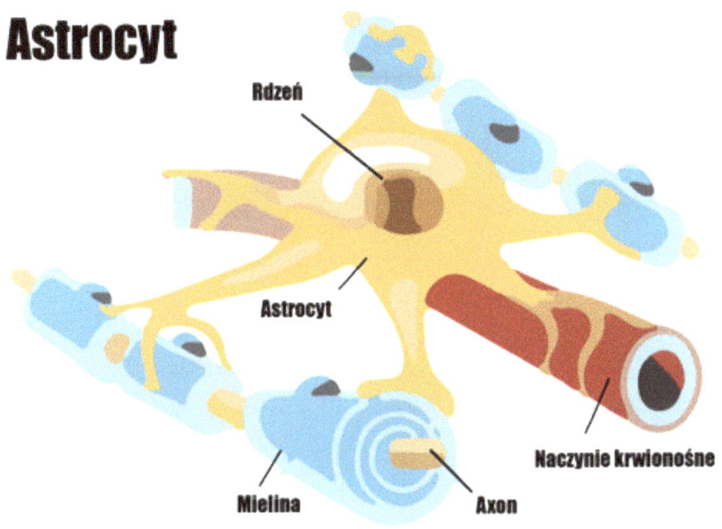

Fig. 113. Astrocyte cells and their connections.

Succinates accumulating in cells are an indicator of inflammation. They inform about the ischemic situation in the human body system. They activate dendritic cells that are related to the immune functions in the body. Succinates can affect the body's immune functions and increase blood flow [Gilissen et al. 2016].

It has been shown that amber compounds rebuild neurons, reverse the damage to the morphology of the neuronal cell. The damaged cortex and hippocampal neurons are repaired. Amber also influences gamma-aminobutyric acid (GABA) in the cortex of the brain. Amber compounds can regulate the level of this acid [Zhu et al. 2019].

Amber containing succinates affects the lipid metabolism in human cells. It is pointed out that amber extract regulates the glucose uptake into cells and, at the same time, promotes the secretion of glycerol. Amber extract can regulate the secretion of adipokines (substances that maintain energy balance in the cells of the adipose tissue), related to diseases such as atherosclerosis and diabetes [Achari, Jain 2017]. Fat cells treated with amber extract caused an increase in secreted adiponectin – a polypeptic hormone secreted by mature fat cells. Therefore, it is possible that amber extract has a positive effect on the correct level of adiponectin, which is important in preventing atherosclerosis and diabetes [Sogo et al. 2021].

The effect of amber extract on peptides in the brains of patients suffering from Alzheimer's' disease has been also investigated. It was found that amber extract has a positive effect on the process of cell apoptosis in the body, protects neurons against the harmful effects of excessively secreted peptide Amyloid β. Deposition of Amyloid β can lead to the destruction of neurons.

Apoptosis (Ancient Greek apóptōsis – falling off) is a process of programmed cell death. It is a completely natural phenomenon that allows the removal of abnormal, damaged and used cells from the body.

It 'was found that amber extract protects neuronal cells against harmful effects of amyloid β. It reduces the destruction of the body's own cells as part of the natural processes occurring in the body and accelerates the removal of unnecessarily degraded cells. Amber extract increased the antioxidant capacity of cells. It also lowered expression of proteins (BACE1 proteins) generated by cells [Luo et al. 2021].

Dimethyl succinate – $C_6H_{10}O_4$ – is one of the succinates that are involved in a number of cell processes. Its presence has been confirmed in Baltic amber. It has been noticed that in ischemic processes of the heart and in stroke, when cells die, reactive oxygen species are produced and succinate is accumulated in

the cells. After the blood flow in the disrupted organ has been restored, succinate is released again in the cell cycle. However, it produces reactive oxygen species and a reverse flow of electrons occurs in cells [Chouchani et al. 2014]. Administration of succinate has proven to be an effective measure for brain damage. Mitochondria, as we have already mentioned, are power plants in the body's cellular system. The ETC (electron transport chain) requires an oxygen supply. Succinate can support endangered mitochondria, thanks to which the synthesis of ATP, a compound that is the main energy carrier in cells, can be restored. The uptake of succinate is mediated by compounds that occur in the brain in astrocytes and neurons [Jalloh et al. 2017]. Under the influence of succinate, beneficial reactions took place in cells, contributing to protection against the ischemia-reperfusion injury in the body [Chouchani et al. 2014].

In mulberries, carrots and rhubarb stalks, the amount of succinic acid is 10,000 times smaller than in Baltic amber.

Other terpenes that affect the cellular "electron transport chain" and occur in Baltic amber are derivatives of pimaric acid. Pimaric acid is a diterpene ($C_{20}H_{30}O_2$). This compound belongs to the group of resin acids and is often found in oleoresins of pine trees. It can also arise through chemical processes of rosin dehydration. Such compounds have been found in the needles of a New Zealand pine Pinus radiata D [Franich et al. 1993].

Chemical analysis of Baltic amber has also revealed derivatives of this diterpene, i.e. pimarate' compounds ($C_{20}H_{36}$). Pimaric acid derivatives play an important role in influencing cellular processes in which electrical exchange takes place. Dihydropimaric acid, dihydroisopimaric acid and dihydroisopimarinol caused a significant polarization of cell membranes, in contrast to abietic acid, which is similar in structure and did not induce such reactions. Electrically charged Ca+2 particles play an important role in regulating processes in nerve cells. It has been noticed that in the case of strokes there is an intense influx of Ca+2 and thus overload of nerve cells, which causes their destruction. The death of nerve cells results, among others, in dementia and Alzheimer's disease. Increasing the activity of channels in nerve cells by adding terpenoids, including succinates, may result in a better Ca+2 flow, a better electrical activity of the brain, and thus alleviate the symptoms of physiological disorders. Some terpenoids have shown channel-opening activity in nerve cells. Already at 40 mV, an effect on channel opening in neurons was found [Imaizumi et al. 2002].

'The processes of cellular respiration supported by succinates and terpenoids, which are electron carriers, are illustrated by experiments carried out on yeast. Yeast are single-celled, so in some way they can be used to depict the processes taking place in the human body at the cellular level.

As part of the control test, which included only water, starch and yeast, the voltage meter showed a very low voltage of only 2 mV. This means that yeast and water alone do not trigger any essential electrical processes. On the other hand, the voltage analysis in an amber extract prepared on the basis of Baltic amber and alcohol (Photo 88) showed a voltage jump to about 35 mV. It lasted for about an hour. The voltage in the amber extract increased significantly compared to the test with yeast alone. As we can see, the amber extract shows electrical activity, and a similar phenomenon of electrical activity under the influence of succinates probably occurs in the human body. Electrical neuronal activity can increase in the human brain. For now, however, there are no human subject studies.

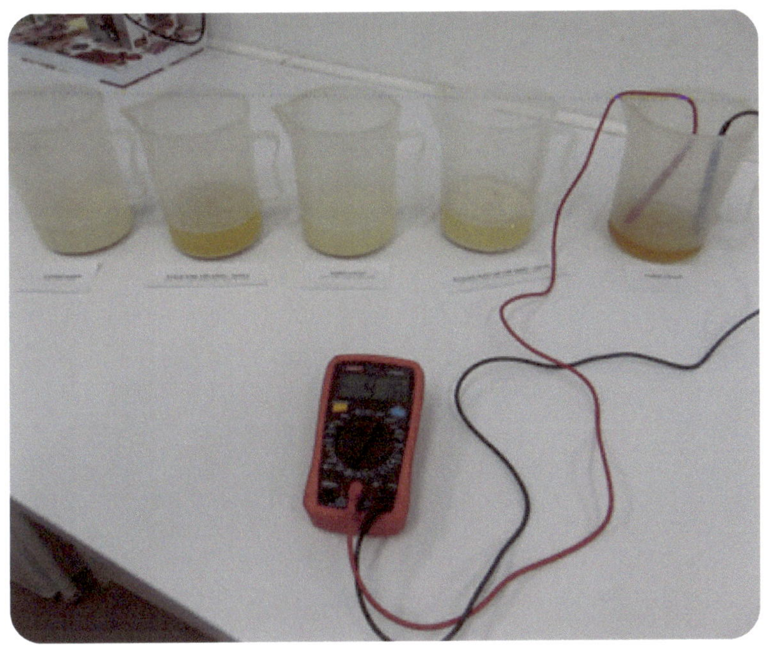

Photo 116. Samples with yeast suspensions and a voltage meter.

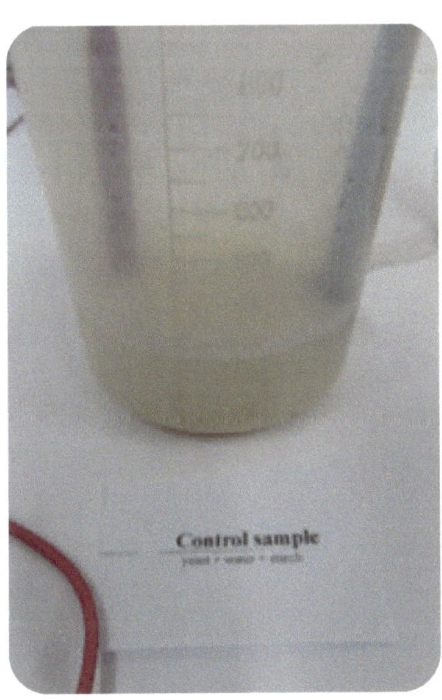

Photo 117. Experimental measurement of the only yeast suspension and almost no voltage.

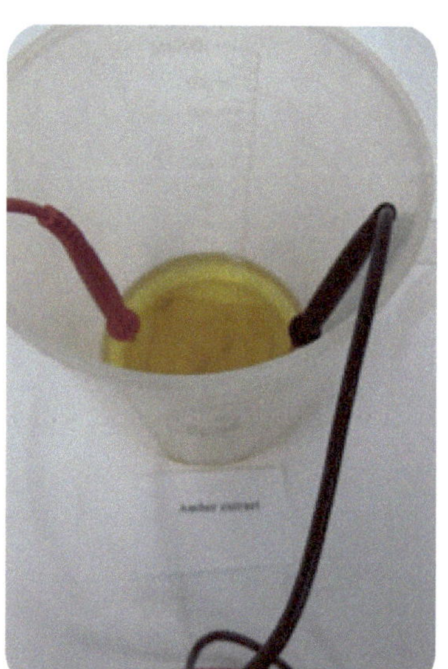

Photo 118. Experimental measurement of the pure amber extract and voltage increase to 35.2 mV.

The electrical action of the amber extract mixed with lemon balm extract has been also investigated. Lemon balm in small concentrations also has a positive effect on the processes of remembering. In this case, the voltage induced in a suspension of yeast, amber and lemon balm was about 27 mV. It was therefore smaller than in the case of the pure amber extract. Lemon balm extract with amber extract causes a calming and antidepressant effect, which may be 'manifested by a decrease in the generated tension in cellular processes in the human nervous system.

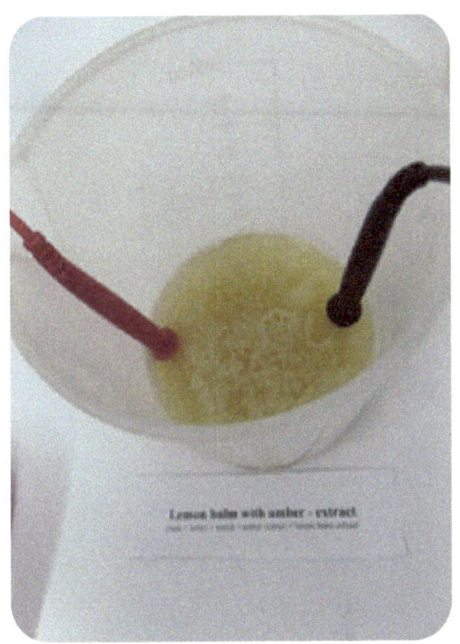

Photo 119. A mixture of yeast, lemon balm and amber extract. Voltage increase to 27.1 mV.

Photo 120. Amber extract in yeast suspension, voltage reduction to 25.7 mV.

A similar voltage drop, compared to the pure amber extract, was observed for a mixture of yeast and amber extract. The maximum voltage was 25.7 mV, which indicates the development of resistance on the yeast membranes. The pure extract has a voltage of 35.2 mV, hence a higher one, as described above. Electric tension in the mixture of yeast and amber extract lasts longer.

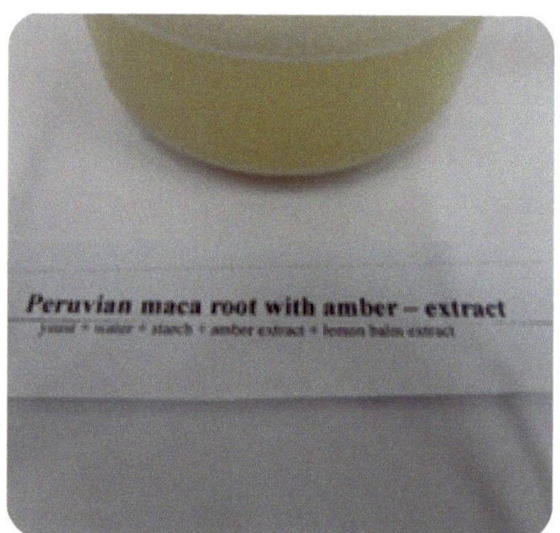

Photo 121. Amber extract, yeast and Peruvian maca extract.

Yet another effect is caused by Peruvian maca extract mixed with amber extract. The voltage in such a mixture is even lower – 17.7 mV – which suggests that Peruvian maca compounds block the energy generated in the amber extract. Peruvian maca, however, has strengthening properties, which may also have a positive effect on patients with neuronal da'mage.

Photo 122. Peruvian maca extract with amber. Photo by P. Barczak.

Experiments with the amber extract in various configurations indicate that the terpenes (succinates) present in it affect yeast cells and thus, through electrical interactions and intense ion flows, this type of chemical structure can improve the functioning of the human brain and body.

PERIOD	EPOCH	AGE Ma	AMBER, SUCCINITE AND OTHER FOSSIL RESINS
QUATE-RNARY	Holocene	11,7 yr	succinite in Europe, redeposited
	Pleistocene	2.58	succinite in Europe, redeposited
NEOGENE	Pliocene	23.03	resin Peru, Duxite Czechia succinite: Saxon amber resin: goitschite, siegburgite Mexican amber Dominican amber
	Miocene		glessite from Malaysia and Indonesia simetite Borneo aamber Kamchatka resin
PALAEOGENE	Oligocene	28.4	Carpathian romanite, Romania Romaniete Turkish Durglessite, Bitterfeld
		32.5	Fushun amber, China
	Eocene	40.4	succinite: Baltic amber Ukrainian amber Somerset a.mber Canada gedanite, glessite, beckerite, stantienite
		53	plaffeiite Schwitzerland krantzite from Saxony- -Anhalt, Germany
		55.8	copalite Great Britain Oise a., France Corbières, France
	Palaeocene	65.5	plaffeiite, Switzerland Sakhalin romanite cedarite, Hanna Basin, WY, USA
CRETACEOUS			cedarite, WY, USA Manitoba, Alberta, Canada Trepcza, Poland ajkaite, Hungary

		70.6-83.5	valchovite, Obara, Czech Republic Chukotka r., Russia Moru amber France
		85.8	Taimyr retinite, resin Piolenc France
		89.3	Caucasian copalite
	Late Cretaceous	93.5	New Jersey resin Burmite, Burma resin Sarthe Department, France
		99,5	Kent copalite Great Britain
		112	Amber Cantabria and Alva, Spain
		125	Archingeay resin, France
		140.2	brasierite, southern Great Britain copalite Gablitz
JURASSIC		155.7	Lebanese amber
TRIAS		201,3	Mutxamiel, Alicante, Spain, Dolomite Italy

Source: based on, B. Kosmowska-Ceranowicz, Amber in Poland and in the World, Warsaw UW, 2017.

CHAPTER
VIII

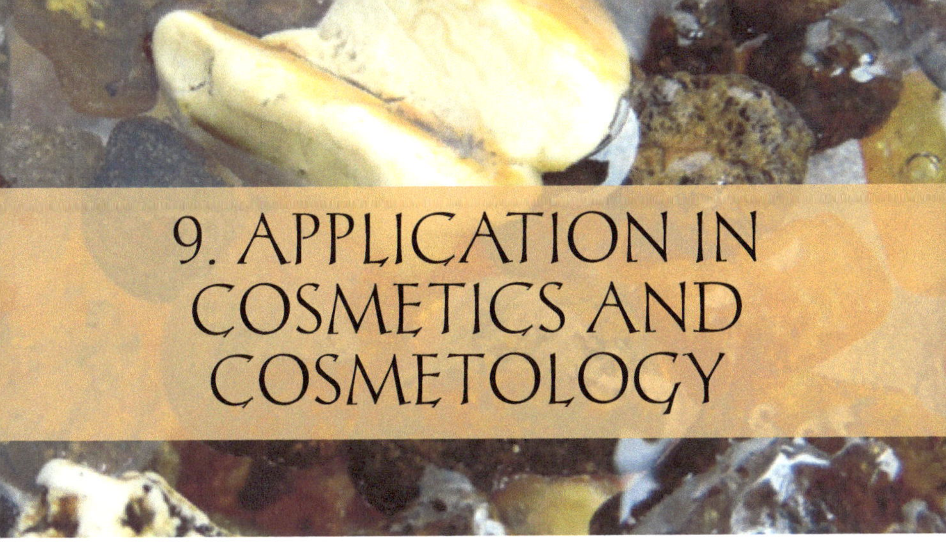

9. APPLICATION IN COSMETICS AND COSMETOLOGY

Amber dust contains about 40 compounds that can absorb water, fats, organic compounds, toxins, microorganisms and viruses on the surface of the skin. Amber is recommended f'or use in inflammatory skin diseases (lumps, vesicles, pustules, abrasions, ulcers, etc.), reddening of the skin, swelling, burning, itching, especially when the lesion is located near folds. Amber powder is highly hygroscopic, therefore it dries and degreases the surface of the skin, absorbs sweat, has anti-inflammatory properties, reduces itching and neutralizes toxins. Several patents have been obtained for cosmetics containing amber. Ground amber is used in peelings or creams. Amber has been tested as a potential agent that can stimulate human fibroblasts, dermis cells responsible for the production of collagen and hyaluronic acid. Glycol extract from Baltic amber had a positive effect on fibroblasts, an increase in skin protein synthesis was noted. Moreover, amber can be considered a component of cosmetic preparations protecting against UV-B radiation. Cosmetics with amber applied to the skin have a smoothing and softening effect, and when applied to hair, they improve its appearance. They accelerate the renewal of skin cells and increase the production of hyaluronic acid. Amber is used in cosmetics for tired skin. It should be noted that people allergic to rosin may also be allergic to the terpenes contained in the extract.

9.1. TEETHERS AND ISSUES OF MOUTH AND TEETH DISINFECTION

During the eruption of milk teeth, parents usually try to help their child so that it suffers as little as possible. Often, eruption is accompanied by gingivitis. Children put various items in their mouth and they are irritable. For understandable reasons, no one wants to subject a little child to painkillers. Parents give their child silicone teethers, frozen fruit and vegetables. For some time, there were texts appearing on the Internet which promoted Baltic amber necklaces. This way of supporting children gained popularity thanks to social media. Scientific research has not been carried out on Baltic amber, therefore it is difficult to find an answer to what extent such teethers help. It is indicated that there is a risk of choking or suffocation by this kind necklaces. These types of warnings have been issued by the Australian Commission for Consumers and Competition (ACCC). In 2011, the Australian Therapeutic Goods Administration (TGA) also issued warnings to two sellers of amber teething necklaces urging companies selling them to justify their marketing claims that the necklaces have analgesic and anti-inflammatory properties [Nissen et al. 2019]. Health Canada also has issued warnings, which recommend that amber necklaces should not be worn around the neck of a child under the age of 3. This is due to one

situation when a child choked on an amber necklace. After being examined,' the toddler never saw a doctor again for this reason. Occasionally, research to explain the mystery of Baltic amber beads is undertaken in the USA and Australia. It has been found that a small amount of succinic acid is released during the crushing of the beads. However, the study found that the secretion of inflammatory cytokines, which play a role in the formation of inflammation and associated pain in the mouth, was reduced at high levels of succinic acid. At the same time, the secretion of substances responsible for the state of apoptosis – as a result of which the body removes unnecessary and diseased cells – increased [Hogquist et al. 1991]. It can therefore be supposed that succinic acid contributes to the faster removal of inflammatory cells during teething.

Generally, monoterpenes have an anti-inflammatory effect. In our analyzes, we found in Baltic amber one of the monoterpenes, i.e. p-cymene. It can be an important pain reliever. Studies in mice have shown a reduction in nociceptive effects, i.e. when a body cell is damaged, pain-minimizing substances are released in the mouth. The actions of p-cymene, by affecting the nervous system, minimize pain [Santana et al. 2011], which may explain the relief of children who gnaw at amber balls.

Another compound found in the amber extract was borneol. This terpene has also been found to have an analgesic effect in mice. The action of this type of monoterpenes on neurotransmitters allowing for pain relief has been indicated [Almeida et al. 2013]. Camphene, another compound found in the Baltic amber extract, has a similar effect. The principle of operation is similar – pain relief through effects on the nervous system [Gadotti et al. 2021].

DL-menthol and eucalyptol were also found in the analyzed Baltic amber extract. The names alone give much food for thought, as both substances are used by the dental and oral hygiene industry. Human studies of menthol fluid containing menthol have shown that menthol mouthwashes are anti-plaque and anti-gingivitis agents [Ali et al. 2015]. Eucalyptol (1.8-cineol) is also a common ingredient in mouthwashes. It has a camphor smell and a refreshing taste. It has been approved by the FDA (USA) for use in food. Its anti-microbial effect is known. It is known to' fight the entire list of bacteria that cause inflammation in the human body.

Baltic amber also includes a number of terpenoids found in trace amounts. A range of such compounds can produce a synergistic effect, which in turn can bring relief to young children who suffer from the eruption of milk teeth. Therefore, amber ethno-necklaces can play their important role in caring for young children. However, each parent must make decisions in their own environment, assessing the pros and cons of using one or another ethno-pharmacological or ethno-stomatological agent.

CHAPTER IX

10. AMBER IN THE WORLD NAMES AND PLACES OF ORIGIN. A SHORT CATALOG OF FOSSIL RESINS

All over the planet there are dozens of places where fossilized resins have been found. Classifying them is beyond the scope of this work. However, it is worth mentioning some of them, which, in a general sense, are related to Baltic amber deposits and can characterize the areas of Baltic amber occurrence or the plants that formed amber. The list has been made on the basis of the atlas "Infrared Spectra of the World's Resins. Holotype Characteristics". The atlas was published in 2015 by the Polish Academy of Sciences, the Earth Museum' in Warsaw, and the resins were cataloged by Barbara Kosmowska-Ceranowicz and Norbert Vávra. In the atlas, the names of resins are presented on the basis of 3 criteria: 1) fossil resins within the system of mineralogical names, 2) fossil resins defined by the geographical name, 3) immature (subfossil) resins referred to as copal. However, the study below omits fossil resins that are no longer available for sale in the world, as the places where they were obtained no longer exist. The resins that are undescribed, their origin and chemical composition are unknown and there are no

descriptions of plants from which they may have arisen are also omitted.

The atlas is based on infra-red spectroscopy. It found its first application in the study of fossil resins in the 1950s. The research on amber is carried out with infrared light, thanks to which the spectrum shown in the device is obtained and, on this basis, we can learn some features of the analyzed substance. The younger the resins, the clearer the recorded bands are. In geological terms, the young ones are between 10,000 years and 5 million years old. The name copal was adopted for such resins.

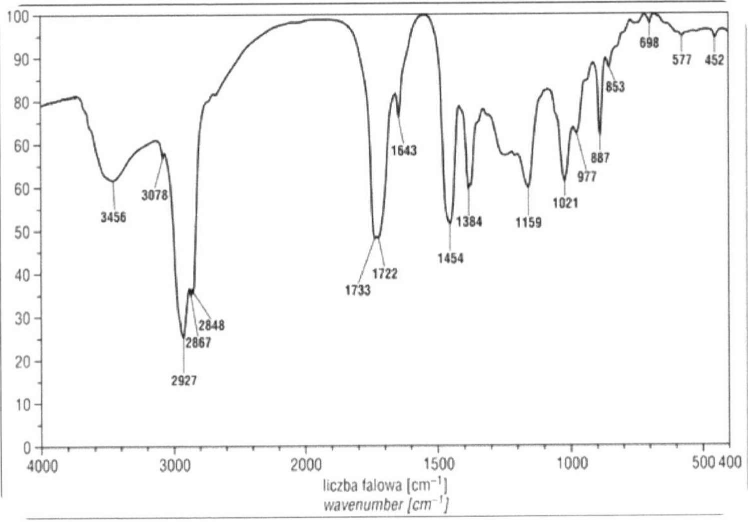

Fig. 114. The spectrum of Baltic amber (succinite) from Mikoszewo near Gdańsk, Poland. Based on: Kosmowska-Ceranowicz, B. (2017). "The Use of Infrared (IR) Spectroscopy of the World's Fossil Resins". Geological Review, 65 (8).

Fossil resins get their status of a given type on the basis of the first description and the name of the organic mineral given by the discoverer. Names of "ambers", according to the alphabetical order, are the following:

Agathocopalite (1953) – occurring in North Island, New Zealand, but also in Dargaville, Kaipara District, Northland Region, New Zealand. The botanical origin from *Agathis australis*. It contains agate acid and is synonymous with Kauri resin, which can now be purchased in the hobbyist market.

Ajkaite / Ajkait (1926) – it occurs in Ajka, district Bakony, Hungary and in France. It probably comes from wood of dicotyledonous plants. This variety of amber is found in the Bakony Mountains in Hungary in lignite and claystones. It comes from the geological period of the Cretaceous (80-90 million years ago). It does not contain succinic acid, and the amount of sulfur in this amber is approx. 1.5%. The IR curves of this resin are almost identical to those of another hobbyist variety of amber – walchovite found in Czechia. Interestingly, the 'arthropods found in this amber are known from Baltic amber, from Cretaceous Burmese amber, from Dominican amber, and even from quite young amber from Madagascar [Szabó et al. 2022].

Allingite (1894) – known from Thonon, France. Its probable botanical source were coniferous trees. Although it was already described in the 18th century, it is still a relatively poorly researched variety. One of the first inventories of French amber occurrences was made by Lacroix (1910) in his treatise on the mineralogy of France. Provence is particularly rich in amber deposits. Amber is found there in lignite sandy marls, especially from the Cretaceous period, but also in lignite deposits in Piolenc. In these deposits you can find pieces of coal stuck together with plants and amber. It has been noticed that the amber grains are the result of bacterial colonization by *Leptotrichites* resinatus [Schmidt, Schäfer 2005]. On the other hand, in the Rustrel region, amber is found in pyrite black marls that contain sulfur compounds [Saint Martin et al. 2021].

Almashite (1924) – known from Piatra Neamt, Romania. Its probable botanical source are coniferous trees. A variety of amber from the Oligocene period, found in 1923 in Romania, Europe. The petrified resin accompanied bituminous charcoal. This amber is green, what is an exception in the world of amber. It is also found in different colors, which makes it a sought-after collectible item in the market. Some argue that almashite is a metamorphosed succinite or rumenite [Cârciumaru et al. 2017].

Ambrite (1861) – discovered in the Hochstetter coal mine, New Zealand, its probable botanical source are coniferous trees, maybe *Agathis australis*. The term "ambrite" refers to amber found in New Zealand lignite mines.

Ambrosine (1870) – Charleston, Carolina, USA.

Amekite (1988) – near Umuahia, Nigeria.

Anthracoxenite (1856) – Slanỳ, Kladno, Bohemia, a kind of asphalt-type resin,

Bathvillite (1863) – the Boghead mine, Scotland, Great Britain,

Beckerite (1880) Baltic coast, the Goitzsche mine in Germany, its probable botanical source is Pinites succinifera. It was popularized by people analyzing amber in the German Goitzsche mine. It does not resemble the quality of jewelry amber, because it is a lump of compact mass connected with fossilized resin. Opaque. Contains traces of succinic acid and wax [Rappsilber, Krumbiegel, 2013].

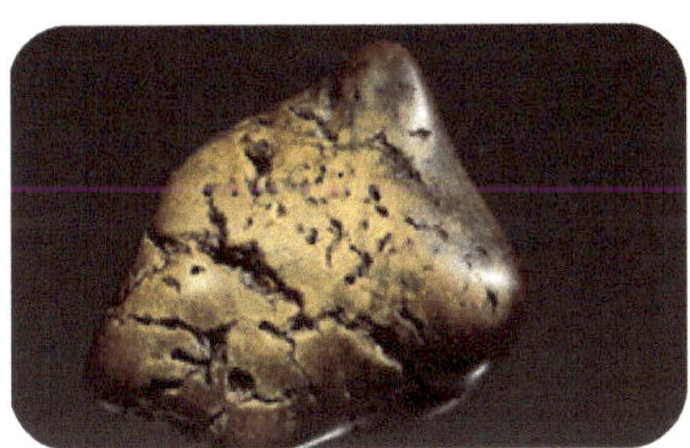

Photo 123 and Photo 124. Beckerit (left photo), 6.8 g, 4 cm. Collection: Goitzsche Bernstein GmbH. Photo by I. Rappsilber. Beckerit (right photo) 1.98 g, 3.5 cm. Collection: Goitzsche Bernstein GmbH. Photo by I. Rappsilber.

Photo 125. Amber from Goitzsche, Germany. Photo by P. Barczak.

Berengelithe (1838) – St. Juan de Berengela, Peru, asphalt resin.
Bielzite (1886) – the Kendeffi coal mine, Romania.
Burmite (1894) – discovered near Mogaung and Hukawng, Burma (Myanmar). It comes from coniferous trees of *Cupressaceae*. The found pollen points to Metasequoia. The resin can also come from plants from both Araucariaceae and Taxodiaceae families. Amber from Burma has been known for 2000 years. Trade contacts between Chinese cities and Burmese amber mines took place in the 1st century AD. There are numerous inclusions of plants, animals and fungi. Recent studies have shown that burmite comes from the upper Cretaceous (99 million years ago), although some deposits may be from the Eocene. A well-known deposit is Noije Bum, Hukawng, in northern Burma. In 1906, it was excavated in the amount of 11 tons. A large number of arachnids are found in the amber, which may suggest the Cretaceous period. Fossilized insects and arachnids found in Burmese amber are also known from Europe and North America [Selden, Rhine, 2017]. Burmite is cracked internally, which causes problems in jewelry processing.

Bitterfeldite (1986) – discovered near Bitterfeld, botanically linked to the *Burseraceae* family.

Chiltonite (2015) – discovered in Chilton Chine, Isle of Wight, botanically linked to Masculostrobus and Araucaricites. Amber was found in a deposit of lignite plant remains from the Cretaceous period. Araucaria fossils as well as those of Pagiophyllum and other extinct coniferous plants were found there [Batten, Austen 2011].

Cedarite / Chemawinite (1896 / 1891) – known from Cedar Lake (Manitoba), Canada. Its botanical sources are *Araucariaceae* (*Agathis*) and *Parataxodium*. Amber nuggets on the shores of Lake Cedar were found by Indians in pre-Columbian times. Amber was collected there in mass quantities and used for the production of varnish. The amber deposit was created as a result of a river drift from an area of about 1000 miles away. It probably dates back to the Cretaceous times. There have been always many different arthropods in amber, which additionally confirms the Cretaceous origin (approximately 79 million years ago). Most of the species found in amber have become extinct. The amber is often found in bituminous coal. In the Grassy Lake deposit, amber is associated with hard coal. It is possible that amber deposited on salt lagoons in the north of the American continent [McKellar 2011].

Copaline (1883) - found near Vienna, Austria. Its other locations are East India and Gablitz, Austria. The resin is sparingly soluble in carbon disulphide. It comes from the Eocene.

Photo 126. Ajkait, Csinger Tal, Ajka, Hungary. Photo by P. Barczak.

Copalite (1847) – known from the Highgate Hill near London, south-eastern coast of Great Britain. Another locations are Gablitz in Austria and Romania. Its botanical source are coniferous woods and *Burseraceae.*

Delatynite (1908) – found in Delyatyn, Ukraine. The amber contains 0.74% of succinic acid. Its botanical source were probably coniferous trees,

Durglessite (1986) – Bitterfeld, Saxony, Germany.

Duxite (1874) – the Duchcov mine in Czechia, Bilina and Vršany in an open-pit mine. Possible botanical source – *Taxodium* dubiu. The name comes from "Dux" – the German name of the town of Duchcov. The resin occurs in combination with coal, and its formation may be connected with volcanic changes and the heat released. As a result of the temperature of 350 degrees C, the resin impregnated the carbon. Duxite infrared radiation analysis showed similarity to Scottish marsh coals. Duxite in the Vršany mine was described in 1998. It appeared in cracks in charred tree trunks and in roots that had traces of fire with charred bark. As a result of the chemical analysis, the following terpenes were found in the amber: drimane, C_{16}-bicyclic sesquitepene, labdane, simonellite, retene and C_{18}-tricyclic diterpene. Duxite is classified as a resin asphalt – a mixture of bitumen, resins and waxes. Labdane and phylocladane have been found in duxite. The trees of the *Cupressaceae* family are a source of natural duxite resin, containing abietane, pimarane and dehydroabietane [Havelcová et al. 2018].

Gedanite (1878) – known from the Gulf of Gdańsk, Bitterfeld and Chatanga in Russia. It contains no succinic acid. Its probable botanical source is *Agathis australis*. Also known as unripe amber. Its composition is completely different from the composition of Baltic amber. It breaks easily, does not contain succinic acid, and when heated in oil, it swells and becomes elastic like rubber. Gedanite dissolves in ether and alcohol. It contains a certain amount of sulfur bound in chemical bonds, it is an insulator. Some fossil resins have the same infrared spectrum as gedanite.

Siberian amber (from the Taymyr Peninsula) is very similar in appearance to gedanite. The amber also has numerous cracks. There are numerous arthropods in this amber. It is supposed that it was created by plants from the cedar family [Rasnitsyn et al. 2016]. Most likely, it was formed in an anaerobic environment [Bogdasarov, Rudko 2018]. Gedanite, cedarite Mexican amber and Colombian copal extracts do not show bactericidal properties towards S. aureus, E. coli, P. aeruginosa, C. albicans [Kaczmarczyk 2016].

Gedan-succinite (1983) – Baltic coast, the botanical origin unknown.

Photo 127. Air bubbles in Baltic amber. Photo by P. Barczak.

Glessite (1883) – defined as opaque red-brown to brown fossil resin with elements of angiosperms. It occurs on Baltic coast; in Bitterfeld and Lusatia in Germany; in Malaysia; Indonesia; at Cold Lake in Canada; in Malvern, Arkansas, USA; in Cape York, Australia. Regarding its botanical origin, it probably comes from *Burseraceae, Dipterocarpaceae* – flowering plants are indicated. Indonesian resins are in the same group as glessite from Bitterfeld, Germany. In Indonesia, they are found in lignite and hard coal mines in deposits from the Miocene period (20-

23 million years ago). Bitterfeld glessite is also connected with coal seams. The resin contains triterpenoids with the structure of oleanane, ursane and lupane, which confirms that flowering plants are its precursor, both in the region of Baltic amber formation and in Indonesia. In glessite there are compounds known from *eucalyptus* oils [Drzewicz et al. 2020].

Photo 128. Glessite from Indonesia. Photo by P. Barczak.

Goitschite (1986) – known from Bitterfeld, Saxony, Germany. It is a whitish, white-green variety of amber. It has been named after the place where it was found. Chemically similar to glessite. Goitschite contains characteristic mono-, sesqui- and diterpenoids proving its origin from flowering plants [Lambert et al. 2015]. The most probable plant sources of goitschite are Picea o Pinus – the most probable producers of Bitterfeld succinites [Vavra 2009].

Photo 129. Goitschite, 3 g x 2.5 cm, Goitzsche Bernstein GmbH collection. Photo by Ivo Rappsilber.

Ionite (1978) – Ione Valley, California USA, probably coming from *Cupressaceae*.

Jaffaite (1953) – near Tel Aviv-Jaffa, Israel, probably coming from Pistacia sp.

Kansasite (1935) – Ellsworth, Kansas, USA.

Kiscellite (1933) – Remete Hill, Budapest, Hungary. Its possible botanical source were coniferous trees.

Kocenite (1868) – Kochenthal near Tels, Tyrol, Austria; Northern Italy. It occurs in the form of fine grains, in limestone formations. It is brittle and is referred to as spruce amber. It comes in the form of drops. Possible Triassic or later origin [Roghi et al. 2013].

Krantzite (1859) – Latdorf near the Nienburg Coal Mine and Schöningen open-cast mine area in Germany. It was formed in a subtropical climate. The resin is often found in the fossil wood of Doliostrobus taxiformis. The coniferous plant occurred in the Paleocene and Eocene periods. The second most popular tree in the creation of this amber could have been Quasisequoia couttsiae and Cunninghamieae, the trees currently growing in China. Over 100 fires have been recorded in the area of lignite deposits [Vahldiek 2015]. The ambers found do not contain succinic acid and sometimes contain a high sulfur content.

Legumcopalite (1953) – Congo, Zanzibar, Gabon, Sierra Leone. Probable botanical source – Copaifera sp., Trachylobium verrucosum.

Newjersite (1937) - Sayreville, New Jersey, USA. Probable botanical sources – *Araucariaceae, Pinaceae, Taxodiaceae, Cupressaceae,*

Plafeiite (1926) – Plaffeien and Schwarzsee, Switzerland. Probable botanical source – angiosperms, which are also known as flowering plants.

Rosthornite (1871) – Sonnenberg near Guttaring, Carinthia, Austria. Probably comes from angiosperm plants, which are also known as flowering plants, possibly *Burseraceae*.

Rumaenite / Rumenite (1891) – known from Valeny di Muntye, Romania. It first of all, occurs in Romana, but also in Burma

(Myanmar), Trepcza in Poland, Sakhalin in Russia, Bolu in Turkey and in Ukraine. It is found in the Neogene sediments of the Carpathian flysch. Rumenite contains 3.2% or even 5.2% of succinic acid. Dark red in color. In the Carpathians, fossil resins, it occurs in the Cretaceous (144–65 million years ago) and Tertiary sediments, in the Eocene sediments (55–32.4 million years) and in the Oligocene flysch (32.5–22.5 million years). It is worth noting that flysch also occurs in Ukraine and Poland. It is often found surrounded by bituminous shales. Rumenite probably comes from coniferous trees. Fragments of Sequoioxylon gypsaceum from the Taxodiaceae family have been found in the amber. Currently, similar trees grow in the USA.

Schibeite (1912) – Golpa near Bitterfeld, Germany, synonymous with glessite.

Schlierseerite (1993) – Leitnernase, Bavaria, Germany, its probable botanical source is *Agathis australis*.

Schraufite (1975) – Vama, Suceava District, Bukovina, Romania; Vienna; Ukraine, Lebanon; northern Italy.

Sieburgite (1875) – Siegburg near Bonn, Germany, probably comes from Liquidambar plants. It lacks amber characteristics, although recognized to be a fossil resin. Also found in New Jersey (USA). Its composition points out that it is a fossilized polystyrene rather than amber.

Photo 130. Siegburgite, Westphalia, Germany. Collection: Piotr Barczak. Photo by Piotr Barczak.

Simetite (1882) – Sicily, Petralia, Italy. It occurs along the Simeto River, from which it takes its name, but also in the town of Enna. This amber is also found on the coast, near the town of Catania. The Mount Etna volcano, known for its sulfur minerals, is active nearby. Simenite is valuable due to its rarity and physicochemical properties known since ancient times. It is red, but also happens to be black. Traces of flowering plants have been found in it. The amber was formed 11-17 million years ago, during the Miocene period. It is younger than Baltic amber. Simetite contains succinic acid. Regarding its botanical origin, it probably comes from coniferous trees.

Stantienite (1880) – known from Baltic coast, Goitsche, Germany. It probably comes from Cupressospermum wood. Its infrared spectrum tells that it is part of Baltic amber, but chemical analysis shows differences between Baltic amber and stantienite, because succinite has the majority of saturated carbon, and stantienite is rather similar to bituminous coal in this respect. It can be compared to the ltu stone found in coal mines in the Turkish province of Erzincan. Commonly known as black amber.

Valchovite (1845) – known from Valchov, Moravia, the Czech Republic. It was formed during the Cretaceous period – a period of exceptionally developed flora as a result of the planet's heat. The subtropical-tropical climate prevailed in the region of Moravia. That is why this amber contains a number of terpenoids, including cycloalkanes. Among them there is naphthalene and compounds similar to sesquiterpenes with cadinene nature. There are no oleans, known from Baltic amber, in it. The more this fossil resin is polymerized, the harder it is and the more insoluble in solvents. Fenchol, camphor, borneol have been identified in valchovite. It is found in flysch, among others from Moravia in the Czech Republic. Spiders, wasps and ants have been discovered in it. Regarding its botanical origin, it probably comes from *Damarophyllum* (*Araucariacea*), but possibly also *Agathis, Pinaceae (Pseudolarix, Pinus)* or *Sciadopityaceae* (the Japanese umbrella pine) [Havelcová et al. 2014].

Photo 131. Valchovite, Obora near Moravská Třebová, Moravia, the Czech Republic. Photo by P. Barczak.

WHY AMBER FLOATS?

It is assumed that the density of Baltic amber succinite is 0.96-1.096 g / cm3. Some ambers, however, have a higher density, up to 2 g / cm3, and others - lower. They were made of the resins of different trees, at different times. In addition, their buoyancy depends on the content of oxygen compounds in the amber structures and the number of bubbles trapped in the fossilized resin.

One of the ways to recognize real amber is to immerse it in salt water, and theoretically it should not sink because its density is comparable to that of salt sea water. To obtain the density of seawater, add 30 g (1.05 oz) of salt to 1 liter of water (0.264 [gal]). Seawater, when it reaches a lower temperature, increases its density, thanks to which some ambers may float to the surface.

In salt water, as seen in the photo, not all ambers float to the surface Some of them sink, and only some float. One of the reasons for this is the pores in amber and the air content of some amber pieces. Interestingly, the fossilized resin is still active. When exposed to water with salt, it emits air bubbles, which indicates that the chemical compounds contained in the resin are still active and react to the changing environment.

Photo 132. French amber, Charente-Maritime (Archingeay), Cretaceous
era.

CHAPTER X

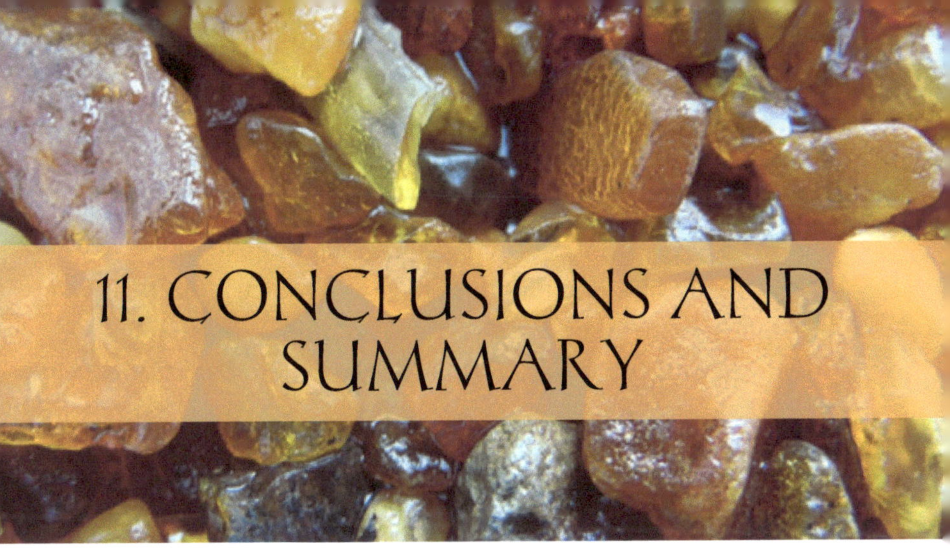

11. CONCLUSIONS AND SUMMARY

Baltic amber – this name may be understood as a brand of a unique resin that refers to abundant deposits in Central Europe. Individual pieces of succinite vary depending on what portion of soil was brought to the Baltic Sea by the glaciers that transported millions of tons of soil and rock during the ice ages. The age of amber is still hypothetical, and the estimation is based on the sediments and theoretical works. An additional difficulty is the fact that amber craftsmen obtain various Baltic amber from different time periods. The succinite deposits at the Baltic Sea have most likely been mixed up as a result of the flow of water and glaciers. With a similarly unclear age of origin we have to do in the case of amber from Saxony in Germany. The deposits of fossil resin in the Goitzsche mine near Bitterfeld most likely date from the Eocene epoch, although on the basis of ancient fauna it is also hypothesized that this amber comes from the middle Cenozoic era [Wimmer et al. 2008], and the discrepancies in origin may concern millions of years. It probably comes from vast rivers with deltas, which later turned into lignite deposits. Some hypotheses point to the age of around 22 million years (early Miocene), others to the Eocene. The climate on Earth cooled from an average temperature of 17 degrees C in the Palaeocene to 12 degrees C in the Oligocene, which shaped the vegetation in this area. In the first texts it was assumed that the trees

Cupressospermum saxonicum (*Cupressaceae*) and *Geinitzia formosa* (*Geinitziaceae*) grew there, but on the basis of pollen it was found that they were rather spruces (*Picea*) and pines (*Pinus*) [Yamamoto et al. 2010]. In the Goitzsche mine other amber is also found than in the area of the former Eocene Sea. An additional difficulty in recognizing the kind of amber are errors in interpretation and numerous varieties found, that pose classification problems. Comparing the samples of Baltic amber and German amber, you can clearly see the differences between them. Baltic amber is more varied in color. Goitzsche amber seems to be uniform. In 1986 – on the basis of the finds in Goitzsche – succinite, goitschite and gedanite were combined into one group. The second group included glessite, bitterfeldite and durglessite. The third group – beckerite, siegburgite and crantzite. The last group included stantienite and pseudostantienite, which is very rare in the area of the Vistula delta and the Baltic coast. Many unrecognized fossil resins are found in Goitzsche, which have not been tested to this day [Fuhrmann 2010].¬¬

Photo 133. Comparison of types of amber. On the left – amber from Goitzche (Germany). On the right – amber from the Gdańsk region (Poland). Photo by P. Barczak.

Amber has been tested relatively systematically for over 300 years, but to this day it is not possible to carry out statistical tests. Over 350 species of arthropods have already been described. Attempts were made to test a batch of 200 kg of amber for inclusions [Klebs 1910], another time the sample was 42 kg and 7079 arthropods were found [Sontag 2003], but the statistical tests are questioned [Penney, Preziosi 2013] . The samples differ from each other and have a low comparative value. No wonder, the number of arthropods found in Baltic amber is over 193,000, so samples with several thousand pieces are only a section [Penney, Preziosi 2013].

Photo 134. Comparison of the so-called "Cinder" from Kaliningrad, Russia (the photo on the left) and amber from the Rovno Region, Ukraine (on the right). Photo by P. Barczak.

The places where amber is obtained are a similar ambiguity. In the area of the Baltic delta, Eocene amber was found at a depth of approximately 100 m, and on this basis it was established that Baltic amber is from the Eocene era. The exploited amber deposits, however, in large part do not come from such deep excavations. They are obtained relatively shallowly from almost surface seams. They are mined in Poland, Ukraine, Lithuania, and in the region of Kalinigrad (former Königsberg), in Latvia, southwestern Ukraine, Jutland (Denmark), on the southern coast of Sweden, the eastern coast of the British Isles, in Spitzbergen

(Norway) and even in Axel Heiberg (Canada). The period of the formation of amber deposits is also a problem. We are talking about millions of years, while the civilization in which written records are created goes back, say, 3000-4000 years. Scientific materials present the emergence of Baltic amber in the Cenozoic era, i.e. about 66 million years ago. The Paleogene itself lasted 41.5 million years, or from 65 million to 23.5 million years ago. The areas where amber was formed hanged from land to sea and vice versa, depending on the orogenic processes, sea level, etc. At the end of the Eocene, a warm climate prevailed and perhaps those deposits were formed by the Pinus succinifera pine, which produces a lot of resin, although this type of concept seems to be a great simplification. Scientific materials concerning sediments also classify them into the Eocene, but also the Oligocene period, and they are the basis the age of the amber found is estimated on. Deposits and sediments, however, were mixed up as a result of glaciations. It is estimated that only the Baltic region contains several hundred thousand tons of amber, i.e. 95% of all resources. In Lithuania, deposits extend to the Kuršių Marios (Curonian) region and to Juodkrantė [Grigelis 2001]. However, it is difficult to describe statistical data, since a very large art of amber enters the market as a result of "silent" extraction from deposits.

Photo. 135. Colored resins sold in China. Amber imitation offered at the Guangzhou fair. Photo by P. Barczak.

ERA	PERIOD		AGE (MILLION YEARS FROM THE BEGINNING OF AN ERA / PERIOD)
cenozoic	quaternary	holocene	0,01
	quaternary	pleistocene	2,6
	neogen		23
	paleogen		65

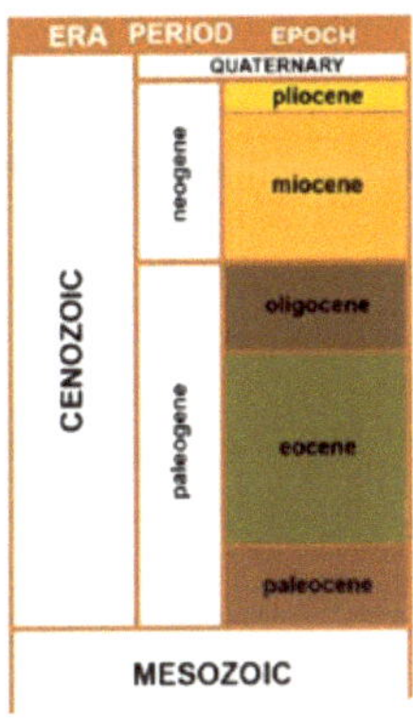

Fig. 115. Division of the geological era into periods and epochs in the hypothetical time of amber formation.

The radiometric tests dating marine glauconite deposits located on the Sambia Peninsula indicated that the lower layer of the so-called blue earth in which amber is located is 44.1± million years old. The "wild earth" lying beneath it is, in turn 47±1.5 million years old [Ritzkowski 1997]. There are also studies showing that amber is 37.7 million years old±3 million years.

There are several concepts where Baltic amber was created. It is assumed that the hypothetical Eridanos River transported resins to the region of Gdańsk and Kaliningrad from the north. However, this concept does not explain why amber is found in many regions of Poland, Belarus, Ukraine and Lithuania. The concept is based on no scientific evidence and is conjectural. Wolfe presented the concept that the Baltic deposits and German Bitterfeld deposits were formed at the time of the disappearance of the sea in the Polish Lowlands and the flow of waters from south to north. The time of creation is estimated at the Miocene. The Bitterfeld deposits were to be created in a separate way, from the local occurrence of trees. The differences in the chemical composition of Bitterfeld amber and Baltic amber were confirmed [Wolfe et al. 2016]. The results can be misleading, however, as the chemical transformations of amber have been taking place for millions of years and transformations of chemical compounds keep taking place also today.

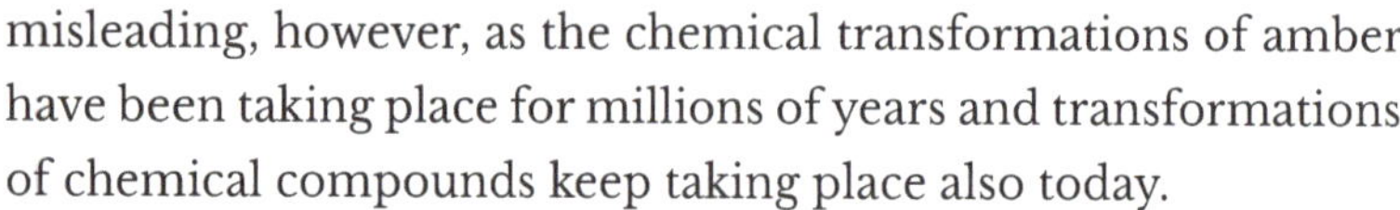

Did representatives of the Myceanean culture travel to the Baltic sea in the 15th century BC?

Nowadays amber artefacts are presented in the Archeological Museum in Athenes, Greece. They come from graves in Myceane from the 16th century BC. A considerable part of the amber has been found in graves in Kakovatos on the Peloponnese Penninsula. More than 220 graves where ancient amber jewellery has been found is known in this area. In the years 1884-1908, when first discoveries were made, there appeared at a theory about Myceaneans' travels to the Baltic sea. It has been adopted by archeologists and it is applied up till now, although there is also a considerable group of scepticists. Profesor Marinatos from Greece pointed out that Greek amber can come from Sicily. And how do ancient ambers presented in museums at the Baltic sea look like? The ecavations made in the town of Elbląg in Poland turn out to be interesting. There used to be an ancient Viking settlement called Truso and Baltic amber was traded there about 890 AD. Today some ancient specimens can be seen in the Archological Museum there. Amber specimens which can be seen there are quite different from those presented in the Archeological Museum in Athenes. They are more refined than those from Myceanean culture from 1600 BC.

Amber artefacts from the Archeological Museum in Elbląg.

They come form 890-900 AD so they are almost „contenporary"
to Myceanen culture products. Amber material and the way of
producing artefacts are different.

Amber artefacts from the Archeological Museum in Athenes, Greece.

The amber forest was also diversified, and it was growing on a huge area from northern Europe to the south of Asia. In addition, it was also transforming, and its resins turned into amber and were transported for millions of years. This complicated mixture of time and changes on Earth is supplemented with climate changes. The Eocene was a period of dynamic weather changes. The beginning of the Eocene was marked by a significant warming that lasted 55.6-55.8 millions of years. Then the temperature on Earth started to cool down, and this process has been going on until the present day. The amber forests were enormous plant complexes – in part with a temperate climate and in part tropical. That is why ambers contain insects known from tropical as well as temperate forests, from deciduous forests as well as coniferous forests [Bogri, Solodovnikov, Żyła 2018]. The nature of the amber forests is evidenced by insects preserved in the resin. The beetle of the Dysanabatium genus was abundant in the Eocene era in what today is Europe, but currently it is present in Southeast Asia, Indonesia, the Himalayas and China. Hemiptera bugs present in amber have been found in Africa [Emeljanov, Shcherbakov 2011]. The analysis is complicated by the fact that the amber forest fauna is a mixture of insects from temperate and tropical forests. The insect known from European amber, Sierola Cameron, is doing well in Australia. Only 3 species out of 206 in this genus do not currently live there [Perkosky 2018].

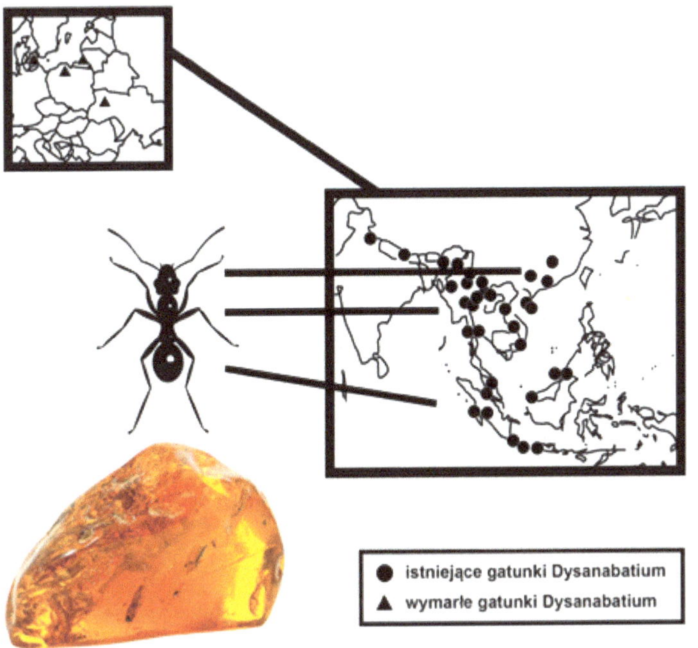

Fig. 116 Based on: Bogri, A., Solodovnikov, A., & Żyła, D. (2018). "Baltic Amber Impact on Historical Biogeography and Palaeoclimate Research: Oriental Rove Beetle Dysanabatium Found in the Eocene of Europe *Coleoptera, Staphylinidae, Paederinae)*". Papers in Palaeontology, 4 (3), 433-452.

Modern analytical techniques make it possible to look deep into fossil resins and identify micro-fungi and molds contained in them. One of them is the black *Metacapnodium* fungus that settles on tree branches and bark. It is also known from today's trees from Australia, although its ancestors were found on leaves fixed in amber from the Miocene and Eocene periods. It also settles on today's *Cupressaceae* trees, traces of it have been found on the *Calocedrus cedar, the Chamaecyparis cypress* and the juniper, and the trees of these kinds are known from Baltic amber.

Black molds and fungi are abundant in the South Pacific region and especially in New Caledonia, which is a resource of flora and fauna documenting the times of creation of Baltic amber [Rikkinen et al. 2003]. The situation is similar in New Zealand. Molds willingly settle on southern beeches (*Fagaceae*: *Nothofagus*), tea trees (*Myrtaceae*: *Leptospermum*), and they are a breeding ground for insects, which in turn produce honeydew, and this is a breeding ground for fungi. The beetles of the *Cyclaxyridae* family, found in Burmese amber (estimated Middle Cretaceous period – 145-66 million years ago), but also in Baltic amber, feed on them in turn. Mold has been also found in Baltic amber dating from the early Cretaceous, which may indicate the diversity of the temporal origin of succinite [Gimmel et al. 2019]. The consternation about the age of Baltic amber may be caused by spider specimens flooded with fossil resin. Some of them come from the Cretaceous era, which is much earlier than the Eocene. This type of spiders is found in Baltic amber and Dominican amber. The fossil record of the Nephila spiders dates back to the Jurassic period. For unknown reasons, some of the spiders do not live in New Zealand and Madagascar (*Theridiidae*). In turn, spiders of the genus Synotaxids genus, known from Baltic amber and Bitterfeld amber. live today mainly in temperate parts of South America, Australia and New Zealand. On the other hand, *Cyatholipidae* spiders living today in Australia and New Zealand belong to the *Cyatholipidae* family whose some species lineages, known from Baltic amber, died out millions of years ago [Hormiga, Griswold 2014].

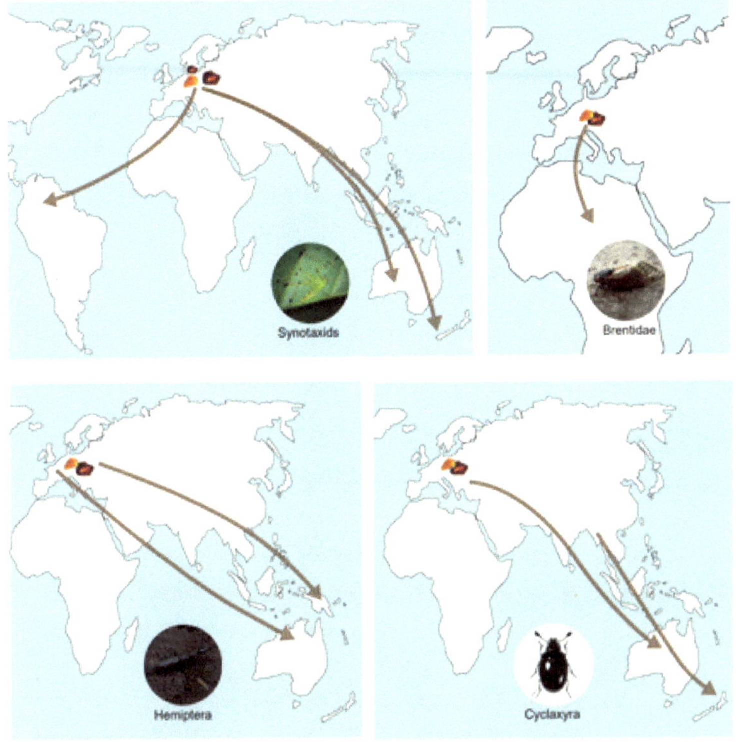

Fig. 117. Transferred plants and insects during the connection of continents. The world of flora and fauna preserved today on particular continents and in Baltic amber.

Endemic *Myrmeciinae* ants, today known from Australia and New Caledonia, used to occupy a much larger range in the past and their fossils from the early Eocene to the Miocene are known from South America, North America, Europe and Asia. On the other hand, ants of the Oecophylla genus are now found in parts of sub-Saharan Africa, southern India, Southeast Asia and northern regions of Australia, and fossils from Europe indicate that the insect lived in the north of the planet from 7 million to 56 million years ago [Barden 2017]. Brentidae beetles currently live in Australia and New Guinea, but are known from fossilized amber from France and the Baltic region. Similarly, the beetle Baltocyba electrinus sp. N., currently known from New Caledonia since 1889, has its "relatives" known from Baltic amber. The Notapionini beetle tribe, found in Baltic amber, feeds on plants in Eastern Australia, New Guinea, New Caledonia and the Melanesian Islands [Legalov 2018]. Insects from the *Psocoptera* family, mined in New Zealand are similar to families found in Baltic amber and Miocene Mexican amber. The *Orthocladiinae* subfamily of flies and the *Cecidomyiidae* family found in New Zealand are also known from Baltic amber and Rovno amber [Baranov, Andersen, Hagenlund 2015]. The situation is similar in the case of the *Forcipomyia* insect. It is an important pollinator of the cocoa tee, which may also indicate trees that grew in the amber rain forest [Schmidt et al. 2018].

Photo 136. A piece of petrified branch with preserved resin inside. French amber, Charente-Maritime (*Archingeay*), Cretaceous era.

The common origin of families, subfamilies and genera of flora and fauna jointly shaped in one continent of Gondwana is also evidenced by the Sciadoceridae flies, found in Australia, New Zealand and Chile, which are also found in Baltic amber [Austin et al. 2004]. The common natural and climatic environment of Gondwana is also evidenced by fossilized ant specimens preserved in Baltic and Australian amber. The discoveries of amber in Australia and New Zealand broadened this view with further evidence of existence of ants in the past. Their domination took place in the Eocene, and fossilized ants are very often found in Baltic, Indian and Dominican amber. Subsequent discoveries in New Zealand Foulden Maar lake show the fauna of that period. It is only slightly different than that known from the forest growing in today's Europe. However, it was dominated by plants of the Lauraceae family, known for their pollen contained in amber. The subtropical forest that prevailed during the Eocene, now known for its fossils in New Zealand, also consisted of occasional *Podocarpus* and *Prumnopitys conifer*, palms, orchids, trees currently known from Peru, Argentina, New Zealand – *Laurelia* and cocoa trees [Mildenhall et al. 2014].

Photo 137. Redwood. The Botanical Garden of the Polish Academy of Sciences, Powsin near Warsaw (Poland). Photo by P. Barczak.

To this day, it is not clear which tree species made Baltic amber. As a result of the joint storage of resins for millions of years in Central Europe, we have obtained a terpene cocktail conained in amber on the basis of which we are to determine the trees that formed the fossilized resin. It is a difficult task. Amber is found all over the planet and was created in various places more than 200 million years ago. In the Dolomites in the Alps, amber dating back 230 million years from the late Triassic period is found. The age of "Chinle amber", appearing in Arizona, USA, is similar. Thai amber (Khlong Min amber) is younger, while the Jurassic period is the time of the formation of Lebanese amber found in northern Lebanon. This type of amber was made up of the extinct *Araucarioxylon* and *Protocarpoxylon* trees. About 100 million years ago, amber was created in Spain and France (Charentes region), and trees of the *Cupressaceae* and *Araucariaceae* families contributed to their creation. Approximately 98 million years ago an amber deposit from Burma was formed, and a little later (89.8–93.9 million years ago) amber from La Garnache, Vendée (France), which comes from resinous trees of the *Cupressaceae* and *Pinaceae* families. During this period (93.9-66 million years ago), resins were fixed in the present regions of the USA (Middlesex, Harding, E. Alabama) and Canada (Grassy Lake, Cedar Lake). And finally, the era of relatively young ambers from the Eocene era came. Amber was prouced in Oise in France, Cambay amber in India, Tiger amber in the USA, Foshun amber in China and Russian Amber from Sakhalin. 37.8–41.2 million years ago amber is formed in theRovno region (Ukraine) – Eocenian Central European succinite, which was formed by cedars, trees of the Araucariaceae family and possibly by *eucalyptus* in areas of this part of Europe undergoing progressing steppe-formation. In the period 23.5–25.5 million years ago, Bitterfeld (German) amber, coming from *Picea*, *Pinus* (*Pinaceae*) and *Cupressaceae* trees, was deposited. In modern times, new deposits of amber and fossil resins are being discovered. About 15-25 million years ago, New Zealand ambers were produced, mostly by Agathis and

Araucariaceae trees. In a similar period (16-18 million years ago), resin was produced in the Dominica Republic, where it came from extinct trees of the Hymenaea protera species (Fabaceae). Trees of the same Fabaceae family are resinous in Mexico. Approximately 14 million years ago, "Zhangpu amber" was created in China, in the Fujian province. Finally in the Miocene, Australian Cape York amber was created [Seyfullah et al. 2018].

The fact that continents millions of years ago were connected into one Pangaea somewhat explains the fact that the same trees are found throughout the planet. Plants' ranges of occurrence interpenetrated from south to north and by the preserved traces of nature as if transferred unchanged from the prehistoric period to today's New Zealand, New Caledonia and Australia.

Amber in New Zealand and Australia was formed quite recently (Cenozoic), which makes it possible to compare sediments, rocks and fossils with Baltic amber. Some of the lignite and coal deposits there were formed by araucarias [Lambert et al. 2008]. In Australia, amber occurs in Miocene coals [Hand et al. 2010]; in New Zealand – in Eocene, Oligocene and Miocee sediments, and generally the deposits were formed by araucaria trees [Lyons et al. 2009]. In Australia, research also shows a trace after the *Dipterocarpaceae* family trees, which do not exist on this continent at present, but have spread to the tropical zones of Southeast Asia [Sonibare et al. 2014]. They are heavily resinous trees. They could also have had an influence in the Eocene on the formation of succinite in what today is Europe. In turn, on the basis of ambers from the Eocene Brunner Coal Measures bituminous coals from the Reefton Coalfield in New Zealand, it has been established that they were produced by Agathis trees. Traces of the *Podocarpaceae* family trees, which today contain resin mainly in the leaves, have also been discovered in the amber found there. In this region, trees of this type existed as early as in the Cretaceous era [Parrish et al. 1998]. Pollen records indicate that, for example, *Araucariacites australis* has been present in Australia and New Zealand since the Cretaceous [Raine, Mildenhall, and Kennedy 2006]. The wollemia, on the other hand, was present from the Jurassic to the early Miocene [MacPhail, Carpenter 2013].

The fossilized resins found in Europe today may even come from the Carboniferous, 320 million years ago. However, trees from this period were resinous to a minimum extent, and the amber specimens are scant. On the other hand, large amber deposits were formed in the Cretaceous period and in the Eocene and Miocene sediments. All over the world there was a mass resin production by forests and trees at that time. Mesozoic tree species of the Araucariaceae family and the extinct Cheirolepidiaceae family [Nohra et al. 2015] were abundant and secreted resins. From the Late Cretaceous to the Early Cenozoic, angiosperm trees, uch as *Dipterocarpaceae* [Rust et al. 2010] and Fabaceae [Langenheim 1995, 2003] also were contributing significantly to the amber fossil record. Today's Pinaceae and Araucariaceae produce large amounts of resin, with the Araucariaceae occurring predominantly in the middle and southern latitudes of Borneo, the Philippines, Chile, Argentina, southern Brazil, New Caledonia, New Zealand, Norfolk Island, Australia and New Guinea. The most abundant Araucariaceae grow in New Caledonia. In the past, trees of this type were probably the main component of the Mesozoic forest ecosystems in both hemispheres. The nutritional web of the soils existing 100 million years ago in the north of the planet was functionally very similar to the soils of today [Adl et al. 2011]. During a fire a comparison of *Araucaria columnaris* trees growing on a limestone substrate in New Caledonia today showed the phenomenon of resin-forming by trees. The fire caused a massive release of resin that fell into the salt water and was then transported by waves over long distances [Girard et al. 2009]. The *Araucaria columnaris* resin is primarily white or light yellow,

and changes its color to red-brown under the influence of the temperature of fire, whichmay be some hint during examination of the ambers currently obtained [Seyfullah et al. 2018]. Similar phenomena could have taken place in the seas of Central Europe.

Some ambers can be identified by analyzing fossilized plants preserved in coal. For example, New Zealand outcrop ambers are recognized on the basis of fossil trees. However, Baltic amber is a mixture. Bitterfeld amber consists of 9 types of fossilized resins, while Baltic amber is a composition of ambers brought from all over Central Europe from south to north, from north to south, by glaciers and gigantic post-glacial rivers created as a result of the influence of ice. Their conservation are the current amber deposits. The accompanying sediments consisted of hydrocarbons, sulfur, sulfur compounds, including gypsum and glauconite. To this day, traces of sulfur compounds and glauconite sediments are found in the vicinity of Baltic amber. Thanks to chemical and geological processes, amber resins have survived to this day. Similar processes could take place in the Bulgarian amber deposits in the Sozopol region. The ambers from there found in an underwater petrified forest in the Black Sea show pyrite, which is known to be made of sulfur compounds, in the cracks. The fillings also include sulfates [Yossifova, Eskenazy, Valčeva 2011], which may have contributed to the preservation of resins.

Gypsum, which crystallized in the trunks of petrified trees in Bulgaria, also participated in the creation of fossils, which may be another indication of how the Baltic amber deposits were formed in the area of the Carpathian foothills in Poland and Ukraine, where there are also deposits of gypsum, sulfur, bitumen, coal and oil.

The Baltic amber was produced by trees and other plants which produce much resin, what can be exemplified by kauri resins, which were mined in New Zealand at the beginning of the 20th century in the total amount of 450,000 tons [Haywood 1989]. Those deposits were created as a result of climate change. However, kauri resins are around 100,000 – not millions of years – old [Boswijk G. 2005]. In New Caledonia, on the other hand, where there is an environment of amber-bearing trees similar to those from the Cenozoic, there are no amber deposits. Therefore, this fact confirms the thesis that what counts is the way of preserving resins for millions of years – and not so much the type of plants from which amber was made. Of course, the chemical composition of amber is important, but the way in which it is made is also important.

Interesting experiments have been carried out in New Caledonia, where the world of plants is similar to that of amber forest. The resin formed is lifted by the seawater and settles, as a rule, elsewhere than where it emerged from the tree [Seyfullah et al. 2018]. The environment of the Paratethys Sea, however, may have been slightly different from that of New Caledonia. This is indicated by the Bulgarian ambers lying on the remnants of the former Paratetys Sea, i.e. the Black Sea. The petrified forest from the Upper Cretaceous or the Miocene (in the area of present-day Bulgaria) formed in that region consisted of marsh cypresses – *Taxodium* – and so far no trees of the Agathis genus have been found. Traces in coal indicate fires, and ancient lagoons coexisted with a sea of very low salinity [Yossifova, Eskenazy, Valčeva 2011]. Many elements of the flora and fauna of present-day Australia are also known from traces preserved in amber. One of the trees so far not described in scientific studies and probably contributing to the development of the chemical composition of Baltic amber is the *eucalyptus*, to which we have devoted a separate chapter. The chemical compounds found in the analyzed amber sample indicate the influence of this type of trees on the production of amber-bearing resin. Since "Baltic amber" could have been formed over a long time and is a compilation of various

geological eras, at a certain stage of the development of present-day Europe, *eucalyptus* could form resins in peat masses. Perhaps some of the oils contained in *eucalyptus* remains could have an effect on the proto-amber.

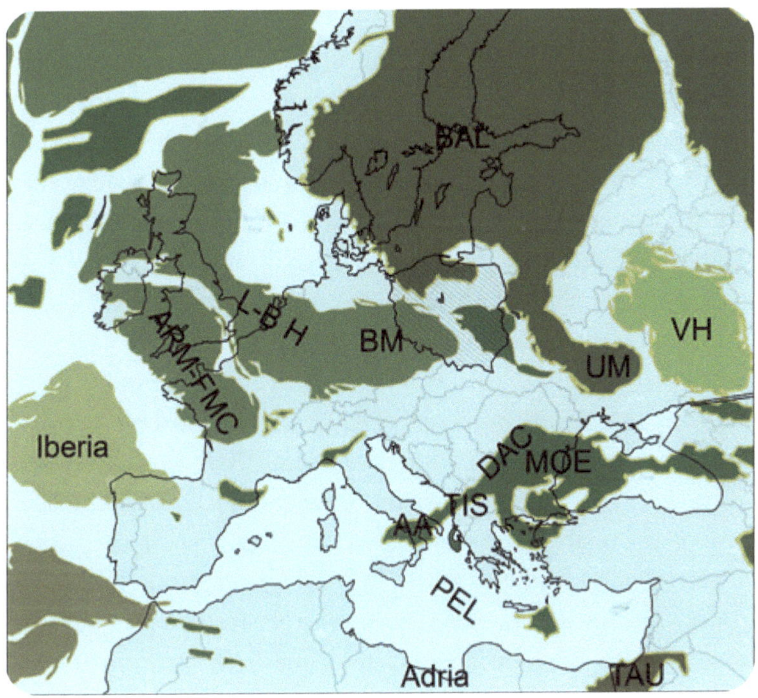

Fig. 118. A hypothetical map of the continent reaching Gdańsk and the Baltic region in the Cretaceous period, revealing the possibilities of creating Baltic amber. Adapted from: Csiki-Sava, Z., Buffetaut, E., Ősi, A., Pereda-Suberbiola, X., & Brusatte, SL (2015). "Island Life in the Cretaceous-Faunal Composition, Biogeography, Evolution, and Extinction of Land-Living Vertebrates on the Late Cretaceous European Archipelago". ZooKeys, (469), 1.

During the Cretaceous period (from ~ 145.0 to 66.0 million years ago), as in Bulgaria, there were also created numerous ambers all over the world. In Austriain Golling and Hauterivian, we find retinites that are more than 130 million years old. In Great Britain

375

there re fossilized resins from the same period [Jarzembowski 1995]. In Japan, on the island of Honshu, amber dating back to around 120 million years has been found. In northern Spain, amber from the Cretaceous period has been found in more than 120 different places [Peñalver et al. 2012].

In Siberia, on the Taymyr peninsula, fossil resins have been known for a long time, they are found in carbon layers, and were probably produced by plants such as Pinus (Haploxylon), Cedrus, Sciadopytis, *Taxodium*, *Cupressaceae*, Taxodiaceae, Cupressacites [Rasnitsyn et al. 2016]. Resins are also found in Alaska (USA) and most likely are the result of secreting resin by trees of the Araucariaceae family. Another place in the US where amber from the Cretaceous era occurs is the Neuse River near Goldsboro (North Carolina). Cretaceous amber has also been found at the Bone Butte site at Hell Creek, Harding County, North Dakota. Amber deposits in France (Charentese, Piolenc, Belcodène, Vendée and Oise) and Burma (Hukawng) come also from the Cretaceous period, although the deposit in Burma was originally estimated at the Miocene, later the Eocene, and as a result, it was considered to be mixed amber from various eological eras, including the Cretaceous – the scientists were convinced about it because of specific arthropods found there. In Azerbaijan, in Agdzhakend, chalk amber has been found in coal inserts. In Armenia, amber from the Cretaceous period has been documented in the Tavush province near Shavarshavan and Koti (Armenian Կոտի). The ambers found there are reddish or brownish and rather dark [Rasnitsyn et al. 2016]. In Canada, Alberta and Manitoba, Cretaceous amber has been discovered at Cedar Lake, and throughout Canada. 30 Cretaceous mber sites have been documented [McAlpine, Martin 1966].

A review of amber from the Cretaceous period proves that this type of resins have survived all over the world and could also be formed in Central Europe, although most of them do not have a jewelry value such as Baltic amber. The illustrative map presented above indicates that in the Cretaceous period in the area where the so-called Baltic amber occurs, there was a seashore and fossilized deposits may have formed proto-resins. Most likely, they were formed then - in the Cretaceous - also in this area, although today they are obtained as a mixture of amber from several geological epochs and described as Eocene deposits. During millions of years of transformation of the planet, amber deposits have moved and mixed up as a result of geological changes, especially during the ice age that lasted thousands of years. The amber mixture was given the name "Baltic amber". Therefore, a simplified marketing name has been adopted to describe all fossil resins found on the Baltic coast, in Poland, Denmark, Sweden, Germany, Russia and Ukraine.

A health-promoting effect of Baltic amber is the result of the synergy of active compounds. It has antioxidant properties due to the presence of succinic acid, its anti-rheumatic ffect may be due to the content of terpene compounds. Antiviral and antibacterial effects of terpenes are a well-known and appreciated feature. Amber also has an immunostimulating effect. Unfortunately, there are few animal studies and no clinical studies of succinite to confirm its health-promoting effects. The knowledge about this type of its features is based on the tradition of use, it comes mainly from folk medicine. For this reason, contemporary research focuses on confirming the validity of the ethnopharmacological use of amber by means of new concepts and analytical techniques.

Amber extracts and amber tinctures affect the human condition. Its effects on the circulatory system are particularly known. Even medics during the World War II appreciated amber vessels in which the stored blood did not condense. A lot of studies and signals indicate the positive effect of Baltic amber terpenes on the circulatory system. The mechanisms of their action are unknown, but the described reactions of the human body

indicate very positive effects of stimulating the heart and the circulatory system by amber extracts. University papers describe ts positive influence on the vascular system, including stabilizing hemorrhoid-related ailments.

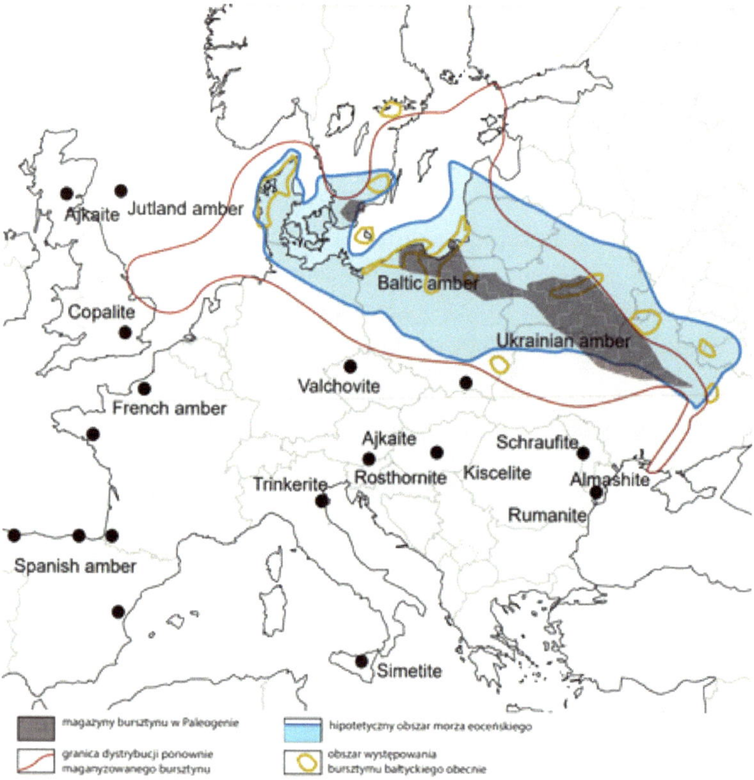

Fig. 119. Amber deposits known from literature. Based on: Czebreszuk, J. (2007). "The Role of the Sambian Center in Creating Cultural Meaning of Amber in the Third and Second Millennium BC. The Outline of Major Problems". Gaigalas, A., & Halas, S. (2009). "Stable isotopes (H, C, S) and the Origin of Baltic Amber". Geochronometry, 33 (1), 33-36.Katinas V, 1971. "Amber and Amber-Bearing Deposits of the Southern Pribaltic". Mintis, Vilnius: 156pp (in Russian).

Baltic amber is a well-known specific that affects the nervous system, especially neurons in the human brain. The book presents the effects of amber on electrical flows in unicellular yeast, as a result of which the interaction of electrons between cell membranes increased. We were the first to perform this type of research, documenting the electrical activity of amber in the human body. The conclusions from this experience may be significant especially at a time when the number of neurodegenerative diseases such as dementia, Parkinson's disease, Alzheimer's disease, especially in the richest countries, is constantly increasing. Stimulating neurons in the brain neurological system may contribute to improving the functioning of the entire synergistic system of the human body. This type of "wake up" of the neuronal system of the brain can help the body recognize damaged nerve pathways in dementia-related diseases and stimulate it to regenerate damaged parts of the brain. This is a hypothesis that requires scientific confirmation and very expensive medical research.

At the experimental stage, we created dietary supplements based on amber extracts by adding the ginkgo, Peruvian maca and ginseng extracts to them and compiling an atomizer that can be helpful in oral application of amber.

The properties of Baltic amber have been known for centuries – beginning from antiquity, when its properties of attracting grass particles were noticed and when the present-day word "electricity" was invented by the ancient Greeks, who saw the phenomenon of electric attraction but could not define it yet. Baltic amber was appreciated in the Roman Empire. The "miracle stone" from the north was an expensive medicine that was used to heal emperors, including Nero. A small amber figurine reached the price of a slave then.

Medieval Teutonic Knights were also interested in amber – they originally used it for making rosaries and for medical purposes. Many years later, the Prince of Prussia, who was quite active in business, promoted Baltic amber by commissioning medical books. It was then that amber tinctures became known throughout civilized Europe, in which Italian pharmacists were in the lead.

The book also presents experiments explaining the electrical properties of amber. After all, this polymerized resin was created from plants and has stored the terpenes derived from them for millions of years to this day. Today trees and plants function in the same way as they did millions of years ago. They operate on the basis of the generated electricity, which drives all life mechanisms of trees and other plants. In the book, we present one of the experiments documenting the generation of electricity in a tree. The compounds that drive electrochemical transformations are the terpenes and terpenoids, to which we have devoted much of the work.

The amount of known terpenes exceeds 40,000 species. These are the basic chemicals secreted by plants. Sometimes they have a protective function, other times they are given off by trees to encourage an insect to fly in and pollinate young buds. Some terpenes are poisons that are non-toxic to humans, others are aromas. Their impact is the subject of ongoing scientific research. Undoubtedly, however, many of them have a bactericidal, antiviral and antifungal effect, because trees, defending themselves against external stress factors, release such chemical compounds. They are used more and more often by pharmacy to protect people from disease threats.

Our civilization began the third millennium, and the secrets of the fossilized resin found in the Baltic region, in Poland, Ukraine, Denmark, Great Britain, Germany and Sweden are still quite a

mystery. That is why it is worth rediscovering succinite, looking at it from the perspective of the millions of years in which the planet has evolved, and taking advantage of the opportunity for a wider analysis than that which was available to previous generations.

ACKNOWLEDGMENTS

The book has been written thanks to the support of several people. I discussed a number of issues with scientists in the field and practitioners. I received a great deal of support from Ms. Alicja Pielińska from the Polish Academy of Sciences Museum of the Earth in Warsaw, who provided me with books, articles, and scientific materials on amber, as well as pointed out certain deficiencies and proofreading errors in the work. During the several years of writing the work I also benefited from a huge collection of literature on amber, which   is assembled at the Museum of the Earth in Warsaw. I was also supported by Ludomir Jastrzębski, a geologist and practitioner, especially on practical geological issues concerning amber. He provided me with the first samples of ground amber and put me in touch with many amber museums in Poland. Thanks to him, I was able to buy several unique ambers, which I used in the book. The draings of geological epochs painted by Katarzyna Doszla, were reviewed by Professor Jerzy Dzik from the Institute of Paleobiology of the Polish Academy of Sciences in Warsaw. Professor Ewa Wagner-Wysiecka of the Gdansk University of Technology, who has been working on the problems of Baltic amber for many years, kindly sent me one of her articles. I received comments in the field of pharmacy and medicine from Dr. Daniel Załuski, whom I would like to thank in particular. He was always willing to discuss my concerns about pharmacy and medicine. Important information on the methodology of writing the book was provided to me by Prof. Dr. Ewa Osińska from the Warsaw University of Life Sciences. Prof. Osinska always supported me in the research process, while we were implementing the research program on amber financed by the National Center for Research and Development in Warsaw based on European Union funds, but also after the project. She also supported me in removing methodological flaws from the book. Research on iodine and amber was supported by Prof. Marek Gajewski, PhD, of the Warsaw University of Life Sciences. I was also assisted by Dr. Ivo Rappsilber from Germany, who sent me photos of amber and scientific articles on amber from Bitterfeld, and Mr. Doug Lundberg, who specializes in identifying insects in Baltic and Dominican amber from Colorado Springs, USA. Thanks to the latter, I obtained amber samples from the USA.

GEOLOGICAL TIME SCALE

ERA Duration in mln of years	PERIOD Boundary of age in millions of years		EPOCH	AGE
CENOZOIC	**QUATERNARY** 2,58	2,58	HOLOCENE	
			PLEISTOCENE	LATE
				MIDDLE
				CALABRIAN
				GELASIAN
	NEOGENE 23,03	20,45	PLIOCENE	PIACENZIAN
				ZANCLEAN
			MIOCENE	MESSINIAN
				TORTONIAN
				SERRAVALLIAN
				LANGHIAN
				BURDIGALIAN
				AQUITANIAN
	PALEOGENE 66,0	42,97	OLIGOCENE	CHATTIAN
				RUPELIAN
			EOCENE	PRIABONIAN
				BARTONIAN
				LUTETIAN
				YPRESIAN
			PALEOCENE	THANETIAN
				SELANDIAN
				DANIAN
MESOZOIC	**CRETACEOUS** 145,0	79,0	LATE	MAASTRICHTIAN
				CAMPANIAN
				SANTONIAN
				CONIAN
				TURONIAN
				CENOMANIAN
			EARLY	ALBIAN
				APTIAN
				BARREMIAN
				HAUTERIVIAN
				VALANGINIAN
				BERRIASIAN
	JURASSIC 201,03	56,3	LATE	TITHONIAN
				KIMMERIDGIAN
				OXFORDIAN
			MIDDLE	CALLOVIAN
				BATHONIAN
				BAJOCIAN
				AALENIAN
			EARLY	TOARCIAN
				PLIENSBACHIAN
				SINEMURIAN
				HETTANGIAN

Era	Period	Epoch		Stage
	TRIASSIC 252,2 50,9	LATE		RHAETIAN
				NORIAN
				CARNIAN
		MIDDLE		LADINIAN
				ANISIAN
		EARLY		OLENEKIAN
				INDUAN
PALEOZOIC	**PERMIAN** 298,9 46,7	LOPINGIAN		CHANGHSINGIAN
				WUCHIAPINGIAN
		GUADALUPIAN		CAPITANIAN
				WORDIAN
				ROADIAN
		CISURALIAN		KUNGURIAN
				ARTINSKIAN
				SAKMARIAN
				ASSELIAN
	CARBONI-FEROUS 358,9 60,0	PENSYLVAN.	LATE	GZHELIAN
				KASIMOVIAN
			MIDDLE	MOSKOVIAN
			EARLY	BASHKIRIAN
		MISSISIP.	LATE	SERPUKHOVIAN
			MIDDLE	VISEAN
			EARLY	TOURNAISIAN
	DEVONIAN 419,2 60,3	LATE		FAMENNIAN
				FRASNIAN
		MIDDLE		GIVETIAN
				EIFELIAN
		EARLY		EMSIAN
				PRAGIAN
				LOCHKOVIAN
	SYLURIAN 443,8 24,6	PRIDOLI		
		LUDLOW		LUDFORDIAN
				GORSTIAN
		WENLOCK		HOMERIAN
				SHEINWOOODIAN
		LLANDOVERY		TELYCHIAN
				AERONIAN
				RHUDDANIAN
	ORDOVICIAN 485,4 41,6	LATE		HIRNANTIAN
				KATIAN
				SANDBIAN
		MIDDLE		DARRIWILIAN
				DAPINGIAN
		EARLY		FLOIAN
				TREMADOCIAN
	CAMBRIAN 541,0 55,6	FURONGIAN		AGE 10
				JIANGSHANIAN
				PAIBIAN
		EPOCH 3		GUZHANGIAN
				DRUMIAN
				AGE 5
		EPOCH 2		AGE 4
				AGE 3
		TERRENEUVIAN		AGE 2
				FORTUNIAN

12. BIBLIOGRAPHY

» Abdelgaleil S. A. (2010). Molluscicidal and insecticidal potential of monoterpenes on the white garden snail, Theba pisana (Muller) and the cotton leafworm, Spodoptera littoralis (Boisduval). Applied Entomology and Zoology, 45(3), 425-433.

» Abd-ElGawad, A. M., El Gendy, A. E. N. G., Assaeed, A. M., Al-Rowaily, S. L., Alharthi, A. S., Mohamed, T. A., ... & Elshamy, A. I. (2020). Phytotoxic effects of plant essential oils: A systematic review and structure-activity relationship based on chemometric analyses. Plants, 10(1), 36.

» Abdel-Sattara E., Azza R. Abdel Monemb, Shahira M. Ezzatb ,Ali M. El-Halawany, Samar M. Mouneird, Chemical and Biological Investigation of Araucaria heterophylla Salisb. Resin, 2009 Verlag der Zeitschrift für Naturforschung.

» Achari A.E., Jain, S.K. Adiponectin, a Therapeutic Target for Obesity, Diabetes, and Endothelial Dysfunction. Int. J. Mol. Sci. 2017.

» Adl S., Girard, V., Breton, G., Lak, M., Maharning, A., Mills, A., ... & Néraudeau, D. (2011). Reconstructing the soil food web of a 100 million-year-old forest: the case of the mid-Cretaceous fossils in the amber of Charentes (SW France). Soil Biology and Biochemistry, 43(4), 726-735.

» Ahlskog J. E., Geda Y. E., Graff-Radford N. R., Pettersen R. C., Physical Exercise as a Preventive of Disease-Modifying Treatment of Dementia and Brain Aging, Mayo Clinic Proceedings, Vol. 86, Issue 9, pp. 876-88, 2011.

» Akermi, S., Smaoui, S., Elhadef, K., Fourati, M., Louhichi, N., Chaari, M., ... & Mellouli, L. (2022). Essential oil from Cupressus sempervirens: testing antimicrobial multi-target mechanisms, predicting chemical toxicity and assessing safety in zebrafish embryos. Molecules, 27 (9), 2630.

» Alekseev V. I., & Grzymala, T. L. (2015). New Aderidae (Coleoptera: Tenebrionoidea) from Baltic and Bitterfeld amber. Zootaxa, 3956(2), 239-257.

» Ali N. A., Abbas, M. J., & Al-Bayaty, F. H. (2015). Evaluation of potential effect of menthol solution on oral hygiene status of dental students in a university in Iraq. Tropical Journal of Pharmaceutical Research, 14(4), 687-692.

» Almeida J. R. G. D. S., Souza, G. R., Silva, J. C., Saraiva, S. R. G. D. L., Júnior, R. G. D. O., Quintans, J. D. S. S., ... & Junior, L. J. Q. (2013). Borneol, a bicyclic monoterpene alcohol, reduces nociceptive behavior and inflammatory response in mice. The Scientific World Journal, 2013.

» Al-Tamimi W.H., Sahar A.A. Malik Al-Saadi, Ahmed A. Burghal, Antibacterial activity and GC-MS analysis of baltic amber against pathogenic bacteria, International Journal of Advanced Science and Technology Vol. 29, No. 11s, (2020).

» Ámbar E. (2006). SELENOPIDAE Y THOMISIDAE (ARACHNIDA: ARANEAE). Boletín Sociedad Entomológica Aragonesa, 1(38), 209-212.

» Andernaci I.G., De medicina veteri et nova faciunda commentarius secundus: Volume 2, Officina Henricpetrina, 1571.

» Anderson, K. B., & LePage, B. A. (1995). The nature and fate of natural resins in the geosphere VI. Analysis of fossil resins from Axel Heiberg Island Canadian Arctic (No. ANL/CHM/CP-85869; CONF-940813-41). Argonne National Lab.(ANL), Argonne, IL (United States).

» Anwar T., Qureshi, H., Mahnashi, M. H., Kabir, F., Parveen, N., Ahmed, D., ... & Alhaider, H. A. (2021). Bioherbicidal ability and weed management of allelopathic methyl esters from Lantana camara. Saudi Journal of Biological Sciences, 28(8), 4365-4374.

» Apetz N., Münch G., Govindaraghavan S., Gyengesi E., Natural Compounds and Plant Extracts as Therapeutics Against Chronic Inflammation in Alzheimer's Disease – A Translational Perspective. CNS & Neurological Disorders – Drug Targets, 2014.

» Aranda F. J., & Villalain, J. (1997). The interaction of abietic acid with phospholipid membranes. Biochimica et Biophysica Acta (BBA)-Biomembranes, 1327(2), 171-180.

» Arya A., Chahal R., Habibur R., Rao R., Rahman M.H., Kaushik D., Akhtar M.F., Saleem A., Shaden M.A. K., El-Seedi H.R.,Kamel M., Albadrani G.M., Abdel-Daim M.M., Mittal V., Acetylcholinesterase Inhibitory Potential of Various Sesquiterpene Analogues for Alzheimer's Disease Therapy, Biomolecules 2021.

» Armstrong D. W., Zhou, E. Y., Zukowski, J., & Kosmowska□ Ceranowicz, B. (1996). Enantiomeric composition and prevalence of some bicyclic monoterpenoids in amber. Chirality, 8(1), 39-48.

» Asad M., Meshal Alhumoud, Hepatoprotective Effect and GC-MS Analysis of Traditionally Used Boswellia Sacra Oleo Gum Resin (Frankincense) extract in rats, African Journal of Traditional, Complementary and Alternative Medicines, 2015.

» Aurifaber A., (1572) Succini historia. Kurzer und gründlicher Bericht, woher der Agtstein oder Börnstein ursprünglich komme, dass er kein Baumharz sey, sondern ein Geschlecht des Bergwachs, und wie man jenen manigfaltiglich in Arzneyen möge gebrauchen, Königsberg 1572.

» Austin A. D., Yeates, D. K., Cassis, G., Fletcher, M. J., La Salle, J., Lawrence, J. F., ... & Taylor, G. S. (2004). Insects 'Down Under'– Diversity, endemism and evolution of the Australian insect fauna: examples from select orders. Australian Journal of Entomology, 43, 216-234.

» Alya M., Ismat, N., Zeb, S., & Niaz, A. (2011). Analysis of total flavonoids and phenolics in different fractions of bark and needle extracts of Pinus roxburghii and Pinus wallichiana. Journal of Medicinal Plants Research, 5(21), 5216-5220.

» Bachir R.G., Benali M. Antibacterial activity of the essential oils from the leaves of *Eucalyptus* globulus against *Escherichia coli* and *Staphylococcus aureus*. Asian Pac J Trop Biomed, 2012.

» Baranov V., Andersen, T., Hagenlund, L.K., 2015. A new species of Bryophaenocladius Thienemann, 1934 (Diptera, Chironomidae, Orthocladiinae) from Baltic amber. Norwegian Journal of Entomology 62, 53–56.

» Barazani, O., Dudai, N., & Golan-Goldhirsh, A. (2003). Comparison of Mediterranean Pistacia lentiscus genotypes by random amplified polymorphic DNA, chemical, and morphological analyses. Journal of Chemical Ecology, 29(8), 1939-1952.

» Barczak P., Amber. Elixir of immortality. Succinite solution for Alzheimer's disease and dementia. Scientific study, Warsaw, 2022.

» Barczak P., Załuski D., Amber Elixir of Immortality: Health properties of compounds with pharmacological action, Warsaw 2021.

» Barden P. (2017). Fossil ants (Hymenoptera: Formicidae): ancient diversity and the rise of modern lineages. Myrmecological News, 24(1), 30.

» Bartus, R.T., Dean R.L., Beer B., Lippa A.S., The cholinergic hypothesis of geriatric memory dysfunction. Science 1982.

» Batish D.R., Singh HP, Setia N, Kaur S, Kohli RK. (2006). Chemical composition and inhibitory activity of essential oil from decaying leaves of *Eucalyptus citriodora*. Z Naturforsch, C, J Biosci, 61, 52–56.

» Batten D. J., & Austen, P. A. (2011). The Wealden of south-east England. Palaeontological Association: London, UK, 15-51.

» Beimforde C., Leyla J. Seyfullah, Vincent Perrichot, Kerstin Schmidt, Jouko Rikkinen, Alexander R. Schmidt, Resin exudation and resinicolous communities on Araucaria humboldtensis in New Caledonia,Arthropod-Plant, 2017.

» Bellien J, Joannides R. (Mar 2013). "Epoxyeicosatrienoic acid pathway in human health and diseases". Journal of Cardiovascular Pharmacology. 61 (3): 188–96.

» Bogdasarov M. A. (2005). Fossil resins of northern Eurasia. Dissertation. 2005.

» Bogdasarov, M. A. & Rudko, G. I. (2018). The Geological-Evolutional Concept of Resin Genesis. Journal of Geological Resource and Engineering, 6, 112-123.

» Bogri A., Solodovnikov, A., & Żyła, D. (2018). Baltic amber impact on historical biogeography and palaeoclimate research: oriental rove beetle Dysanabatium found in the Eocene of Europe (Coleoptera, Staphylinidae, Paederinae). Papers in Palaeontology, 4(3), 433-452.

» Boneti T.D.M., Medicina Septentrionalis Collatitia, Geneve, LXXV.

» Bose, M. N., & Manum, S. B. (1990). Mesozoic conifer leaves with'Sciadopitys-like'stomatal distribution: a re-evaluation based on fossils from Spitsbergen, Greenland and Baffin Island.

» Boswijk G. (2005). A history of kauri. In Australia and New Zealand Forest Histories. Araucarian Forests (ed. J. Dargavel), pp. 19–26. Australian Forest History Society, Kingston.

» Brody, R. H., Edwards, H. G., & Pollard, A. M. (2002). Fourier transform-Raman spectroscopic study of natural resins of archaeological interest. Biopolymers: Original Research on Biomolecules, 67(2), 129-141.

» Brophy, J. J., Goldsack, R. J., Wu, M. Z., Fookes, C. J., & Forster, P. I. (2000). The steam volatile oil of Wollemia nobilis and its comparison with other members of the Araucariaceae (Agathis and Araucaria). Biochemical Systematics and Ecology, 28(6), 563-578.

» Buku, A., Soeparman, S., Wardana, I. N. G., Wahyudi, S., & Roreng, P. P. (2006). LIQUID BROWN COAL ANALYSIS USING GAS CHROMATOGRAPHY-MASS SPECTROMETRY.

» Burks R.A., Mottern JL, Pownall NG, Waterworth R, Paine TD. First record of Closteroceruschamaeleon, parasitoid of the *Eucalyptus* Gall Wasp Ophelimusmaskelli (Hymenoptera, Chalcidoidea, Eulophidae), in the New World. Zookeys, 2015.

» Caron L., Deslauriers, A., Mshvildadze, V., & Pichette, A. (2013). Volatile compounds in the foliage of balsam fir analyzed by static headspace gas chromatography (HS-GC): an example of the spruce budworm defoliation effect in the boreal forest of Quebec, Canada. Microchemical Journal, 110, 587-590.

» Cârciumaru M., Niţu, E. C., Miclăuş, C., Ion, R., Cîrstina, O., Ionuţ Lupu, F., ... & Grigore, S. (2017). Amber Deposits in Romania, with Particular Emphasis on Those Located on the Eastern Side of the Carpathians (Bibliographical Considerations and a few Field Investigations). Annales d'Université" Valahia" Târgovişte. Section d'Archéologie et d'Histoire, 19(1), 33-56.

» Chalchat J. C., J. L. Chabard, M. S. Gorunovic, V. Djermanovic & V. Bulatovic (1995) Chemical Composition of *Eucalyptus* globulus Oils from the Montenegro Coast and East Coast of Spain, Journal of Essential Oil Research.

» Checker R., Sharma, D., Sandur, S. K., Subrahmanyam, G., Krishnan, S., Poduval, T. B., & Sainis, K. B. (2010).

» Chan P., Tomlinson, B., Chen, Y.-J., Liu, J.-C., Hsieh, M.-H., Cheng, J.-T., 2000. A double-blind placebo-controlled study pf the effectiveness and tolerability of oral stevioside in human hypertension. Br. J. Clin. Pharmacol. 50, 215-220.

» Chen S.W. et al., Anxiolytic-like effect of succinic acid in mice, Life Sciences 73, 2003.

» Chen Q., Xiaoge Zhao , Tingya Lu , Yao Yang , Yi Hong , Minyi Tian , Ying Zhou, Chemical Composition, Antibacterial, and Anti-Inflammatory Activities of Essential Oils from Flower, Leaf, and Stem of *Rhynchanthus beesianus*, BioMed Research International, 2021.

» Chhagan Lal , et al. Juglans nigra: Chemical constitution and its application on Pashmina (Cashmere) fabric as a dye. J. Nat. Prod.

Plant Resour, 2011, 1.4: 13-19.Plumbagin inhibits proliferative and inflammatory responses of T cells independent of ROS generation but by modulating intracellular thiols. Journal of cellular biochemistry, 110(5), 1082-1093.

» Chelliah, D. A. (2008). Biological activity prediction of an ethno medicinal plant Cinnamomum camphora through bioinformatics. Ethnobotanical leaflets, 2008(1), 22.

» Chętnik A.,- Przemysł i sztuka bursztynowa nad Narwią, Ludt. 39, 1952, s. 357.

» Chen C-N,Shing-Hwa Liu,Shoei-Yn Lin-Shiau,Honokiol, a Neuroprotectant against Mouse Cerebral Ischaemia, Mediated by Preserving Na+, K+-ATPase Activity and Mitochondrial Functions, Volume101, 2007.

» Chouchani E.T., Pell VR, Gaude E, Aksentijević D, Sundier SY, Robb EL, Logan A, Nadtochiy SM, Ord ENJ, Smith AC, Eyassu F, Shirley R, Hu CH, Dare AJ, James AM, Rogatti S, Hartley RC, Eaton S, Costa ASH, Brookes PS, Davidson SM, Duchen MR, Saeb-Parsy K, Shattock MJ, Robinson AJ, Work LM, Frezza C, Krieg T, Murphy MP. Ischaemic accumulation of succinate controls reperfusion injury through mitochondrial ROS.

» Nature. 2014 Nov 20;515(7527):431-435. doi: 10.1038/nature13909. Epub 2014.

» Clarke J.H. , The Homeopathic World: A Monthly Journal of Medical, Social, and Sanitary Science. Ed. by J.H.Clarke, M.D. London. The Homeopathic Publishing Company, 1896.

» Conwentz H. Monographie der baltischen Bernsteinbäume / H. Conwentz. – Danzig, 1890. – 151 p.

» Culpeper N. Culpeper's Complete Herbal & English Physician. London, Applewood Press; 2006: 276.

» Czebreszuk J. (2007). The Role of the Sambian Centre in Creating Cultural Meaning of Amber in the Third and Second Millennium BC. The Outline of Major Problems.

» Dahlström Å, Bros L., The amber book, Geoscience Press, Tuscon, 1996, za Szwedo, J. (2002). Amber and amber inclusions of planthoppers, leafhoppers and their relatives (Hemiptera, Archaeorrhyncha et Clypaeorrhyncha) (pp. 37-56).

» D'Costa,D.,Boswijk,G.&Ogden,J.(2009). Holocene vegetation andenvironmental reconstructions from swamp deposits in the Dargaville region ofthe North Island, New Zealand: implications for the history of Kauri (Agathisaustralis).The Holocene19.

» Deka D. C., & Basumatary, S. (2011). High quality biodiesel from yellow oleander (Thevetia peruviana) seed oil. Biomass and Bioenergy, 35(5), 1797-1803.

» Denk, T., & Bouchal, J. M. (2021). Dispersed pollen and calyx remains of Diospyros (Ebenaceae) from the middle Miocene "Plant beds" of Søby, Denmark. GFF, 143(2-3), 292-304.

» Devi K.S., Sruthy.P.B, Anjana.J.C, J. Rathinamala, S. Jayashree,International Journal of Applied Biology and Pharmaceutical Technology, www.ijabpt.com.

» Devièse T., Ribechini, E., Castex, D., Stuart, B., Regert, M., & Colombini, M. P. (2017). A multi-analytical approach using FTIR, GC/MS and Py-GC/MS revealed early evidence of embalming practices in Roman catacombs. Microchemical Journal, 133, 49-59.

» Devkota K.P., Ratnayake, R., Colburn, NH, Wilson, JA, Henrich, CJ, McMahon, JB i Beutler, JA (2011). Inhibitory onkogennego czynnika transkrypcyjnego AP-1 z *Podocarpus* latifolius. Dziennik produktów naturalnych , 74 (3), 374-377.

» Dillhoff R.M., Thomas A. Dillhoff, David R. Greenwood, Melanie L. DeVore, and Kathleen B. Pigg, The Eocene Thomas Ranch flora, Allenby Formation, Princeton, British Columbia, Canada, Botany, 2013.

» Dimpfel W., I. Pischel, R. Lehnfeld, Effects of lozenge containing lavender oil, extracts from hops, lemon balm and oat on electrical brain activity of volunteers, European Journal Medical Research (2004).

» Ding Z., A research note studies on extraction and isolation of flavonoids from ginkgo leaves,

» Food & Nutrition Press, 1999.

» Domagała J. (1982). Polihalit w cechsztynie rejonu Zatoki Puckiej w świetle nowych danych. Przegląd Geologiczny, 30(9), 463-466.

» Dominiguez-Gonzalez M.R. et al., Evaluation of iodine bioavailability in seaweed using in vitro methods, J Agric Food Chem., 2017.

» Dong T., Nian Chen, Xiao Ma, Jian Wang, Jing Wen, Qian Xie,

Rong Ma, The protective roles of L-borneolum, D-borneolum and synthetic borneol in cerebral ischaemia via modulation of the neurovascular unit, Biomedicine & Pharmacotherapy, 2018.

» Donzelli G., Teatro Farmaceutico, Dogmatico, e Spagirico del Dottor Giuseppe Donzelli. – Venezia: Andrea Poletti, 1737.

» Doskočil L., Grasset, L., Válková, D., & Pekař, M. (2014). Hydrogen peroxide oxidation of humic acids and lignite. Fuel, 134, 406-413.

» Drabkin I.E. Caelius Aurelianus. On Acute Diseases and on Chronic Diseases. Chicago : University of Chicago Press, 1950: 166; Riddle, J.M., 1973: 9.

» Drzewicz P., Naglik, B., Natkaniec-Nowak, L., Dumańska-Słowik, M., Stach, P., Kwaśny, M., ... & Kubica-Bąk, D. (2020). Chemical and spectroscopic signatures of resins from Sumatra (Sarolangun mine, Jambi Province) and Germany (Bitterfeld, Saxony-Anhalt). Scientific Reports, 10(1), 1-14.

» Duffin C. J. (2013). History of the external pharmaceutical use of amber. Pharm Hist (Lond), 43(3).

» Duffin C.J. & Pymm, R.J. A Survey of Pharmaceutical 'Stones' – Part 1. "Pharmaceutical Historian", 45 (1); 2015: 2-8.

» Eberle J.J., David R. Greenwood, Life at the top of the greenhouse Eocene world—A review of the Eocene flora and vertebrate fauna from Canada's High Arctic, INVITED REVIEW ARTICLE, 2012.

» Ebadollahi A. , Jalal Jalali Sendi , Mostafa Maroufpoor, Mehdi Rahimi-Nasrabadi, Acaricidal Potentials of the Terpene-rich Essential Oils of Two Iranian *Eucalyptus* Species against Tetranychus urticae Koch, Journal of Oleo Science, 2017.

» Eckert A. (2012). Mitochondrial effects of *Ginkgo biloba* extract. International psychogeriatrics, 24(S1), S18-S20.

» Edelfrawi M.E., Edelfrawi, A.T., 1997. Comparative molecular and pharmacological properties of cholinergic receptors in insects and mammals. ACS Symposium Series 658, 327–338.

» Ehrmann, R. (2011). Mantodea from Turkey and Cyprus (Dictyoptera: Mantodea). Articulata, 26(1), 1-42.

» Ellis L., Singh R.K., Alexander R., Kagi R.I., Identification and occurrence of dihydro-ar-curcumene in crude oils and sediments, Organic Geochemistry, 1995.

» El Tahir KEH, Al-Ajmi MF, Al-Bekairi AM 2003. Some cardiovascular effects of the dethymoquinonated Nigella sativa volatile oil and its major components []-pinene and p-cymene in rats. Saudi Pharm J 113: 104-110.

» Elditt H. L., Bernstein -Regal in Preussen , Alt preussische Monatsschrift, 1868, s. 580.

» Elshamy, A. I., Ammar, N. M., Hassan, H. A., Al-Rowaily, S. L., Ragab, T. I., El Gendy, A. E. N. G., & Abd-ElGawad, A. M. (2020). Essential oil and its nanoemulsion of Araucaria heterophylla resin: Chemical characterization, anti-inflammatory, and antipyretic activities. Industrial crops and products, 148, 112272.

» Fall, P. L. (2012). Modern vegetation, pollen and climate relationships on the Mediterranean island of Cyprus. Review of Palaeobotany and Palynology, 185, 79-92.

» Fernandez X., Louisette Lizzani-Cuvelier,1 André-Michel Loiseau,1 Christine Perichet,2 Claire Delbecque2 and Jean-François Arnaudo, Chemical composition of the essential oils from Turkish and Honduras Styrax, Flavour and Fragrance Journal, 2005.

» Ferro A. et al., Short-term succinic acid treatment mitigates cerebellar mitochondrial OXPHOS dysfunction, neurodegeneration and ataxia in a Purkinje-specific spinocerebellar ataxia type 1 (SCA1) mouse model, https://doi.org/10.1371/journal.pone.0188425, 2017.

» Fijoł-Adach E., Feledyn-Szewczyk B., Kazimierczak R., Stalenga J. 2016: Wpływ systemu produkcji rolnej

» na występowanie substancji bioaktywnych w owocach truskawki. Postępy Techniki Przetwórstwa

» Spożywczego, 1, 78-86.

» Emeljanov A. F., & Shcherbakov, D. E. (2011). A new genus and species of Dictyopharidae (Homoptera) from Rovno and Baltic amber based on nymphs. ZooKeys, (130), 175.

» Engel M. S. (2005). An Eocene ectoparasite of bees: The oldest definitive record of phoretic meloid triungulins (Coleoptera: Meloidae; Hymenoptera: Megachilidae). Acta zoologica cracoviensia, 48(3-4), 43-48.

» Feyroz R. H., First record of the *eucalyptus* gall wasp, Leptocybe invasa Fisher and La Salle (Hymenoptera: Eulophidae), in Iraq, ACTA AGROBOTANICA, 2012.

» Fleisher, A., & Fleisher, Z. (2000). The volatiles of the leaves and wood of Lebanon Cedar (Cedrus libani A. Rich) aromatic plants of the Holy Land and the Sinai. Part XIV. Journal of Essential Oil Research, 12(6), 763-765.

» Francis P.T., Palmer AM, Snape M, et al. The cholinergic hypothesis of Alzheimer's disease: a review of progress. Journal Neurol Neurosurg Psychiatry 1999;54:137–47.

» Franich R. A., Kroese, H. W., Jakobsson, E., Jensen, S., & Kylin, H. E. N. R. I. K. (1993). Trace constituents of natural and anthropogenic origin from New Zealand Pinus radiata needle epicuticular wax. New Zealand Journal of Forestry Science, 23(1), 101-9.

» Frezza C., Alessandro Venditti, Daniela De Vita, Chiara Toniolo, Marco Franceschin, Antonio Ventrone, Lamberto Tomassini, Sebastiano Foddai, Marcella Guiso, Marcello Nicoletti, Armandodoriano Bianco, Mauro Serafini, Phytochemistry, Chemotaxonomy, and Biological Activities of the Araucariaceae Family—A Review, Plants 2020.

» Flávia A., Santos, Regilane M. Silva, Adriana R. Tomè, Vietla S. N. Rao, Margarida M. L. Pompeu, Maria J. Teixeira, Luiz A. R. De Freitas and Valderes L. De Souza, 1,8-Cineole protects against liver failure in an in-vivo murine model of endotoxemic shock, Journal of Pharmacy and Pharmacology, 2000.

» Fonseca M. C., Aguiar C.J., Rocha Franco J.A., Gingold R.N., M. Leite M.F., Expanding the frontiers of krebs cycle intermediates, cell communication and signaling, Volume 14, 2016.

» Franus M. (2010). Zastosowanie glaukonitu do usuwania śladowych ilości metali ciężkich. Politechnika Lubelska.

» Francis P.T., Palmer A.M., Snape M., et al., The cholinergic hypothesis of Alzheimer's disease: a review of progress, Journal Neurology Neurosurgery Psychiatry, 1999.

» Fuhrmann R. (2010). Die Bitterfelder Bernsteinarten. Mauritiana, 21, 13-58.

» Fujita, E. (1974). The chemistry on diterpenoids in 1971. Bulletin of the Institute for Chemical Research, Kyoto University, 52(3), 519-560.

» Fujiwara M., Yagi N., Miyazawa M., Acetylcholinesterase inhibitory activity of volatile oil from Peltophorum dasyrachis Kurz ex

Bakar (yellow batai) and bisabolane-type sesquiterpenoids, Journal Agricultural Food Chem., 2010.

» Gadotti V. M., Huang, S., & Zamponi, G. W. (2021). The terpenes camphene and alpha-bisabolol inhibit inflammatory and neuropathic pain via Cav3. 2 T-type calcium channels. Molecular brain, 14(1), 1-10.

» Ganapaty, S., Thomas, P. S., Karagianis, G., Waterman, P. G., & Brun, R. (2006). Antiprotozoal and cytotoxic naphthalene derivatives from Diospyros assimilis. Phytochemistry, 67(17), 1950-1956.

» García-Risco M.R., Lamia Mouhid, Lilia Salas-Pérez, Alexis López-Padilla, Susana Santoyo, Laura Jaime, Ana Ramírez de Molina, Guillermo Reglero, Tiziana Fornari, Biological Activities of Asteraceae (Achillea millefolium and Calendula officinalis) and Lamiaceae (Melissa officinalis and Origanum majorana) Plant Extracts, Plant Foods Hum Nutr, 2017.

» Garneau F.-X.; Collin, Guy; Gagnon, Helena; Pichette, André (2012). „Skład chemiczny hydrozolu i olejku eterycznego trzech różnych gatunków z rodziny Pinaceae: Picea glauca (Moench) Voss., Picea mariana (Mill.) BSP i *Abies balsamea* (L.) Mill". Journal of Essential Oil Bearing Plants . 15 (2): 227.

» Garrison M.S., Anthony K. Irvine, William N. Setzer, Chemical composition of the resin essential oil from Agathis atropurpurea from North Queensland, Australia, American Journal of Essential Oils and Natural Products 2016.

» Gąsiewicz A. (2006). Struktury mikroorganiczne w siarce rodzimej zapadliska przedkarpackiego. Przegląd Geologiczny, 54(12), 1057-1059.

» Gąsiewicz A., Jasionowski, M., & Poberzhskyy, A. (2012). Wpływ eksploatacji siarki na cechy geochemiczne środowiska powierzchniowego złóż siarki z pogranicza polsko-ukraińskiego. Biuletyn Państwowego Instytutu Geologicznego.

» Gelnitius A. Disquisitio philosophica de succini natura / Pauli A. (praes.). – Dantisci; ex Molibdographia Hünefeldiana, 1614.; Nothnagel E. Manuale di Materia Medica e Terapia. – Napoli: Giovanni Jovene Libraio Editore, 1876.

» Gheribi E. 2011: Związki polifenolowe w owocach i warzywach. Medycyna Rodzinna, 4, 111-115.

» Ghesini, S., & Marini, M. (2015). Description of a new termite species from Cyprus and the Aegean area: Reticulitermes aegeus sp. nov. Bulletin of Insectology, 68(2), 207-210.

» Gilissen J., François Jouret, Bernard Pirotte, Julien Hanson, Insight into SUCNR1 (GPR91) structure and function, Pharmacology & Therapeutics (2016).

» Gimmel M.L., Szawaryn K, Cai C, Leschen RAB. 2019 Mesozoic sooty mould beetles as living relicts in New Zealand. Proc. R. Soc. B 286: 20192176. http://dx.doi.org/10.1098/rspb.2019.

» Girard, V., Sschmidt, A. R., Struve, S., Perrichot, V., Breton, G., & Nerauedau, D. (2009). Taphonomy and palaeoecology of mid-Cretaceous amber-preserved microorganisms from southwestern France. GEODIVERSITAS, 31(1).

» Goeppert H. K. Dir Flora des Bernsteins/H. K. Goeppert, A. Menge. – Danzig, 1883. – 128 p.

» González M.A., Aromatic Abietane Diterpenoids: Their Biological Activity and Synthesis, Natural Product Reports, 2015

» González-Burgos E., Mindaugas Liaudanskas, Jonas Viškelis, Vaidotas Žvikas, Valdimaras Janulis, M. Pilar Gómez-Serranillos, Antioxidant activity, neuroprotective properties and bioactive constituents analysis of varying polarity extracts from *Eucalyptus* globulus leaves, Journal of Food and Drug Analysis, Volume 26, Issue 4, 2018.

» Goudarzi S., M. Rafieirad, Evaluating the effect of α-pinene on motor activity, avoidance memory and lipid peroxidation in animal model of Parkinson disease in adult male rats, Research Journal of Pharmacognosy, 2017.

» Gough L. J. The composition of succinite (Baltic amber)/ L. J. Gough, J. S. Mills//Nature. – 1972. – Vol. 239. – P. 527–528.

» Góra J., Lis A., Najcenneiejsze olejki eteryczne. Część I. Monografie Politechniki Łodzkiej, Łódź 2019.

» Granger R.E., Erica L. Campbell, Graham AR Johnston, (+)- And (–)-borneol: efficacious positive modulators of GABA action at human recombinant α1β2γ2L GABAA receptors, Biochemical Pharmacology, 2005.

» Grein, M., Roth-Nebelsick, A., Wilde, V., 2010. Carbon isotope composition of middle Eocene leaves from the Messel Pit, Germany. Palaeodiversity 3.

» Grigelis A. (2001). Outline on geology of amber-bearing deposits in the Sambian peninsula. Acta Academiae Artium Vilnensis, 22, 35-40.

» Grímsson F., Zetter, R., Grimm, G.W. et al. *Fagaceae* pollen from the early Cenozoic of West Greenland: revisiting Engler's and Chaney's Arcto-Tertiary hypotheses. Plant Syst Evol 301, 809–832 (2015).

» Gucwa I., Poprawa D., Dystrybucja mikroelementów i bituminów w skałach fliszowych polskiej części Karpat, Przegląd Geologiczny, vol. 44, nr 5, 1996.

» Guilbert E. R. I. C., Chazeau, J., & De Larbogne, L. B. (1994). Canopy arthropod diversity of New Caledonian forests sampled by fogging: preliminary results. Memoirs of the Queensland Museum. Brisbane, 36(1), 77-85.

» Habtemariam S., (2018). Iridoids and other monoterpenes in the Alzheimer's brain: Recent development and future prospects. Molecules, 23(1), 117.

» Haklıdır M., & Kapkın, Ş. (2005). Black Sea, a hydrogen source. In Proceedings International Hydrogen Energy Congress and Exhibition IHEC.

» Heiss E. (2020). Juvenile spider riding a bug—phoresy in Baltic amber?. Palaeoentomology, 3(6), 561-563.

» Halamski A.T., Latest Cretaceous leaf floras from southern Poland and western Ukraine, Acta Palaeontologica Polonica 58, 2013.

» Hamel D., Melanie Sanchez, François Duhamel, Olivier Roy, Jean-Claude Honoré, Baraa Noueihed, Tianwei Zhou, Mathieu Nadeau-Vallée, Xin Hou, Jean-Claude Lavoie, Grant Mitchell, Orval A. Mamer, and Sylvain Chemtob, G-Protein–Coupled Receptor 91 and Succinate Are Key Contributors in Neonatal Postcerebral Hypoxia-Ischemia Recovery, Arteriosclerosis, Thrombosis, and Vascular Biology, 2013.

» Hamsarekha R., Gopinath, K., Srikanth, J., Sivakumar, M., Reddy, C. U. M., & Maheswara, U. (2017). In silico and in vitro xanthine oxidase inhibitory activity of embilica officinalis (amla). International Journal of Pharmaceutical Sciences and Research, 8(11), 4614-4623.

» Han M., Y. Liu, B. Zhang, J. Qiao, W. Lu, Y. Zhu, Y. Wang, C. Zhao, Salvianic borneol ester reduces β-amyloid oligomers and prevents cytotoxicity, Pharmaceutical Biology, 2011.

» Hand S., Archer, M., Bickel, D., Creaser, P., Dettmann, M., Godthelp,

H., ... & Penney, D. (2010). Australian cape york amber. Biodiversity of fossils in amber from the major world deposits, 69-79.

» Hartmann P.J., Succini Prussici physica et civilis historia cum demonstratione, etc, 1677.

» Havelcová M., Machovič, V., Sýkorová, I., Lapčák, L., Špaldoňová, A., Mach, K., & Dvořák, Z. (2018). Duxite–Fossil resin of Miocene age. Organic Geochemistry, 124, 190-204.

» Havelcová M., Sýkorová I., Karel Mach, Zdeněk Dvořák, Organic geochemistry of fossil resins from the Czech Republic, Procedia Earth and Planetary Science 10 (2014).

» Haywood B. W. (1989). Kauri Gum and the Gumdiggers: A Pictorial History of the Kauri Gum Industry in New Zealand (Pictures from the Past), Second Edition. Gordon Ell, Bush Press, Auckland.

» He W., Miao, F.J. -P., Lin, DC -H., Schwandner, RT, Wang, Z, Gao.J., Chen, J. -L, Tian, H., & Ling, L (2004).

» Hellemond, A. Fossil spiders in Baltic amber.2019.

» He W., Miao, F. J. P., Lin, D. C. H., Schwandner, R. T., Wang, Z., Gao, J., ... & Ling, L. (2004).Citric acid cycle intermediates as ligands for orphan G-protein-coupled receptors. Nature 429,188-193.

» He Y., Chen Z., Gong G., Evans A.C., Neuronal Networks in Alzheimer's Disease, The Neuroscientist, Volume 15, Number 4, 2009.

» Health Canada. Is Your Child Safe? Play Time. Ottawa: Health Canada; 2012. Available at: http://www.hc-sc.gc.ca/ cps-spc/pubs/ cons/child-enfant/play-jeu-eng.php (accessed November 30, 2015).

» Hekimi S., Lapointe J., Wen J., Taking a "good" look at free radicals in the aging process, Trends Cell Biol, 2011.

» Hinrichs K., Bernstein, das "Preußische Gold" in Kunst- und Naturalienkammern und Museen des 16. – 20. Jahrhunderts, Dissertation, Humboldt-Universität, Berlin,2007.

» Hernández Vázquez L., Javier Palazon, Arturo Navarro-Ocaña, The Pentacyclic Triterpenes [], []-amyrins: A Review of Sources and Biological Activities, Phytochemicals 23, 2012.

» Hill R.S., Yelarney K. Beer, Kathryn E. Hill , Elizabeth Maciunas, Myall A. Tarran, Carmine C. Wainman, Evolution of the eucalypts – an interpretation from the macrofossil record, Australian Journal of Botany 2016.

» Hogquist, K. A., Nett, M. A., Unanue, E. R., & Chaplin, D. D. (1991). Interleukin 1 is processed and released during apoptosis. Proceedings of the National Academy of Sciences, 88(19), 8485-8489.

» Hormiga G., & Griswold, C. E. (2014). Systematics, phylogeny, and evolution of orb-weaving spiders. Annual Review of Entomology, 59, 487-512.

» Horváthová E.; Kozics, K.; Srancíková, A.; Hunáková, L.; Gálová, E.; Sevcovicová, A.; Slamenová, D. Borneol administration protects primary rat hepatocytes against exogenous oxidative DNA damage. Mutagenesis 2012, doi:10.1093/mutage/ges023.

» Hsieh C.-L., M.-H. Tseng, R.-N. Pan, J.-Y. Chang, C.-C. Kuo, T.-H. Lee and Y.-H. Kuo,Novel Terpenoids from Calocedrus macrolepis var. formosana , Chem. Biodiv., 2011.

» Hur J., Pak, S. C., Koo, B. S., & Jeon, S. (2013). Borneol alleviates oxidative stress via upregulation of Nrf2 and Bcl-2 in SH-SY5Y cells. Pharmaceutical Biology, 51(1), 30-35.

» Imaizumi Y., Sakamoto, K., Yamada, A., Hotta, A., Ohya, S., Muraki, K., ... & Ohwada, T. (2002). Molecular basis of pimarane compounds as novel activators of large-conductance Ca2+-activated K+ channel α-subunit. Molecular Pharmacology, 62(4), 836-846.

» Iqbal M. et al., Antifungal and antibacterial activities of substituted benzyl 4-ketohexanoates. Indian Journal Science and Technology, 2016.

» Ivic L., Sands, T. T., Fishkin, N., Nakanishi, K., Kriegstein, A. R., & Strømgaard, K. (2003). Terpene trilactones from *Ginkgo biloba* are antagonists of cortical glycine and GABAA receptors. Journal of Biological Chemistry, 278(49), 49279-49285.

» Jalloh I., Helmy A, Howe DJ, Shannon RJ, Grice P, Mason A, Gallagher CN, Stovell MG, van der Heide S, Murphy MP, Pickard JD, Menon DK, Carpenter TA, Hutchinson PJ, Carpenter KL. Focally perfused succinate potentiates brain metabolism in head injury patients. J Cereb Blood Flow Metab. 2017

» James R. M.D., Pharmacopoeia Universalis: Or, A New Universal English Dispensatory, London, 1747.

» Jarzembowski E.A., 1995. The first insects in Cretaceous (Wealden) amber from the UK. Geology Today 11, 42.

» Jeong H. U., Kwon, S. S., Kong, T. Y., Kim, J. H., & Lee, H. S. (2014). Inhibitory effects of cedrol, β-cedrene, and thujopsene on

cytochrome P450 enzyme activities in human liver microsomes. Journal of Toxicology and Environmental Health, Part A, 77(22-24), 1522-1532.

» Jeznach A., 2012: Znaczenie jodu dla heterotrofów i autotrofów. Sochaczew.

» John J.F., Naturgeschichte des succins, oder des sogenannten bernsteins; nebst theorie der bildung aller fossilen, bituminösen inflammabilien des organischen reichs und den analysen derselben, Köln, 1813.

» Juergens U.R., (2014). Właściwości przeciwzapalne monoterpenu 1,8-cyneolu: aktualne dowody na jednoczesne stosowanie leków w chorobach zapalnych dróg oddechowych. Badania nad lekami , 64 (12), 638-646.

» Juergens L.J., Worth, H. & Juergens, U.R. New Perspectives for Mucolytic, Anti-inflammatory and Adjunctive Therapy with 1,8-Cineole in COPD and Asthma: Review on the New Therapeutic Approach. Adv Ther 37, 1737–1753 (2020). https://doi.org/10.1007/s12325-020-01279-0.

» Kaczmarczyk I. Bursztyn bałtycki w medycynie–badania aktywności biologicznej bursztynu. AMBERIF 2016, 35.

» Kaczmarczyk I., Baltic amber as a potential source of active agents against selected microorganisms, International Symposium Amber, Science and Art, 2018.

» Kafetzidou A. L. "A case study of fossil leaves from the Early Miocene Petrified Forest at Akrocheiras site in Sigri area Lesvos island, 2020.

» Kałwa K., Wyrostek J. 2018: Ocena zawartości związków biologicznie aktywnych w herbacie zielonej i

» czarnej. Inżynieria Przetwórstwa Spożywczego, 2, 15-21.

» Kamarudin N.A. , Nik Nur Hakimah Nik Salleh, Suat Cheng Tan,v Gallotannin-Enriched Fraction from Quercus infectoria Galls as an Antioxidant and Inhibitory Agent against Human Glioblastoma Multiforme, Plants 2021.

» Kampen Van J, Robertson H, Hagg T, Drobitch R. Neuroprotective actions of the ginseng extract G115 in two rodent models of Parkinson's disease. Exp Neurol. 2003; 184(1): 521-529.

» Kand Y-H., Luke R. Howard, Phenolic Composition and Antioxidant Activities of Different Solvent Extracts from Pine Needles in Pinus Species, J Food Science and Nutrition, 2010.

» Kang G., Mishyna, M., Appiah, K. S., Yamada, M., Takano, A., Prokhorov, V., & Fujii, Y. (2019). Screening for plant volatile emissions with allelopathic activity and the identification of L-Fenchone and 1, 8-Cineole from star anise (*Illicium verum*) leaves. Plants, 8(11), 457.

» Karnkowski P. (1962). Uwagi o roponośności i gazonośności polskich Karpat fliszowych i ich przedgórza. Przegląd Geologiczny, 10(7), 333.

» Kasprzyk A. (2005). Modele genetyczne badeńskich anhydrytów w zapadlisku przedkarpackim na obszarze Polski. Przegląd Geologiczny, 53(1), 47-54.

» Katinas V.I. Amber and amber-bearing deposits of the South Baltic states / V.I. Katinas // Sat. scientific tr. / LitNIGRI. – Vilnius, 1971. - Issue. 20: Amber and amber-bearing deposits of the Southern Baltic. – 150 s./ Katinas V., 1987: On the origin of amber. "The Pravda newspaper". Moscow.

» Kennedy D.O., Scholey A.B., Tildesley N.T., Perry E.K., Wesnes K.A., Modulation of mood and cognitive performance following acute administration of Melissa officinalis (lemon balm), Pharmacology Biochemistry Behavior, 2002.

» Keskin I., Y Gunal, S Ayla, B Kolbasi, A Sakul, U Kilic, O Gok, K Koroglu, H Ozbek, Effects of Foeniculum vulgare essential oil compounds, fenchone and limonene, on experimental wound healing, Biotech Histochem 2017).

» Khan A., Vaibhav, K.; Javed, H.; Tabassum, R.; Ahmed, M.E.; Khan, M.M.; Khan, M.B.; Shrivastava, P.; Islam, F.; Siddiqui, M.S.; et al. 1,8-cineole (eucalyptol) mitigates inflammation in amyloid beta toxicated PC12 cells: Relevance to Alzheimer's disease. Neurochem. Res. 2014.

» Khan-Mohammadi-Khorrami MK, Asle-Rousta M, Rahnema, Amini R, The Effect of Alpha-pinene on Amyloid-beta-induced Neuronal Death and Depression in Male Wistar Rats, Journal of Ardabil University of Medical Sciences, Vol. 20, No. 4, Winter 2021.

» Khazanov V.A., Vengerovsky A.I., Effect of Silimarin, Succinic Acid, and Their Combination on Bioener-getics of the Brain in Experimental Encephalopathy, Bulletin of Experimental Biology and Medicine, Vol. 144, No. 6, 2007.

» Kennedy D.O., Scholey AB, Tildesley NT, Perry EK, Wesnes KA (2002) Modulation of mood and cognitive performance following

acute administration of Melissa officinalis (lemon balm). Pharmacol Biochem Behav 72: 953- 64.

» Kijowska-Oberc J., Ratajczak E., Fuchs H., Alarm energetyczny nasion, Magazyn Polskiej Akademii Nauk, 1/65/2021.

» Kim D-S., Young-Min Goo, Jinju Cho, Jookyeong Lee, Dong Yeol Lee, Seung Mi Sin,Young Sook Kil, Won Min Jeong, Keon Hee Ko, Ki Jeung Yang,Yun Geun Kim, Sang Gon Kim, Kiseong Kim , Young Jun Kim, Jae Kyeom Kim, Eui-Cheol Shin, Effect of Volatile Organic Chemicals in Chrysanthemum indicum Linné on Blood Pressure and Electroencephalogram, Molecules 2018.

» Klebs R. 1910. Über Bersteineinschlüsse im allgemeinen und die Coleopteren meiner Bernsteinsammlung. Schriften der Physikalisch-Ökonomischen Gesellschaft zu Königsberg 51: 217–242.

» Knight Terrell K., et al. "A new Upper Cretaceous (Santonian) amber deposit from the Eutaw Formation of eastern Alabama, USA." Cretaceous Research 31.1 (2010): 85-93.

» Kobert R., Historische Studien aus dem Pharmakologischen Institute der Kaiserlichen Universität Dorpat: Band 3, 1893.

» Kohlmünzer, S. (2007). Farmakognozja: podręcznik dla studentów farmacji. Państwowy Zakład Wydawnictw Lekarskich.

» Kong D., Meng, Y., & McKenna, G. B. (2022). Determination of the molecular weight between cross-links for different ambers: Viscoelastic measurements of the rubbery plateau. Polymer Engineering & Science, 62(4), 1023-1040.

» Kosmowska-Ceranowicz B., Bursztyn w Polsce i na świecie , Wydawnictwo Uniwersytetu Warszawskiego, Warszawa 2017.

» Koszarski L., Wieser T., Nowe horyzonty tufowe w starszym paleogenie Karpat fliszowych, Geological Quarterly, 1960.

» Kościuk M., Tarasiuk, I., Czurak, A., Szydlik, J., Perłowski, J., Torbicz, G., Naliwajko S., Markiewicz-

» Żukowska R., Bartosiuk E., Borawska, M. 2015: Aktywność przeciwutleniająca wybranych owoców

» egzotycznych. Bromatologia i Chemia Toksykologiczna, 3, 407-411

» Kouipou Toghueo R.M., Endophytes from Gingko biloba: the current status, Phytochem Rev,2020.

» Koul S., Arshad Ahmad, Anu Chaudhary, A. Pandurangan, A Mini Review on Chemistry and Biology of Araucaria Angustifolia (Araucariaceae), /International Journal of Pharmaceutical Sciences Letters 2015 Vol. 5.

» Kühn C.G., Claudii Galeni Opera Omnia. Volume 13. Georg Olms, Hildesheim; 1965: 86.

» Kowczyk-Sadowy M., Obidziński S., Joka M., Piekut J. 2016: Ocena zawartości związków fenolowych i

» aktywności wody w wybranych warzywach poddanych suszeniu. Inżynieria Przetwórstwa Spożywczego, 2,

» 19-23

» Koziorowska L., Badania nieorganicznego składu chemicznego bursztynu, Archeologia Polski, 1984.

» Kumar K.J., Sonnathi, S., Anitha, C., & Santhoshkumar, M. (2015). *Eucalyptus* oil poisoning. Toxicol Int, 22(1), 170-1.

» Kumar M.S., Subramaniyan Kumar, Boobalan Raja, Antihypertensive and Antioxidant Potential of Borneol-A Natural Terpene in LNAME – Induced Hypertensive Rats, International Journal of Pharmaceutical & Biological Archives 2010.

» Kumar R., Bukowski MJ, Wider JM, Reynalds CA, Calo L, Bradley L, et al. Mitochondrial dynamics following global cerebral ischemia. Mol Cell Neurosci, 2016.

» Kusiak A., Kędzia, A., Molęda-Ciszewska, B., Kędzia, A. W., Maciejewska, K., Włodarkiewicz, A., & Kwapisz, E. (2010). Działanie olejku z mięty pieprzowej na bakterie beztlenowe. Dent. Med. Probl, 47(3), 334-338.

» Kvaček Z., (2010) Forest flora and vegetation of the European early Palaeogene—a review. Bull Geosci 85:3–16.

» Ladiges P. Y., Michael J. Bayly, and Gareth J. Nelson. "East-west continental vicariance in *Eucalyptus* subgenus *Eucalyptus*." Beyond cladistics: the branching of a paradigm, 2010.

» Lahlou S., André Fernandes Figueiredo, Pedro Jorge Caldas Magalhães, and José Henrique Leal-Cardoso, Cardiovascular effects of 1,8-cineole, a terpenoid oxide present in many plant essential oils, in normotensive rats, Can. J. Physiol. Pharmacol. Vol. 80, 2002.

» Lambert J. B., Santiago-Blay, J. A., & Anderson, K. B. (2008). Chemical signatures of fossilized resins and recent plant exudates. Angewandte Chemie International Edition, 47(50), 9608-9616.

» Lambert J. B., Santiago-Blay, J. A., Wu, Y., & Levy, A. J. (2015). The History and Structure of Stantienite. Bulletin for the History of Chemistry.

» Langenheim J.H., Biology of Amber-Producing Trees: Focus on Case Studies of Hymenaea and Agathis, Amber, Resinite, and Fossil Resins ACS Symposium Series; American Chemical Society: Washington, DC, 1995.

» Langenheim, J. H., & Beck, C. W. (1968). Catalogue of infrared spectra of fossil resins (ambers) I North and South America. Botanical Museum Leaflets, Harvard University, 22(3), 65-120.

» Larson, S.G., 1978. Baltic amber—a paleobiological study, Entomonograph Volume 1, Scandinavian Science Press, Klampenborg, 192 pp.

» Lázaro J. J., Jiménez, A., Camejo, D., Iglesias-Baena, I., Martí, M. D. C., Lázaro-Payo, A., ... & Sevilla, F. (2013). Dissecting the integrative antioxidant and redox systems in plant mitochondria. Effect of stress and S-nitrosylation. Frontiers in plant science, 4, 460.

» Lee C. K., Cheng Y. S. (2001), Diterpenoids from the leaves of Juniperus chinensis var. kaizuka. J. Nat. Prod. 64, 511 – 514.

» Lee K.H., Roh H.J., Kang K.S., Abietic acid isolated from pine resin (Resina Pini) enhances angiogenesis in HUVECs and accelerates cutaneous wound healing in mice, Journal of Ethnopharmacology, 2017.

» Lee S-J., Seung Yuan Lee, Sun Jin Hur, Young-Il Bae, Chang-Ho Jeong, Neuroprotective and antioxidant effects of *Metasequoia glyptostroboides* leaf extract, Current Topics in Nutraceutical Research, 2016.

» Lei C., Jianyu, S., Lin, L., Bing, L., & Wang, L. (2011). A new source of natural D-borneol and its characteristic. Journal of Medicinal Plants Research, 5(15), 3440-3447.

» Legalov A. A. (2018). Annotated key to weevils of the world. Part 1. Families Nemonychidae, Anthribidae, Belidae, Ithyceridae, Rhynchitidae, Brachyceridae and Brentidae. Ukrainian Journal of Ecology, 8(1), 780-831.

» Legalov A. A., Nazarenko, V. Y., & Perkovsky, E. E. (2018). A new genus of fungus weevils (Coleoptera: Anthribidae) in Rovno amber. Fossil Record, 21(2), 207-212.

» Lenda M., Skórka P. Orzech włoski *Juglans regia*–nowy, potencjalnie inwazyjny gatunek w rodzimej florze. Chrońmy Przyrodę Ojczystą, 2009, 65.4: 261-270.

» Levchyk N.V. et al., Biological activity of aqueous solution of amber, Biotechnolog. Act., 2017.

» Levin B. E., Dunn-Meynell, A. A., & Routh, V. H. (1999). Brain glucose sensing and body energy homeostasis: role in obesity and diabetes. American Journal of Physiology-Regulatory, Integrative and Comparative Physiology, 276(5), R1223-R1231.

» Lewis W., An experimental history of the materia medica, or of the natural and artificial substances made use of in medicine, etc, London, 1761.

» Limchoowong N. et al., A green extraction of trace iodine in table salts, vegetables, and food products prior to analysis by inductively coupled plasma optical emission spectrometry, Journal of the Brazilian Chemical Society, 2017.

» Liu X., Cai, W., Niu, M. et al. Plumbagin induces growth inhibition of human glioma cells by downregulating the expression and activity of FOXM1. J Neurooncol 121, 469–477 (2015).

» Liu Y., Li, P., Yang, J., Wang, F., Kim, E., ... & Tu, Y. (2021). Chemical characterization of Wuyi rock tea with different roasting degrees and their discrimination based on volatile profiles. RSC advances, 11(20), 12074-12085.

» Liu Y., Shenshen Yang,Kailong Wang,Jia Lu,Xiaomei Bao,Rui Wang,Yuling Qiu,Tao Wang,Haiyang Yu Cellular senescence and cancer: Focusing on traditional Chinese medicine and natural products, Cell Proliferation, 2020;53.

» Lu Y., Y. Hautevelle, R. Michels, Determination of the molecular signature of fossil conifers by experimental palaeochemotaxonomy – Part 1: The Araucariaceae family, Biogeosciences, 2013.

» Luo Y., Siqi Zhou, Haruna Haeiwa, ReikoTakeda, Kazuma Okazaki, Marie Sekit, TakuyaYamamoto, MikioYamano, Kazuichi Sakamoto, Role of amber extract in protecting SHSY5Y cells against amyloid []1-42-induced neurotoxicity, Biomedicine & Pharmacotherapy Volume 141, September 2021.

» Lyons Paul C., Maria Mastalerz, and William H. Orem. "Organic geochemistry of resins from modern *Agathis australis* and Eocene

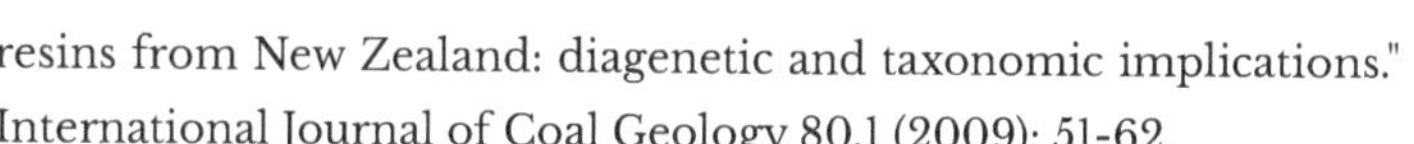

resins from New Zealand: diagenetic and taxonomic implications." International Journal of Coal Geology 80.1 (2009): 51-62.

» Łozińska-Stępień, H., Rytel, A. and Saliński, P., 1986. Szczegółowe objaśnienia do mapy geologicznej Polski, Arkusz Leszkowice. Wydawnictwo Geologiczne, Warszawa.

» MacPhail M, Carpenter RJ. 2013. New potential nearest living relatives for Araucariaceae producing fossil Wollemi Pine-type pollen. Alcheringa 38:135-139.

» Maidanovich I. A. Geology and genesis of amber-bearing deposits of the Ukrainian Polissya / I. A. Maidanovich, Makarenko D. E.. - Kyiv: Naukova Dumka, 1988. - 84 p.

» Malm A., Olejki eteryczne w profilaktyce I leczeniu chorób infekcyjnych, PZWL, Warszawa 2021.

» Małka A., Kramarska R. (2013). The mining of Baltic amber deposits in Poland-an overview. In International Amber Researcher Symposium. Gdańsk, Poland (pp. 22-23).

» Manlio G. G., Luminare majus, opus eximium quod luminare dicitur Medicis et Aromatariis perque necessarium. – Lugduni: de Gabiano, 1536.

» Mannapova R.T., Rapiev R.A., Correction of the biochemical status of the organism airions of amber, UDC, 2013.

» Marreiros B.C., Filipa Calisto, Paulo J. Castro, Afonso M. Duarte, Filipa V. Sena, Andreia F. Silva, Filipe M. Sousa, Miguel Teixeira, Patrícia N. Refojo, Manuela M. Pereira, Exploring membrane respiratory chains, Biochimica et Biophysica Acta (BBA) -Bioenergetics, Volume 1857, Issue 8, 2016.

» Martín-González A., Wierzchos, J., Gutiérrez, J. C., Alonso, J., & Ascaso, C. (2009). Double fossilization in eukaryotic microorganisms from Lower Cretaceous amber. BMC biology, 7(1), 1-11.

» Martins C. B. C., Paulo H. G. Zarbin, Volatile Organic Compounds of Conspecific-Damaged *Eucalyptus benthamii* Influence Responses of Mated Females of Thaumastocoris peregrinus, J Chem Ecol (2013).

» Marusik Y., & Penney, D. (2003, August). A survey of Baltic amber Theridiidae (Araneae) inclusions, with descriptions of six new species. In Proceedings of the 21st European Colloquium of Arachnology (Vol. 4, p. 9).

» Marusik Y. M., & Wunderlich, J. (2008). A survey of fossil Oonopidae (Arachnida: Aranei). Arthropoda Selecta. 17(1-2), 65-79.

» Matsui W. M. From oleoresin-coniferous resin to amber-succinite. Bulletin of the National Natural Science Museum, 2010.

» Matsui V. M. (2013). Geological nature and parent primary sources of the European placers of amber-succinite. Bulletin of the National Science and Nature Museum, (11), 49-54.

» Matsui W.M.,Naumenko U.Z., Transition of plant resins from the biosphere to the lithosphere, GEO&BIO, 2020.

» Matuszewska A., Bursztyn bałtycki i inne żywice kopalne w świetle badań fizykochemicznych, Przegląd Geologiczny 57, 2002.

» Matuszewska A., Bursztyn (sukcynit), inne żywice kopalne, subfosylne i współczesne, Oficyna Wydawnicza WW, Uniwersytet Śląski, 2010.

» Matuszewska A., Chemiczna budowa bursztynu (sukcynitu), Amberif, Badania inkluzji i innych właściwości bursztynu, 2012.

» Matuszewska A., Characteristics of chemical composition of amber distillation products-preliminary results of investigations, Prace Muzeum Ziemi Nr 47, Polska Akademia Nauk, 2004.

» Matuszewska, A. Chemiczna budowa bursztynu (sukcynitu). AMBERIF 2012, 3.

» Matuszewska A., Zarys historii badań kwasu bursztynowego w bursztynie bałtyckim, Amberif, Gdańsk 2016.

» Matuszewska A. Pochodzenie żywic kopalnych The origin of fossil resins. AMBERIF 2017, 11.

» Matuszewska A., Czaja M., Aromatic compounds in molecular phase of Baltic amber synchronous luminescence analysis, Talanta, 2002.

» McAlpine J.F., Martin, J.E.H., 1966. Systematics of Sciadoceridae and relatives with descriptions of two new fenera and species from Canadian amber and erection of family Ironomyiidae (Diptera: Phoroidea). The Canadian Entomologist 98, 527–544.

» doi:10.4039/Ent98527-5

» McGilevery C., and Reed, J. (1993). Aroma Therapy. Ultimate Editions, London.

» McKellar R. C., Chatterton, B. D., Wolfe, A. P., & Currie, P. J. (2011). A diverse assemblage of Late Cretaceous dinosaur and bird feathers from Canadian amber. Science, 333(6049), 1619-1622.

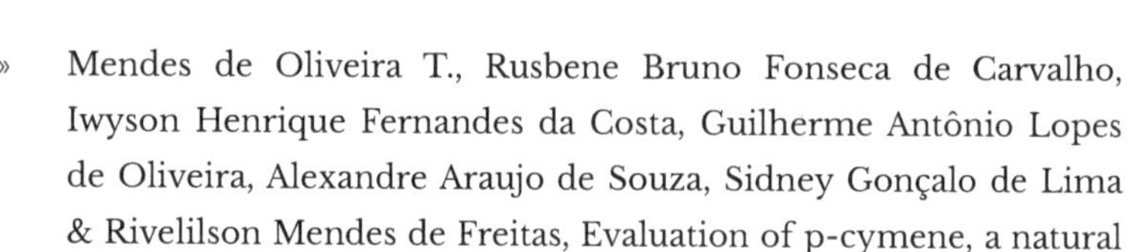

» Mendes de Oliveira T., Rusbene Bruno Fonseca de Carvalho, Iwyson Henrique Fernandes da Costa, Guilherme Antônio Lopes de Oliveira, Alexandre Araujo de Souza, Sidney Gonçalo de Lima & Rivelilson Mendes de Freitas, Evaluation of p-cymene, a natural antioxidant, Pharmaceutical Biology 2015.

» Mesue, Masawaih al-Mardini Yahya ibn; ok. 925–1015.

» Mikucioniene D. et al., Preliminary investigation into the antimicrobial activity of an electrospun polyamide nanofibrous web with micro particles of baltic amber, FIBRES & TEXTILES in Eastern Europe, 2016.

» Mildenhall D. C., Kennedy, E. M., Lee, D. E., Kaulfuss, U., Bannister, J. M., Fox, B., & Conran, J. G. (2014). Palynology of the early Miocene Foulden Maar, Otago, New Zealand: Diversity following destruction. Review of Palaeobotany and Palynology, 204, 27-42.

» Millar A.H., Valentina Mittova, Guy Kiddle, Joshua L. Heazlewood, Carlos G. Bartoli, Frederica L. Theodoulou, Christine H. Foyer, Control of Ascorbate Synthesis by Respiration and Its Implications for Stress Responses, Plant Physiology, Volume 133, Issue 2, October 2003, Pages 443–447.

» Mills J. S., White, R., & Gough, L. J. (1984). The chemical composition of Baltic amber. Chemical Geology, 47(1-2), 15-39.

» Moharrek F., Sanmartin, I., Kazempour-Osaloo, S., & Nieto Feliner, G. (2019). Morphological innovations and vast extensions of mountain habitats triggered rapid diversification within the species-rich Irano-Turanian genus Acantholimon (Plumbaginaceae). Frontiers in genetics, 9, 698.

» Moshkov N.N. The unknown about the known. Amber is a source of beauty, health and longevity. – Kaliningrad, 2007.

» Mosini V., Forcellese, M. L., & Nicoletti, R. (1980). Presence and origin of volatile terpenes in succinite. Phytochemistry, 19(4), 679-680.

» Mullaicharam A.R., A Review on Evidence Based Practice of *Ginkgo biloba* in Brain Health, 2013.

» Murata S., Risa Shiragami, Chihiro Kosugi, Tohru Tezuka, Masato Yamazaki, Atsushi Hirano, Yukino Yoshimura, Masato Suzuki, Kiyohiko Shuto, Nobuhiro Ohkohchi, Keiji Koda, Antitumor effect of 1, 8-cineole against colon cancer, Oncology Reports, December 2013 Volume 30 Issue 6.

» Nanko M., "Definitions and categories of hybrid materials" AZojomo, 6 (2009), 1–8.

» Naumenko U. Z., & Matsui, V. M. (2020). Stages of geological devepment and genetic type of the original primary source in amber-succinite plasters of Ukraine. Geological journal, (4), 76-85.

» Nissen M. D., Lau, E. T., Cabot, P. J., & Steadman, K. J. (2019). Baltic amber teething necklaces: could succinic acid leaching from beads provide anti-inflammatory effects?. BMC complementary and alternative medicine, 19(1), 1-9.

» Nosova N., & Kiritchkova, A. (2015). New data on the Mesozoic conifer genus Sciadopityoides Sveshnikova (Miroviaceae). Review of Palaeobotany and Palynology, 221, 1-21.

» Nowak G., G. L. Clifton, D. Bakajsova, Succinate ameliorates energy deficits and prevents dysfunction of complex I in injured renal proximal tubular cells, J Pharmacol Exp Ther. 2008 Mar; 324(3): 1155–1162.

» Ogunlesi M, Okiei W, Osibote EA. Analysis of the essential oil from the leaves of Sesamum radiatum, a potential medication for male infertility factor, by gas Chromatography – mass spectrometry. Afr J Biotechnol 2010.

» Ogunsuyi H. O., & Daramola, B. M. (2013). Evaluation of almond (Prunus amygdalus) seed oil as a viable feedstock for biodiesel fuel. Mercy Derkyi Esi Awuah Daniel Obeng-Ofori Nana Sarfo Agyemang Derkyi Fred Owusu-Ansah, 18(4), 90.

» Ogunwande, I. A., Flamini, G., Adefuye, A. E., Lawal, N. O., Moradeyo, S., & Avoseh, N. O. (2011). Chemical compositions of *Casuarina* equisetifolia L., *Eucalyptus* toreliana L. and Ficus elastica Roxb. ex Hornem cultivated in Nigeria. South African Journal of Botany, 77(3), 645-649.

» Ono M., M. Yamamoto, C. Masuoka, Y. Ito, M. Yamashita, and T. Nohara, J. Nat. Prod., 1999, 62, 1532-1537; (b) S.-Y. Wang, J.-H. Wu, L.-F. Shyur, Y.-H. Kuo and S.-T. Chang, Holzforschung, 2002.

» Oskolski A.A., X.X. Feng, J.H. Jin Myrtineoxylon gen. nov.: The first fossil wood record of the tribe Myrteae (*Myrtaceae*) in eastern Asia Taxon, 62 (2013), pp. 771-778.

» Osmólski T. (1963). Związek procesu powstawania złóż siarki w miocenie zapadliska przedkarpackiego z litologią ich podłoża. Geological Quarterly, 7(3), 439-445.

» Otto A., & Simoneit, B. R. (2001). Chemosystematics and diagenesis of terpenoids in fossil conifer species and sediment from the Eocene Zeitz formation, Saxony, Germany. Geochimica et Cosmochimica Acta, 65(20), 3505-3527.

» Owemidu I.O., Abayomi Abraham Taiwo, Elizabet Ojochebo Dangana, Caleb Achile Jonah, Muhammed Nafiu Negedu, Haruna Ozovehe Suleiman, Saidi Odoma, Anti-nociceptive activity of the ethanol extract of *Eucalyptus* globulus leaf in experimental animals, Journal of Pharmacognosy and Phytochemistry 2020.

» Öz, M., Deniz, İ., Okan, O. T., & Fidan, M. S. (2015). Chemical Composition of Oleoresin and Larvae Gallery Resin of Pinus Brutia Attacked by Dioryc tria Sylvestrella Ratz. Drvna industrija, 66(3), 179-188.

» Panda M, Tripathi SK, Biswal BK. Plumbagin promotes mitochondrial mediated apoptosis in gefitinib sensitive and resistant A549 lung cancer cell line through enhancing reactive oxygen species generation. Mol Biol Rep. 2020.

» Park J, Kim WJ, Kim W, Park C, Choi CY, Cho JH, Kim SJ, Cheong H, Antihypertensive Effects of Dehydroabietic and 4-Epi-Trans-Communic Acid Isolated from Pinus densiflora, Journal of Medicinal Food, 01 Jan 2021.

» Parrish J. T., Daniel, I. L., Kennedy, E. M., & Spicer, R. A. (1998). Paleoclimatic significance of mid-Cretaceous floras from the middle Clarence Valley, New Zealand. Palaios, 13(2), 149-159.

» Pasaribu G., Winarni, I., Gusti, R. E. P., Maharani, R., Fernandes, A., Harianja, A. H., ... & Kholibrina, C. R. (2021). Current Challenges and Prospects of Indonesian Non-Timber Forest Products (NTFPs): A Review. Forests, 12(12), 1743.

» Passera F.,Il Nuovo Tesoro degl' Arcani Farmacologici Galenici, & Chimici, ò Spagirici. – Venezia: Giovanni Parè, 1688.

» Pastorelli, G., Shashoua, Y., & Richter, J. (2013). Hydrolysis of Baltic amber during thermal ageing–An infrared spectroscopic approach. Spectrochimica Acta Part A: Molecular and Biomolecular Spectroscopy, 106, 124-128.

» Pattanayak P., Behera, P., Das, D., and Panda, S. K. (2010). Ocimum sanctum Linn. A reservoir plant for therapeutic applications: an overview. Pharmacogn. Rev.

» Pederneiras, L. C., Gaglioti, A. L., Romaniuc-Neto, S., & Mansano, V. D. F. (2018). The role of biogeographical barriers and bridges in determining divergent lineages in Ficus (Moraceae). botanical Journal of the Linnean Society, 187(4), 594-613.

» Peñalver E., Labandeira, C.C., Barrón, E., Delclòs, X., Nel, P., Nel, A., Tafforeau, P., Soriano, C., 2012. Thrips pollination of Mesozoic gymnosperms. Proceedings of the National Academy of Sciences 109, 8623–8628.

» Penney D., & Preziosi, R. F. (2013). Estimating fossil ant species richness in Eocene Baltic amber. Acta Palaeontologica Polonica, 59(4), 927-929.

» Pereira J., The Elements of materia medica and therapeutics v.2, 1854: Volume 2, Blanchard and Lea, 1854.

» Perkovsky E.E .(2018). Only a half of Species of Hymenoptera in Rovno Amber Fauna is Common with Baltic Amber. Vestnik Zoologii 52: 353–360. https://doi.org/10.2478/vzoo2018-0037.

» Perkovsky E.E., Zosimovich, V.Yu. & Vlaskin, A.P. 2010. Rovno Amber. In: Penney, D. (ed.), Biodiversity of fossils in amber from the major world deposits. Siri Scientific Press, Manchester: 116–136.

» Pharmacopœia Officinalis & Extemporanea. Complete English Dispensatory, John Quincy, London, 1733.

» Philips R., Martyn Rix: The Botanical Garden. Vol. 1. Trees and shrubs. London: Macmillan, 2002, s. 24.

» Pichersky E., & Raguso, R. A. (2018). Why do plants produce so many terpenoid compounds?. New Phytologist, 220(3), 692-702.

» Pillon Y. (2012). Time and tempo of diversification in the flora of New Caledonia. Botanical Journal of the Linnean Society, 170(3), 288-298.

» Pisulewska E., Z Janeczko, Krajowe rośliny olejkowe, Wydawnictwo Know- How Piotr Kaczmarczyk, Kraków, 2008.

» Plinii C. Secundi, Naturalis Historiae tomus Primus, Apud Hackios, 1669.

» Plößer J., Martin Lucas, Peter Claus, Highly selective menthol synthesis by one-pot transformation of citronellal using Ru/H-BEA catalysts, Journal of Catalysis, Volume 320, 2014.

» Pluznick J.L., Renal and cardiovascular sensory receptors and blood pressure regulation, American Journal of Physiology Renal Physiology, 2013.

» Poinar G. O. (1992). Life in amber. Stanford University Press.

» Poinar G.O., Milki, R. K. (2001): Lebanese Amber. The Oldest Insect Ecosystem in Fossilized Resin. – 96 pp., Corvallis ,Oregon State University Press.

» Porres-Martínez M.; González-Burgos, E.; Carretero, M.E.; Gómez-Serranillos, M.P. In vitro neuroprotective potential of the monoterpenes α-pinene and 1,8-cineolc against H2O2 -induced oxidative stress in PC12 cells. Z. Naturforsch. C. 2016, 71, 191–199.

» Poulin J., Helwig K., Inside Amber: The structural role of succinic acid in class Ia and class Id resinite, Analytical Chemistry, 2014.

» Powolny A.A., Singh, S.V. Plumbagin-induced Apoptosis in Human Prostate Cancer Cells is Associated with Modulation of Cellular Redox Status and Generation of Reactive Oxygen Species. Pharm Res 25, 2171–2180 (2008).

» Pozdnyakov D.I., Denis S., Zolotych D.S.,Larsky M.V., Correction of mitochondrial dysfunction by succinic acid derivatives under experimental cerebral ischemia conditions, 2021.

» Quintans-Júnior L., José C. F. Moreira, Matheus A. B. Pasquali, Soheyla M. S. Rabie, André S. Pires, Rafael Schröder, Thallita K. Rabelo, João P. A. Santos, Pollyana S. S. Lima, Sócrates C. H. Cavalcanti, Adriano A. S. Araújo, Jullyana S. S. Quintans, and Daniel P. Gelain, Antinociceptive Activity and Redox Profile ofthe Monoterpenes (+)-Camphene, p-Cymene, and Geranyl Acetate in Experimental Models, Hindawi Publishing Corporation ISRN Toxicology, Volume 2013.

» Quintans-Júnior L. J., A. G. Guimarães, B. E. S. Araújo et al., "Carvacrol, (−)-borneol and citral reduce convulsant activity in rodents," African Journal of Biotechnology, vol. 9, no. 39, pp. 6566–6572, 2010.

» Rad M.N., Iraj Jafari Anarkooli, Alireza Abdanipour, Neuroprotective Effects of 1,8-Cineole on Apoptosis Inhibition and Bcl-2/Caspase-3 mRNA Expression in the Hippocampus of Epileptic Pilocarpine Model Rats, J Adv Med Biomed Res 2022.

» Radchenko, A. G., & Perkovsky, E. E. (2020). New finds of the fossil ant genus *Prionomyrmex* Mayr (Hymenoptera, Formicidae, *Myrmeciinae*) in late Eocene European amber. Paleontological Journal, 54(6), 617-626.

» Ragazzi E., Amber, a Stone of Sun for Ancient Medicines, Acta medico-historica Rigensia (2016) X:208-234 DOI: 10.25143/amhr.2016.X.11.

» Raine JI, Mildenhall DC, Kennedy EM. 2006. New Zealand fossil spores and pollen: an illustrated catalogue, GNS science miscellaneous series (2nd ed). vol. 4.

» Rao V., Flavia Almeida Santos, Antiinflammatory and antinociceptive effects of 1,8-cineole a terpenoid oxide present in many plant essential oils, PhytotherapyResearch, July 2000.

» Rana N., Bais S., Neuroprotective effect of J. *communis* in Parkinson disease induced animal models, MS thesis in 6 International Scholarly Research Notices Pharmacy, Pharmacology Department, Punjab Technical University, Punjab, India, 2014.

» Rasnitsyn A. P., Bashkuev, A. S., Kopylov, D. S., Lukashevich, E. D., Ponomarenko, A. G., Popov, Y. A., ... & Vorontsov, D. D. (2016). Sequence and scale of changes in the terrestrial biota during the Cretaceous (based on materials from fossil resins). Cretaceous Research, 61, 234-255.

» Rehman N. U., Alsabahi, J. N., Alam, T., Khan, A., Rafiq, K., Khan, M., & Al-Harrasi, A. (2021). chemical constituents and carbonic anhydrase II activity of essential oil of Acridocarpus orientalis A. Juss. in comparison with stem and leaves. Journal of Essential Oil Bearing Plants, 24(1), 68-74.

» Rappsilber I., Krumbiegel G., & Wimmer R., Overview of Bitterfeld amber. In The International Amber Researcher Symposium, 2013.

» Richter G.A., Ausführliche Arzneimittellehre: Handbuch für praktische Aerzte, Band 2, 1827.

» Riddle J.M. Amber in ancient Pharmacy. The Transmission of Information About a Single Drug. A Case Study. "Pharmacy in History" 15 (1); 1973: 3-17.

» Rikkinen J., Dörfelt, H., Schmidt, A. R., & Wunderlich, J. (2003). Sooty moulds from European Tertiary amber, with notes on the systematic position of Rosaria ('Cyanobacteria'). Mycological Research, 107(2), 251-256.

» Ritzkowski S. (1997). K-ar-altersbestimmungen der bernsteinführenden sedimente des samlandes (paläogen, bezirk kaliningrad). Metalla (Sonderheft), 66, 19-23.

» Roghi G., Ragazzi, E., Gianolla, P., Coppellotti, O., & Fedele, P. (2013). Sull'ambra triassica della dolomite a 100 anni dalla segnalazione di Ernst Koken, (1913-2013). Frammenti–Conoscere e tutelare la natura Bellunese, 5, 53-64.

» Rolim T. L., Meireles, D. R. P., Batista, T. M., De Sousa, T. K. G., Mangueira, V. M., De Abrantes, R. A., ... & Sobral, M. V. (2017). Toxicity and antitumor potential of Mesosphaerum sidifolium (Lamiaceae) oil and fenchone, its major component. BMC complementary and alternative medicine, 17(1), 1-12.

» Romero-Estrada A., Maldonado-Magaña, A., González-Christen, J. et al. Anti-inflammatory and antioxidative effects of six pentacyclic triterpenes isolated from the Mexican copal resin of Bursera copallifera. BMC Complement Altern Med 16, 422 (2016).

» Rowell M. V., Jordan G. J., & Barnes R. W. (2001). An in situ, late Pleistocene Melaleuca fossil forest at Coal Head, western Tasmania, Australia. Australian Journal of Botany, 49(2), 235-244.

» Rudloff E. V. (1961). Gas–liquid chromatography of terpenes: part IV. The analysis of the volatile oil of the leaves of eastern white cedar. Canadian Journal of Chemistry, 39(6), 1200-1206.

» Różański S., Z dziejów wschodnio-pruskich regaliów bursztynowych, Komunikaty Mazursko-Warmińskie nr 2, 180-197, 1959.

» Ryka W., Problem powstania tarnobrzeskiego złoża siarki rodzimej w świetle badań petrograficznych, Kwartalnik Geologiczny, t. 30, nr 3/4, 1986 r., str. 473 -498.

» Sadgrove NJ, Senbill H, Van Wyk BE, Greatrex BW. New Labdanes with Antimicrobial and Acaricidal Activity: Terpenes of Callitris and Widdringtonia (*Cupressaceae*). Antibiotics (Basel). 2020.

» Sadowska, A., Świderski, F., Kromołowska, R. 2011: Polifenole - źródło naturalnych przeciwutleniaczy.

» Postępy Techniki Przetwórstwa Spożywczego, 108-111.

» Sadowski E.-M., Schmidt A.R., Seyfullah L.J., Kunzmann L. 2017a. Conifers of the 'Baltic amber forest' and their palaeoecological significance. Stapfia 106, 1–73.

» Sadowski E.-M., Schmidt A. R. & Denk T. 2020: Staminate inflorescences with in situ pollen from Eocene Baltic amber reveal high diversity in *Fagaceae* (oak family). – Willdenowia 50: 405–517. doi: https://doi.org/10.3372/wi.50.50303

» Sadowski E. M., Seyfullah, L. J., Regalado, L., Skadell, L. E., Gehler, A., Gröhn, C., ... & Schmidt, A. R. (2019). How diverse were ferns in the Baltic amber forest?. Journal of Systematics and Evolution, 57(4), 305-328.

» Sánchez-Navas A., et al. Color, mineralogy and composition of Upper Jurassic West Siberian glauconite: useful indicators of paleoenvironment. The Canadian Mineralogist, 2008, 46.5: 1249-1268.

» Saint Martin J. P., Dutour, Y., Ebbo, L., Frau, C., Mazière, B., Néraudeau, D., ... & Valentin, X. (2021). Reassessment of amber-bearing deposits of Provence, southeastern France. BSGF-Earth Sciences Bulletin, 192(1), 5.

» Santana M. F., Quintans-Júnior, L. J., Cavalcanti, S. C., Oliveira, M. G., Guimarães, A. G., Cunha, E. S., ... & Bonjardim, L. R. (2011). p-Cymene reduces orofacial nociceptive response in mice. Revista Brasileira de Farmacognosia, 21, 1138-1143.

» Santos M.R.V., Flávia V. Moreira, Byanka P. Fraga, Damião P. De Sousa, Leonardo R. Bonjardim, Lucindo J. Quintans-Junior, Cardiovascular effects of monoterpenes:

» a review, Revista Brasileira de Farmacognosia Brazilian Journal of Pharmacognosy 21(4): 764-771, Jul./Aug. 2011.

» Savkevich S. S., 1983. Processes of amber and some amber-like resins transformation depending on conditions of their formation and natural environment. Proceedings of the USSR Academy of Sciences, Series Geology, 12: 96–106.

» Schmidt A., Sadowski, E. M., Kaasalainen, U., & Rikkinen, J. K. (2019). A botanical view of the 'Baltic amber forest': new evidence from seed plants, lichens and fungi. Baltic amber (succinite): an intriguing resin.

» Schmidt A. R., & Schäfer, U. (2005). Leptotrichites resinatus new genus and species: a fossil sheathed bacterium in Alpine Cretaceous amber. Journal of Paleontology, 79(1), 175-184.

» Schmidt A. R., Kaulfuss, U., Bannister, J. M., Baranov, V., Beimforde, C., Bleile, N., ... & Lee, D. E. (2018). Amber inclusions from New Zealand. Gondwana Research, 56, 135-146.

» Sebti M., Lahouel, M., & Zellagui, A. (2020). Phytochemical and pharmacological study of four aromatic plants growing in northeast of Algeria. Algerian Journal of Environmental Science and Technology, 6(3).

» Selden P.A. & Nudds, J.R. 2004. Evolution of Fossil Ecosystems. Manson Publishing, London, 160 pp.

» Selden P.A., & Ren, D. (2017). A review of Burmese amber arachnids. The Journal of Arachnology, 45(3), 324-343.

» Sembratowicz I., Rusinek-Prystupa E. 2015: Zawartość substancji bioaktywnych w owocach

» pozyskiwanych z upraw ekologicznych i konwencjonalnych. Problemy Higieny i Epidemiologii, 96 (1),

» 259-263.

» Seyfullah L., Christina Beimforde, Jacopo Dal Corso, Alexander R. Schmidt, Jouko Rikkinen, Vincent Perrichot, Production and preservation of resins - past and present, May 2018, Biological Reviews.

» Shimizu E. , Shimoda, N. , Kawamura, T. , Ueda, N. and Kimura, K. (2020) Comparison of the Biological Activity and Constituents in Japanese Ambers. Advances in Biological Chemistry.

» Sichone K. (2013). Pyrolysis of sawdust (Doctoral dissertation, University of Waikato.

» Siedow J. N., & Umbach, A. L. (1995). Plant mitochondrial electron transfer and molecular biology. The Plant Cell, 7(7), 821.

» Simutnik S.A., Evgeny E. Perkovsky, Dmitry V. Vasilenko, First record of Leptoomus janzeni Gibson (Hymenoptera, Chalcidoidea) from Rovno amber, Journal of Hymenoptera Research 80: 137–145, 2020.

» Słodkowska, B., & Kasiński, J. R. (2016). Paleogen i neogen: czas dynamicznych zmian klimatycznych. Przegląd Geologiczny, 64(1).

» Słodkowska B., Kramarska R., Kasiński J. R., (2013, March). The Eocene climatic optimum and the formation of the Baltic amber deposits. In The international amber researcher symposium."Amber. Deposits-Collections-The Market". Gdansk, Poland: Gdansk International fair Co. Amberif (pp. 28-32).

» Sogo E., Zhou, S.; Haeiwa, H.; Takeda, R.; Okazaki, K.; Sekita, M.; Yamamoto, T.; Yamano, M.; Sakamoto, K. Amber Extract Reduces Lipid Content in Mature 3T3-L1 Adipocytes by Activating the Lipolysis Pathway. Molecules 2021.

» Sokoloff D. D., Ignatov, M. S., Remizowa, M. V., Nuraliev, M. S., Blagoderov, V., Garbout, A., & Perkovsky, E. E. Staminate flower of Prunus s. l.(Rosaceae) from Eocene Rovno amber (Ukraine), 2018.

» Son K.-H., H.-M. Oh, S.-K. Choi, D. C. Han and B.-M. Kwon, Bioorg. Med. Chem. Lett., 2005, 15, 2019-2021.

» Sonibare O. O., Agbaje, O. B., Jacob, D. E., Faithfull, J., Hoffmann, T., & Foley, S. F. (2014). Terpenoid composition and origin of amber from the Cape York Peninsula, Australia. Australian Journal of Earth Sciences, 61(7), 979-985.

» Sontag E. 2003. Animal inclusions in a sample of unselected Baltic amber. Acta Zoologica Cracoviensia 46 (Supplement: Fossil Insects): 431–440.

» Sontag E., Szadziewski R. 2011. Biting midges (Diptera: Ceratopogonidae) in Eocene Baltic amber from the Rovno region (Ukraine). Polish Journal of Entomology 80: 779-800.

» Soodi M, Naghdi N, Hajimehdipoor H, Choopani S, Sahraei E. Memory-improving activity of Melissa officinalis extract in naïve and scopolamine-treated rats. Res Pharm Sci. 2014;9(2):107-114.

» Sousa, V., Ferreira, J., Miranda, I., Quilhó, T., & Pereira, H. (2021). Quercus rotundifolia Bark as a Source of Polar Extracts: Structural and Chemical Characterization. Forests, 12(9), 1160.

» Staccioli G., Mellerio G., Alberti M. B. (1993) Investigation on terpene-related hydrocarbons from a Pliocenic fossil wood. Holzforschung 47, 339–342.

» Stanić I. Fosili Medvednice. Diss. University of Zagreb. Faculty of Science. Department of Biology, 2018.

» Stilwell J. D., Langendam, A., Mays, C., Sutherland, L. J., Arillo, A., Bickel, D. J., ... & Peñalver, E. (2020). Amber from the Triassic to Paleogene of Australia and New Zealand as exceptional preservation of poorly known terrestrial ecosystems. Scientific Reports, 10(1), 1-11.

» Stillé A.,M.D., Therapeutics and Materia Medica: A Systematic Treatise on the Action ..., Tom 2, Philadelphia, 1868.

» Stollberg R. (2013). Groundwater contaminant source zone identification at an industrial and abandoned mining site-a forensic backward-in-time modelling approach.

» Sumsakul W., Plengsuriyakarn, T., Chaijaroenkul, W. et al. Antimalarial activity of plumbagin in vitro and in animal models. BMC Complement Altern Med 14, 15 (2014).

» Sylwan, Zbiór Nauk i Urządzeń Leśnych i Łowieckich, Tom XIV, Warszawa,1838.

» Synoradzki L., Arct J., Safarzyński S., Hajmowicz H., Sobiecka A., Dankowska E., Charakterystyka i zastosowanie bursztynu bałtyckiego w przemyśle farmaceutycznym i kosmetycznym, Przemysł Chemiczny, 91/1/2012.

» Szabó M., Hammel, J. U., Harms, D., Kotthoff, U., Bodor, E., Novák, J., ... & Ősi, A. (2022). First record of the spider family Hersiliidae (Araneae) from the Mesozoic of Europe (Bakony Mts, Hungary). Cretaceous Research, 131, 105097.

» Szajdek, A., Borowska, J. 2004: Właściwości przeciwutleniające żywności pochodzenia roślinnego.

» Żywność Nauka Technologia Jakość, 4 (41), 5-28.

» Szwedo, J., & Kania, I. (2015). Rekonstrukcje klimatyczne na podstawie inkluzji, kzbp.biol.ug.edu.pl.

» Tabarraei H., Hassan, J., Parvizi, M. R., & Golshahi, H. (2019). Evaluation of the acute and sub-acute toxicity of the black caraway seed essential oil in Wistar rats. Toxicology reports, 6, 869-874.

» Tahar S., Bendif H., Zedam A., Flamini G., Maggi F., A new chemotype with high tricyclene content from the essential oil of Salvia aegyptiaca L. growing in Algerian Pre-Sahara, Natural Product Research Formerly Natural Product Letters, 2021.

» Takahashia N., Teruo Kawada, Tsuyoshi Goto, Chu-Sook Kim, Aki Taimatsu, Kahori Egawa, Takayuki Yamamoto, Mitsuo Jisaka, Koji Nishimura, Kazushige Yokota, Rina Yu, Tohru Fushiki, Abietic acid activates peroxisome proliferator-activated receptor-Q (PPARQ) in RAW264.7 macrophages and 3T3-L1 adipocytes to regulate gene expression involved in infammation and lipid metabolism, FEBS Letters 550 (2003).

» Takei M., Tachikawa E., Umeyama A., Dendritic Cells Promoted by Ginseng Saponins Drive a Potent Th1 Polarization, Biomarker Insights, 2008.

» Takei M., Umeyama A., Arihara S., T-cadinol and calamenene induce dendritic cells from human monocytes and drive Th1 polarization, European Journal of Pharmacology, 2006.

» Tappert R., Wolfe, A. P., McKellar, R. C., Tappert, M. C., & Muehlenbachs, K. (2011). Characterizing modern and fossil gymnosperm exudates using micro-Fourier transform infrared spectroscopy. International Journal of Plant Sciences, 172(1), 120-138.

» Tarasevich V. F., & Alekseev, P. I. (2017). Pollen inculusions of flowering plants in Baltic amber.(Kaliningrad region, late Eocene). Silvae Genetica, 41(105), 109.

» Tholl D. (2015). Biosynthesis and biological functions of terpenoids in plants. Biotechnology of isoprenoids, 63-106.

» The College Journal of Medical Science: Volume 4, Wilstach, Keys & Company, Cincinati, 1859.

» Tibballs J., 1995. Clinical effects and management of *eucalyptus* oil ingestion in infants and young children. Medical Journal of Australia 163:177–180.

» Tildesleya N.T.J., Kennedya D.O., Perryb E.K., Ballardc C.G., Savelevd S., Wesnesa K.A., Scholeya A.B., Salvia lavandulaefolia (Spanish Sage) enhances memory in healthy young volunteers, Pharmacology, Biochemistry and Behavior, 2003.

» Trettin R., Gläser, H. R., Schultze, M., & Strauch, G. (2007). Sulfur isotope studies to quantify sulfate components in water of flooded lignite open pits–Lake Goitsche, Germany. Applied Geochemistry, 22(1), 69-89.

» Trifunovic A., Nils-Göran Larsson, Tissue-Specific Knockout Model for Study of Mitochondrial DNA Mutation Disorders. Chandan K. Sen, Lester Packer, Methods in Enzymology, Academic Press, Volume 353, 2002.

» Troncoso Ch., Claudia Perez, Victor Hernandez, Manuel Sanchez-Olate, Darcy Rios, Aurelio San Martin, José Becerra, Induction of Defensive Response in *Eucalyptus* globulus Plants and its Persistence in Vegetative Propagation, Natural Product Communications, 2013.

» Tumiłowicz P., Ludwik Synoradzki, Agnieszka Sobiecka, Jacek Arct, Katarzyna Pytkowska, Sławomir Safarzyński, Bioactivity of Baltic amber – fossil resin, Polimery 2016.

» Tung Y. T., Huang, C. C.; Ho, S. T.; Kuo, Y. H.; Lin, C. C.; Lin, C. T.; Wu, J. H. (2011). "Bioactive phytochemicals of leaf essential oils of Cinnamomum osmophloeum prevent lipopolysaccharide/D-galactosamine (LPS/D-GalN)-induced acute hepatitis in mice". J. Agric. Food Chem. 59.

» Tschirch A. & Aweng, E., Über den Succinit. – Archiv der Pharmacie, 1894.

» Tschirch A., Aweng E., De Jong C., Hermann S., Über den Bernstein, „Helvetica Chimica Acta", t. 6, s. 214-225, 1923: z Longina Koziorowska, Badania nieorganicznego składu chemicznego bursztynu, Archeologia Polska, 1984.

» Tutkovsky P.A. Amber of Kyiv. In: Southwest Region. Popular natural history essays. Vol. 1. (Kyiv, 1893). P. 12—18).

» Ueno H., Atsumi Shimada, Shunsuke Suemitsu, Shinji Murakami, Naoya Kitamura, Kenta Wani, YuTakahashi, Yosuke Matsumoto, Motoi Okamoto, Takeshi Ishihara, Alpha-pinene and dizocilpine (MK-801) attenuate kindling development and astrocytosis in an experimental mouse model of epilepsy, IBRO Reports, Volume 9, December 2020.

» Ulubelen A., H. Birman, S. Oksuz, G. Topcu, U. Kolak, A. Barla and W. Voelter, Planta Med., 2002.

» Ullah A., Munir, S., Mabkhot, Y., & Badshah, S. L. (2019). Bioactivity profile of the diterpene isosteviol and its derivatives. Molecules, 24(4), 678.

» Untzer M., Hermetica principia, descriptio, ejusdemque per remedia ... probatissima cùm dogmaticorū, tum chymicorum, methodica curatio. Duobus libris comprehensa, etc, 1616.

» Urata G., & Iljima N. (2000). The Extract of *Ginkgo biloba* A New Field for the Application of Flavonoids and Terpenoids. Journal of Home Economics of Japan, 51(8), 735-746.

» Urbański T., Glinka, T., Wesołowska, E. (1976). On chemical composition of Baltic amber. Bulletin de l'Academie Polonaise des Sciences, Série des sciences chimiques, 24(8), 625.

» Utescher T., et al. "Palaeoclimate and vegetation change in Serbia during the last 30 Ma." Palaeogeography, Palaeoclimatology, Palaeoecology 253.1-2 (2007).

» Vahldiek B. W. Bernstein aus dem Tagebau Schöningen, Baufeld Süd,(Niedersachsen, Norddeutschland): Der Mutterbaum von Krantzit ist identifiziert!. 43. Jahrgang 2015 Heft 2, 35.

» Varamini S., Amirhossein Sakhteman, Gholamhosein Yousefi, Masood Sepehri, Pouya Faridi, Mohammad M. Zarshenas, Volatile composition analysis of five different Commiphora mukul (Hook.

» ex Stocks) Engl. gum samples, Trends in Phramaceutical Sciences 2016.

» Wąsik, A. (2019). Gemstones in Latin literary sources: from Pliny the Elder to Isidore of Seville.

» Vavra N., Chemical characterization of fossil resins ("Amber") - A critical review of methods, problems and possibilities: determination of mineral species, botanical sources and geographical attribution, 1993.

» Vávra N., The Chemistry of Amber – Facts, Findings and Opinions, Annalen des Naturhistorischen Museums in Wien 111 A, 2008.

» Vávra N., (2009). The chemistry of amber-facts, findings and opinions. Annalen des Naturhistorischen Museums in Wien. Serie A für Mineralogie und Petrographie, Geologie und Paläontologie, Anthropologie und Prähistorie, 445-473.

» Verdeguer M., Sánchez-Moreiras, A. M., & Araniti, F. (2020). Phytotoxic effects and mechanism of action of essential oils and terpenoids. Plants, 9(11), 1571.

» Vivek K. Bajpai, Ajay Sharma , Sun Chul Kang, Kwang-Hyun Baek, Antioxidant, lipid peroxidation inhibition and free radical scavenging efficacy of a diterpenoid compound sugiol isolated from *Metasequoia glyptostroboides*, Asian Pacific Journal of Tropical Medicine, 2014.

» Vivek K. Bajpai,Sun Chul Kang, A diterpenoid sugiol from *Metasequoia glyptostroboides* with α-glucosidase and tyrosinase inhibitory potential,Bangladesh Journal of Pharmacology, 2014.

» Vivekanandhan P., Ayyakkannu Usha-Raja-Nanthini, Gurusamy Valli, Muthugoundar Subramanian Shivakumar, Comparative efficacy of *Eucalyptus* globulus (Labill) hydrodistilled essential oil and temephus as mosquito larvicide, Natural Product Research Formerly Natural Product Letters,Taylor & Francis, 2020.

» Volkov A. G., & Ranatunga, D. R. A. (2006). Plants as environmental biosensors. Plant signaling & behavior, 1(3), 105-115.

» Wagner M. (2007). Węglowe osady miocenu Kępy Swarzewskiej na wybrzeżu bałtyckim. Geologia/Akademia Górniczo-Hutnicza im. Stanisława Staszica w Krakowie, 33(1), 69-88.

» Wagner-Wysiecka E., Mid-infrared spectroscopy for characterization of Baltic amber (succinite), Spectrochimica Acta Part A: Molecular and Biomolecular Spectroscopy 196 (2018) 418–431.

» Wake G., Court J., Pickering A., Lewis R., Wilkins R., Perry E., CNS acetylcholine receptor activity in European medicinal plants traditionally used to improve failing memory, Journal of Ethnopharmacology 69, 2000.

» Webb N. J. A., & Pitt, W. R. (1993). *Eucalyptus* oil poisoning in childhood: 41 cases in south-east Queensland. Journal of paediatrics and child health, 29(5), 368-371.

» Więcław D., Lytvyniuk, S. F., Kovalevych, V. M., & Peryt, T. M. (2008). Inkluzje fluidalne w halicie oraz bituminy w solach ewaporatów mioceńskich ukraińskiego Przedkarpacia jako wskaźnik występowania nagromadzeń węglowodorów w niżej leżących utworach. Przegląd Geologiczny, 56(9), 837-841.

» Wilde V., Frankenhäuser, H., & Lenz, O. K. (2021). A myricaceous male inflorescence with pollen in situ from the middle Eocene of Europe. Palaeobiodiversity and Palaeoenvironments, 101(4), 873-883.

» Wolfe A. P., McKellar, R. C., Tappert, R., Sodhi, R. N., & Muehlenbachs, K. (2016). Bitterfeld amber is not Baltic amber: Three geochemical tests and further constraints on the botanical affinities of succinite. Review of Palaeobotany and Palynology, 225, 21-32.

» Wolfe A. P., Tappert, R., Muehlenbachs, K., Boudreau, M., McKellar, R. C., Basinger, J. F., & Garrett, A. (2009). A new proposal concerning the botanical origin of Baltic amber. Proceedings of the Royal Society B: Biological Sciences, 276(1672), 3403-3412.

» Wimmer R., L. Pester & L. Eissmann. 2008. Geologie der Bitterfelder Bernsteinlagerstätte unter Berücksichtigung neuer Erkenntnisse. In Rascher, J., Wimmer, R., Krumbiegel, G. & Schmiedel, S. (eds) Bitterfelder Bernstein versus Baltischer Bernstein – Hypothesen, Fakten, Fragen. Exkursionsführer und Veröffentlichungen der Deutschen Gesellschaft für Geowissenschaften 236: 34-45.

» Worth H., Uwe Dethlefsen, Patients with Asthma Benefit from Concomitant Therapy with Cineole: A Placebo-Controlled, Double-Blind Trial, Journal of Asthma Volume 49, 2012 - Issue 8.

» Woo J., Yang, H., Yoon, M., Gadhe, C. G., Pae, A. N., Cho, S., & Lee, C. J. (2019). 3-Carene, a phytoncide from pine tree has a sleep-enhancing effect by targeting the GABAA-benzodiazepine receptors. Experimental neurobiology, 28(5), 593.

» Wood G.B., Franklin Bache Grigg, Elliot & Company, 1849 – 1380.

» Wrzeciono U., & Zaprutko, L. (2001). Chemia związków naturalnych. Chemistry of natural compounds.

» Wu A., Noble E. E., Tyagi E., Ying Z., Zhuang Y., Gomez-Pinilla F., Curcumin boosts DHA in the brain: Implications for the prevention of anxiety disorders, Biochimica Biophysica Acta (BBA), 2015.

» Wunderlich J. (2004). The fossil spiders of the family Linyphiidae in Baltic and Dominican amber (Araneae: Linyphiidae). Beiträge zur Araneologie, 3, 1298-1373.

» Wunderlich J. (2004). The fossil spiders (Araneae) of the families Tetragnathidae and Zygiellidae n. stat. in Baltic and Dominican amber, with notes on higher extant and fossil taxa. Beitrage Araneologie, 3, 899-955.

» Wunderlich J. 2004ak. Members of the family Philodromidae (Araneae) in Baltic amber. Beiträge zur Araneologie, 3, 1689-1693.

» Wunderlich J. 2011. Taxonomy of extant and fossil (Eocene) European ground spiders of the family Gnaphosidae (Araneae), with a key to the genera, and descriptions of new taxa. Beiträge zur Araneologie, 6:19–97.

» Xu D., Zhang, S.; Foster, D.J.; Wang, J. The effects of isosteviol against myocardium injury induced by ischaemia-reperfusion in the isolated guinea pig heart. Clin. Exp. Pharmacol. Physiol. 2007.

» Xu G., Junsheng Guo & Chunming Sun, Eucalyptol ameliorates early brain injury after subarachnoid haemorrhage via antioxidant and anti-inflammatory effects in a rat model, , Pharmaceutical Biology, 59:1, 114-120, 2021.

» Xu J., Hu, Z. Q., Wang, C., Yin, Z. Q., Wei, Q., Zhou, L. J., ... & Yin, L. Z. (2014). Acute and subacute toxicity study of 1, 8-cineole in mice. International journal of clinical and experimental pathology, 7(4), 1495.

» Yamamoto S., Otto, A., Krumbiegel, G., & Simoneit, B. R. (2006). The natural product biomarkers in succinite, glessite and stantienite ambers from Bitterfeld, Germany. Review of Palaeobotany and Palynology, 140(1-2), 27-49.

» Yong L, Ren Mihong, Wang Jiajun, Ma Rong, Chen Hai, Xie Qian, Li Hongyan, Li Jinxiu, Wang Jian, Progress in Borneol Intervention for Ischemic Stroke: A Systematic Review, Frontiers in Pharmacology 2021.

» Yossifova M. G., Eskenazy, G. M., & Valčeva, S. P. (2011). Petrology, mineralogy, and geochemistry of submarine coals and petrified forest in the Sozopol Bay, Bulgaria. International Journal of Coal Geology, 87(3-4), 212-225.

» Yousuf Al M.H., Bashir A.K., Hali B., Mom T., Blunden G., Some effects of Salvia aegyptiaca L. on the central nervous system in mice, Journal of Ethnopharmacology, 2002.

» Yuan J., Guo W., Yang B., Liu P., Wang Q., Yuan H. 116 cases of coronary angina pectoris treated with powder composed of radix ginseng, radix notoginseng and succinum // J. Tradit. Chin. Med., 1997; 17 (1): 14–17.

» Załuski D., Cieśla Ł., Janeczko Z., The Structure-Activity Relationships of Plant Secondary Metabolites with Antimicrobial, Free Radical Scavenging and Inhibitory Activity towards Selected Enzymes, Studies in Natural Product Chemistry (Bioactive Natural Products), editor prof. Atta-ur Rahman, Elsevier Science Publishers, Amsterdam, Holand, 2015.

» Zan J, Zhang H, Lu MY, et al. Isosteviol sodium injection improves outcomes by modulating TLRs/NF-κB-dependent inflammatory responses following experimental traumatic brain injury in rats. Neuroreport, 2018.

» Zarubina I.V., Lukk, M. V., & Shabanov, P. D., Antihypoxic and antioxidant effects of exogenous succinic acid and aminothiol succinate-containing antihypoxants, Bulletin of Experimental Biology and Medicine, 2012.

» Zhang A., Mernitz, K., Wu, C., Xiong, W., He, Y., Wang, G., & Wang, X. (2021). ATP drives efficient terpene biosynthesis in marine thraustochytrids. MBio, 12(3), e00881-21.

» Zhang S., Li, L., Hu, J., Ma, P., & Zhu, H. (2020). Polysaccharide of *Taxus chinensis* var. mairei Cheng et LK Fu attenuates neurotoxicity and cognitive dysfunction in mice with Alzheimer's disease. Pharmaceutical Biology, 58(1), 959-968.

» Zhu Z., Chen C, Zhu Y, Shang E, Zhao M, Guo S, Guo J, Qian D, Tang Z, Yan H, Duan J. Exploratory Cortex Metabolic Profiling Revealed the Sedative Effect of Amber in Pentylenetetrazole-Induced Epilepsy-Like Mice. Molecules. 2019.

» Zinkel D.F., & Magee, T. V. (1987). Diterpene resin acids from the

needle oleoresin of Pinus strobus. Phytochemistry, 26(3), 769-774.

»   Zinkel D. F., & Magee, T. V. (1991). Resin acids of Pinus ponderosa needles. Phytochemistry, 30(3), 845-848.

»   Żabka M. (1988). Fossil Eocene Salticidae (Araneae) from the collection of the Muzeum of Earth in Warsaw. In Annales Zoologici (Vol. 41, No. 13, pp. 415-420). Państwowe Wydawnictwo Naukowe.

425

https://informacje.pan.pl/informacje/nauki-scisle-i-nauki-o-ziemi/2715-zmiany-klimatu-oficjalne-stanowisko-polskiej-akademii-nauk.

THE AUTHOR'S OTHER BOOKS ABOUT AMBER.